Master Optical Techniques

Master Optical Techniques

ARTHUR S. DE VANY

John Wiley & Sons, New York • Chichester • Brisbane • Toronto

6*H -1274 v

PHYSICS

Library of Congress Cataloging in Publication Data:

De Vany, Arthur S
    Master optical techniques.

    (Wiley series in pure and applied optics)
    Includes index.
    1. Optical instruments—Design and construction.
I. Title. [DNLM: 1. Optics. 2. Optics—Instrumentation. QC 357.2 D488m]

TS517.D48   681'.4   80-24442
ISBN 0-471-07720-8

Printed in the United States of America

10 9 8 7 6 5 4 3 2 1

Preface

This book is directed to those who want to learn the techniques of the master optician. Students, teachers, and practicing opticians will find here a practical development of the techniques and operations involved in the fabrication of prisms, lenses, and aspherical optical elements. A sequential approach will carry the optician through each step from the selection and testing of the optical material to the polishing and testing of the finished work. Extensive use of illustrations is made to facilitate understanding and to simplify the text. Modern techniques and recent developments are fully covered in step-by-step method used to describe all the techniques discussed.

The book is divided into four parts: I. Fundamental Operations, II. Fabrication of Prisms and Lenses, III. Telescope Systems, and IV. Optical Testing. Each section is subdivided into the fabrication phases of one particular component and will help the beginner to undertake the production of various elements and to find the answers to a number of questions involved in their fabrication.

Part I deals with the principles of optics and discusses current practices in diamond sawing, curve generating and edging, hand polishing, and testing flat and radius work.

Part II discusses in depth the fabrication of right-angle and dove prisms, penta, deviation roof prisms, and many other systems. The making of lens elements and test plates of glass and crystal is also treated.

Part III describes the fabrication of parabodial mirrors and cameras of the Schmidt-Cassegrain, Cassegrain, Maksutov, and Schmidt types.

Part IV introduces the knife-edge technique and describes single and double wire tests. The use of the laser in simple optical setups in which lateral-sheared interferograms are formed is reviewed. Ronchigrams formed by optical testing of lens systems are given a more thorough treatment than can be found in any other source.

I am indebted to many people, and it is with sincere pleasure that I acknowledge my debt to the late Marcus Brown, Don Hendrix, Phillip Taylor, Harold Hall, Paul Halderman, Otto Hartman, and many others at the Northrop Corporation. Special thanks go to my wife, whose excellent draftsmanship illustrates many pages of this book, and to my son, Arthur De Vany, who revised the manuscript. I cannot forget John Howard, editor of *Applied Optics*, and his staff who wrestled the papers I have published in that journal into shape and which now, in revised form, are part of this book.

Arthur S. DeVany

San Clemente, California
January 1981

Contents

Part I

Fundamental Operations

1

The Optical Technician

The precision optical industry is responsible for the development and construction of research and test equipment in which precise tolerances beyond those obtainable under the best production control methods must be maintained. A precision optician works on optical components that demand a high degree of perfection for installation in tracking and guidance systems in atmospheric and space environments, test and alignment controls, and research and measurement devices. These components are also used in medical and industrial research, and commercial quality control.

The capability of machine to achieve precision in optical components is limited. To carry the precision beyond machine limits requires meticulous and time-consuming handwork. The optical systems involved may be many feet in diameter (Palomar reflector), with tolerances in millionths of an inch (0.000001 in), or no larger than the period at the end of this sentence.

The relation between optical technician and optical technology is well expressed by J. W. Lanner and E. Goldstein.

> It might be of interest here to make note of an apparent paradox in contemporary optical technology. The design, development, and construction of sophisticated satellite-borne physical science experiments often costs millions of dollars in equipment as well as in the efforts of physicists, engineers, and others. Many months of analytical and experimental work often precede final hardware construction. This work is for the most part subject to precise logical approaches. However, ultimately the whole scientific-administration structure of the project is often completely dependent upon the nonscientific individual, the optical artisan, to fabricate and evaluate the quality of a precise optical surface. This concept might appear to be unduly dramatic. Nevertheless, at times it seems that a very large dog, labeled science, is being wagged by a very small tail, labeled *art*.*

Computer-operated polishing machines, available only in the largest laboratories, have to some extent helped the artisan to figure aspheric surfaces

* Quoted with permission from J. W. Lanner and E. Goldstein, "Some Comments upon Current Optical Shop Practices," *Applied Optics*, May 1966, p. 678.

3

precisely. Yet the quality is not acceptable. Astigmatic surfaces often result when aspheric surfaces are so determined. At this point the skilled artisan must apply a full-sized soft polisher, contoured to an astigmatic surface, to iron it out. This is but one example of working the surfaces of optical components when the optical technician, called on to improvise, must be ready to meet demands of the optical engineer.

Who would make an excellent optician or optical technician? One who has hand skills and some experience in carpentry or mechanical crafts in which rotating saws and milling are used would do well. Amateur telescope builders who have learned the fragile nature of glass handling would also do well. The qualities of persistence and ingenuity required are illustrated in Jeff Schroeder's account of his experiences in building an 11-in. (27.6-cm) refractor. Schroeder had had some experience, although limited, in the production of telescope mirrors. The most ingenious aspect of his work was the use of distant (5 miles) sodium street lights as a source and the interpretation of the correction of the spherical aberration with a Ronchi grating. He figured the inner concave surface optically toward a slight ellipsoidal surface with the distant light source and Ronchi test to indicate when he was over- or under-correct. He tested this ellipsoidal curve at the radius of curvature in his work area at home and thereby eliminated the need for setting the refractor over and over again. Of course, after each period of optical refiguring he had to retest. This method is a substitute for an optical flat of high quality.

Some of the skills required of the optical technician are listed:

1.   Maintenance of equipment, work area, records, and schedules and the development of safe work habits.
2.   Operation of equipment such as a generator, glass saw, edger, hand spindle (roughing), and fine grinding, polishing, and coating machines.
3.   Production of precision diameters, radii, thickness, angles, and parallelism on optical components with various grinding and polishing compounds.
4.   Testing of optical flats, radii, physical dimensions, angles, and optical quality with monochromatic light, micrometer, spherometer, lens bench, and autocollimator.
5.   Accurate use of autocollimator, optical test bench and related gear, interferometer, and Ronchi and knife-edge tests in maintaining tolerances and testing for optical properties.
6.   Construction of test set-ups for the inspection of polishing and figuring operations and interpretation of tests performed to guide work toward the production of usable optical components.

7. Operation of the coating machine in the preparation of material for testing purposes.

Knowledge is required in the following areas:

1. Fundamental theory of light.
2. Principles of refraction and diffraction.
3. Characteristics of glass.
4. Procedures for processing glass or optical material from rough parts to finished product.
5. Elementary mathematics.
6. Interpretation of blueprint specifications, tolerances, and tests of components at various stages of development.

In addition, the technician must develop professional pride in the exacting tolerances of his trade and master the basic skills applicable to each phase of his work.

BIBLIOGRAPHY

Lanner, J. W. and Goldstein, E. "Some Comments upon Current Optical Shop Practices," *Appl. Opt.*, **5,** 678 (1966).

2

Basic Operations

The beginning optician must learn to operate a variety of machines which includes the curve generator, edger, diamond saw, band saw, and polishers of various types. He must also learn to mount the glass elements on vacuum or pitched mounting fixtures and plaster blocks. Good safety habits must be observed when working with rotating machinery, for safety can be gained only by a well-established procedure.

1. DIAMOND SAWING

The first machine to be mastered is the diamond saw. A number of types of diamond saw are on the market but all are similarly operated. The one most commonly seen in optical shops is the Felkner, which is also used by construction workers for cutting cement blocks. The rotating diamond blade is mounted on a spindle in a vertical position and two hoses feed a coolant over the glass being cut. Ordinarily the optician will hand-feed small sections into the rotating saw, but for larger sections a hydraulic feed may be needed.

The diamond sawing of tile pieces is described in the following steps:

Step 1. Set the saw blade about 0.040 in. or 1 mm below the level of the table. Anchor the mounting of the saw blade by tightening the hydraulic screw on the mount.

Step 2. Mark the diameters of the sections to be cut with a soft pencil.

Step 3. Turn on the machine and direct the running coolant toward the saw slot in the sliding table.

Step 4. With the tile against a horizontal stop, slide the table and cut along the mark. Experienced opticians will often bypass this step and move the tile into the rotating saw with both hands. They make sure, however, that the fingers are safely out of the way and the sliding table is anchored.

If a purple powder instead of the normal white appears, too much pressure is being applied on the material. A typical mount diamond saw is shown in Fig. 2.1.

Fig. 2.1. A Flecker-Dresser diamond slitting saw with a coolant pump. Model 80-BQ cutoff machine and cabinet mounted.

As the glass or crystal is sawed, the diamond wheels or wire may become clogged and it will be necessary to remove the material with a carborundum or white special stone.

1.1. Cutting Thicker Sections

Some types of diamond saw, which are mounted on hydraulic control spindles, permit thick glass slabs to be positioned on the translating table.

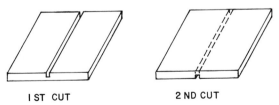

I ST CUT   2 ND CUT

Fig. 2.2. An improvised method of slitting a thick glass slab into two sections by alternate sawing of grooves.

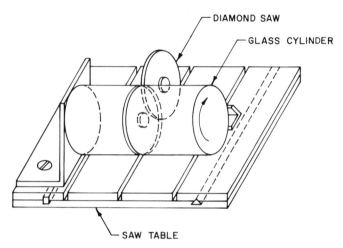

Fig. 2.3. Slitting disks of glass from a larger glass cylinder.

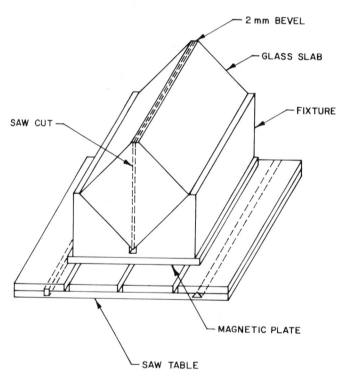

Fig. 2.4. Slitting two tank prisms from one rectangular slab held in the fixture with double-coated tape or pitch.

To saw a slab into several sections one must cut partly through one-half of its thickness, invert it, and make a similar cut. Cutting these glass sections is governed by two limitations—the diameter of the saw blade and its splash guard and the mounting washers that hold the blade to the rotating spindle. This operation is shown in Fig. 2.2.

1.2. Cutting Circular Sections

Often thin circular glass sections must be cut froom one large-diameter cylinder of glass and some improvisation by the operator is necessary. The glass cylinder is mounted on a locked table to which a vee metal stop is anchored. Next a 90° metal upright stand is anchored to the table for the glass cylinder to ride against while it is being rotated by hand. This sawing operation is illustrated in Fig. 2.3. In this operation the saw blade is brought down into the glass cylinder and a cut is made to ⅛ in. (3 mm) of the holding washers. The glass cylinder is then turned and cut until a small core remains in the middle. One should not attempt to break the sections apart by prying if the core has a large diameter. Instead, the core can readily be sawn apart by using a fine-tooth handsaw and 80 carborundum. This procedure works slowly but effectively. Other sections may be cut by placing a rectangular section of wood between the upright stand of the glass cylinder.

1.3. Sawing Out Prisms

Large and small prisms are cut from semipolished plano-parallel elements. These semipolished surfaces must be protected by cementing glass covers to both surfaces. To do this the elements are heated to 212°F (100°C) in an oven and pitch is smeared on each surface. Lens or tissue paper is placed between the polished surface and precut window glass.

It is good practice to use wide masking tape to mark the prisms to be cut from the glass slabs. A cardboard or thin plastic pattern of the prism helps to plan the layout of the sawing. Allow for saw cuts and prism generation: glass is cheaper than labor.

1.4. Beam Splitters

Tank and beam splitters are cut from planoparallels by first sawing rectangular slabs. A difficult diagonal cut must be made through the elongated glass section by taping double-coated slabs of glass to a vee block, as shown in Fig. 2.4. The 90° vee holder is anchored to the sliding table of the saw. A 0.1-in. (2.5-mm) bevel is first hand-ground on one corner of the glass section to serve as a guide line and starting area for the cut, but a number of tests

Fig. 2.5. An elliptical glass flat (42 × 30 cm) band sawed from a large Pyrex disk. The surrounds are used to form a circular disk during fabrication by felt-pitching to the ellipse.

Fig. 2.6. Disk grinding the sides of the sawed ellipse with a carborundum disk grinder. The inner edges of the surrounds were fine-ground with a flex grinder disk to which carborundum papers attached.

Fig. 2.7. The No. 2006 wire saw made by Laser Technology Inc., North Hollywood, CA., for slitting crystals with a running diamond-pregnated wire. Coolants must be carefully chosen.

must be made before cutting. The prism and its bevel are run between the stationary blade to guarantee their alignment. If they are not aligned precisely, the assembly must be moved to the correct position and fastened securely. The saw blade which is set to follow the milled recess in the holder must not cut into the metallic holder. Some caution must be exercised when the saw blade reaches full cut. The glass must not be allowed to break away and fall on the metal table. To prevent chipping a heavy corrugated rubber mat is placed near the holder. The prisms are easily removed from the holder by prying them from the tape with a wood or plastic wedge. Always wipe the holder dry before attempting to install a new prism and renew the double-coated tape. Use only a small piece of tape at each end of the holder because too much will make the prism sections difficult to remove. Bevel all sharp corners on the prisms.

2. BAND OR WIRE SAWING

Glass or crystal sections are often band-sawed. The optical band saw is similar to the conventional product in wood or metal workshops. A common band-saw operation consists of removing the glass surrounds from a large glass disk before it is fabricated into an elliptical optical flat. (See Fig. 2.5.) This method requires a flexible, impregnated metal saw blade and the heavy glass disk is moved on small steel ball bearings pressed into a plywood support. The ellipse to be cut out is drawn on wide masking tape. A coolant is pumped between the running blade and glass to prevent strain. Strains may be removed by sand blasting the disk and surrounds or the glass disk may be fine-ground (see Fig. 2.6); the surrounds are fine-ground with a flexible disk grinder on a drill motor.

It is impractical to use the wire saw method to cut heavy glass because of the pressures required.

In many shops the wire saw or wet-string saw is preferred for cutting crystal elements for infrared (IR) sensors. The diamond impregnated saw is used on the hard crystals and the wet-string method on the softer crystals such as sodium iodide. Coolants must be carefully selected to avoid dissolving the crystal being cut. A wire saw is shown in Fig. 2.7.

3. CURVE OR SURFACE GENERATING

Curve or surface generating on glass elements is a frequent operation. A typical surface generator (often called a grinder is shown in Fig. 2.8. This model, a Strasbaugh 7-M, can handle a glass blank as large as 24 in. (61 cm) and is capable of generating concave, convex, or flat surfaces, bevels, edging, and drilling holes. The glass disks are held by a vacuum holder equipped with a compressible "O" ring. These operations are shown in Fig. 2.9.

Grinding or generating requires a diamond cup wheel. The diamond wheels are set at an angle to generate curved surfaces, flat to generate flat surfaces. The out-diameter of the cutter generates a concave surface; its inner edge is used for convex surfaces.

The angular tilt of the various concave and convex surfaces is based on a simple expression:

$$\sin X = d/2R,$$

where X is the angular tilt of the wheel, d is the diameter of the cutting edge of the diamond wheel, and R is the desired or prescription radius on the glass element. It must be kept in mind that the outer edge of the wheel is applied to concave surfaces and the inner edge, to convex surfaces. These

Fig. 2.8. Strabaugh Model 7-M curve generator and grinder for 24-in. (60 cm) diameter elements. Glass thicknesses of 7 in. (17.8 cm) can be handled.

edges can be roughly measured by a machinist's ruler. Example: The desired radius is 6.125-in., concave. The outer edge of the cutter measures 3.0 in. Then

$$\sin X = 3.0/2\ (6.125) = 0.24490$$

and $X = 14°10'$. The angular tilt of the cutter must then be set at $14°10'$ and the outer edge must be centrally located in the center of the glass element.

One should always choose a diamond wheel larger than the element being generated. The radius will often vary slightly and must be corrected by readjusting the tilt. The radius generated is checked with a spherometer.

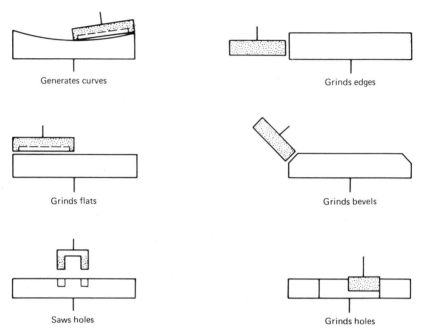

Generates curves

Grinds edges

Grinds flats

Grinds bevels

Saws holes

Grinds holes

Fig. 2.9. Various operations that can be performed with the Strabaugh Model 7-M generator. Special vacuum chucks are required for drilling and sawing central holes.

Fig. 2.10. Strasbaugh Digitron Spherometer Model 18 As for measuring the sagitta of the generated surfaces to an accuracy of 0.001 mm.

14

3.1. The Spherometer

The spherometer measures the degree to which a generated curve departs from a given sphere. The ball type contains three balls pressed against the glass surface. The radius of curvature of the surface r is given by

$$r = (R^2/2h) + (h/2) \pm a \tag{1}$$

where h is the sagittal height, a is the diameter of the balls, and R is the radial distance from the plunger to the center of the contact ball. The sign of a is minus for convex and plus for concave surfaces. Spherometers are available in $\frac{3}{4}$-, 1.5-, 2-, and 3-in. diameters. (See Fig. 2.10.)

Many optical shops use ring-type spherometers instead of the ball type.

3.2. Operating the Strasbaugh Generator

Let us examine the steps required to generate a typical concave surface. First we must determine whether one surface of the glass disk is free of folds. Folds will not permit the vacuum to hold the piece. If folds are present, they must be removed by rough grinding on a spindle with a flat, rotating iron tool. The disk should be edge ground. If it is not, it must be edged on the cylinder grinder or generator.

The rotating spindle is set to turn at 10 rpm and the glass blank must be running true before the dial indicator is set for checking the runout. If the runout is too great, the disk must be moved one-half the amount. Check the runout with a dial indicator, note the total runout, and divide by two. To correct move the disk by one-half the reading. Several adjustments may be required before a zero reading can be achieved. Take a pencil and locate the center of the disk by drawing concentric circles in that area.

The diamond wheel of the generator has now been set to the calculated value. Raise the spindle (stationary) to a point within $\frac{1}{8}$ in. (3 mm) of the glass disk. Translate the generator wheel to the center of the disk, using the pencil marks. Turn on the coolant pump and direct the liquid toward the center. Turn on the spindle and set the disk on the "O" ring at ON. Next, turn on the generator wheel and hand feed it into the glass disk to a depth of 1 mm. Lower the glass disk and determine whether the central area is free of small lands. If a land of glass is evident, move the cutter by translating it into the land. Turn on the spindle with the mounted glass disk and the coolant and raise the disk into the rotating wheel. Turn on the automatic feed and feed the spindle slowly upward into the disk.

When the curve has been generated more than one-half the diameter, lower the disk and turn off all equipment. Check the radius of the concave surface with the spherometer. If the radius is wrong, the tilt of the cutter must be changed. If it is too large, the cutter tilt must be reduced to a smaller angle. If the radius is too small, the cutter tilt must be raised to a wider angle. If the measured value is within 0.020 in. or (0.5 mm), it will not pay to change the angular tilt because the curve can be adjusted during grinding.

Convex-curve generating is similar in all aspects to concave-curve generating. The only important differences are that the inner edge of the cutter is used and the angular tilt is in the opposite direction. Glass lands or humps are more often generated in the center because of the difficulty of locating the exact cutting edge of the wheel. If they do occur, the cutter is always moved toward the inner cutting edge.

Generating flat surfaces is similar to generating curve radii (a flat surface has an infinite radius of curvature). The tilt of the cutter is zero. It is good practice to have all flat surfaces slightly on the concave side, perhaps with 0.002 in. (0.5 mm) overall diameter. This makes it easier to fine grind a convex surface because the edges are ground down and the pits near the edge are more readily removed.

4. EDGING

Glass edging can be done by two methods. The first makes use of a cup diamond wheel set at 5 to 10° and the same tilt angle at which a convex radius is generated. The tilted edge advances downward into the glass blank and the edging is circular. To prevent chipping stop short within 0.25 in. (6.35 mm) of the down edge. This point can be determined by using a dial indicator with a magnetic post near the translating slide of the cutter. Next, move the cutter away from the glass blank and raise the spindle with the glass edge near the cutting edge of the tilted cutter. Move the translating slide with cutter to the preceding setting of the dial indicator and finish the edging of the last 0.25 in. (0.05 mm) by feeding it in the opposite direction. (See Fig. 2.9.)

The second method requires a conventional diamond wheel similar to that used on a cylinder grinder. Here the spindle that carries the glass blank is raised and lowered as the sliding cutter is advanced. *It is very important* that the grinding wheel always have partial contact with the surface of the glass or chipping will occur.

Beveling is done with a cup diamond wheel or the conventional diamond edging wheel. The cutter is preset to the desired beveling angle, generally 45°.

5. DRILLING (SAWHOLES)

Holes are drilled with a core drill which penetrates the glass as it rotates. The glass plate must be carefully centered and pitch-cemented to a glass

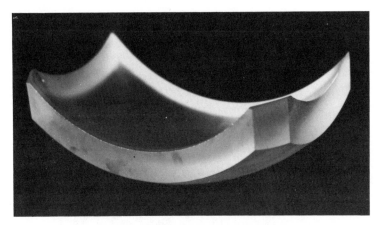

Fig. 2.11. A Maksutov corrector lens sector curve generated by the Strasbaugh generator. Note the bevels and flat edges with steep radii on the two glass surfaces.

backing to prevent the vacuum from being lost as the core drill bores through the element. A special metal holder with a recess can be improvised to allow the core drill to advance freely into the holder. (See Fig. 2.9.)

The combined operations of generating, edging, and sawing are well illustrated by the Maksutov sector correcting element shown in Fig. 2.11. Note the steep concave and convex surfaces of the deep radii, the beveled corners, and the circular edging.

6.   GENERATING PRISM'S ANGULAR FLAT SURFACES

One commonly used generator for prism angular flat surfaces is the LOH (Wutzler) G-6 machine. (See Fig. 2.12.) It has two diamond cup wheels which can generate two angular surfaces in one setting. Its versatility makes it possible to set up precise glass angles with an autocollimator with small aluminized plano-parallels that are wet-wrung on the prism's angular surface. Of course, the master angle of the prism's angular surface must be properly set to control the predetermined angle. These machines are generally used for prisms with 0.75-in. (19 mm) or larger faces. Smaller prisms with smaller base lengths are generated by another type of generator which is discussed later.

When a large number of prisms is the be mass produced, several methods are available for generating the angular surfaces. Many optical firms make use of pressed molded prisms which have rough angular flat surfaces. The first operation generates the two side angles of the prism by pitch-cementing each side to a parallel cast-iron plate and flat generating both surfaces.

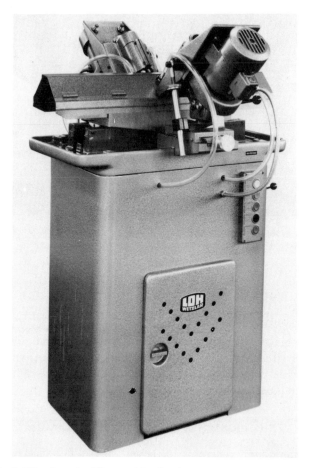

Fig. 2.12. The LOH universal milling machine for simultaneously milling of two right-angle faces on a prism.

Twenty or more prisms may be surfaced at one time, depending on their base diameters. The compound reflecting glass angles may be generated by mounting the prisms on precut cast-iron angular channel blocks before surfacing each angular flat surface. (See also Fig. 13.6.)

Another means is the Blanchard surfacing method which provides individual angular holding fixtures for each glass angle to be generated. Generally, the angular fixture is designed so that one reference side angle of the prism is always cemented to the cast-iron fixture. Many prisms can be flat-surfaced at one time with the Blanchard generator. On the Blanchard machine the prisms are held down by magnetic chucks. These metal holders of cast iron must necessarily be made to extremely close angular tolerances. If the cementing of the prisms to the fixtures is done with care, the pyramidal error on the reflecting glass angles will be minimal. It is always good practice to make all surfaces slightly concave to prevent the prisms from rocking while being assembled during plastering for fine grinding and polishing.

The surfacing of small prisms which measure 0.5 in. (12.5 mm) or less is more easily controlled by attaching a special holding fixture equipped with an autocollimator to one end of the spindle and an octagonal aluminized element to the other. (See Fig. 2.13.) A secondary indexing head is often contained in the fixture for facet surfacing. This fixture can be mounted on a LOH generator or other grinding machine. (See Figs. 2.14a and b.)

Setting this special fixture requires some care, mainly in paralleling the holding fixtures to the autocollimation surface of the octagonal plane. This can be accomplished with a precision dial indicator on a height gauge. Because small prisms have small surfaces and facets, the holding fixtures must be precisely made. (See Fig. 2.15.)

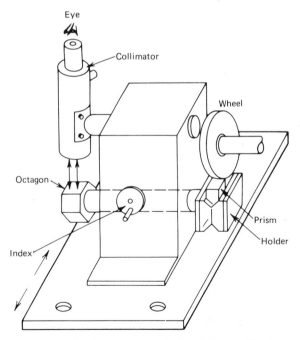

Fig. 2.13. A special collimating fixture that can be attached to machines shown, Figs. 2.14 *a* and *b*. Note that these fixtures are part of the LOH milling machine.

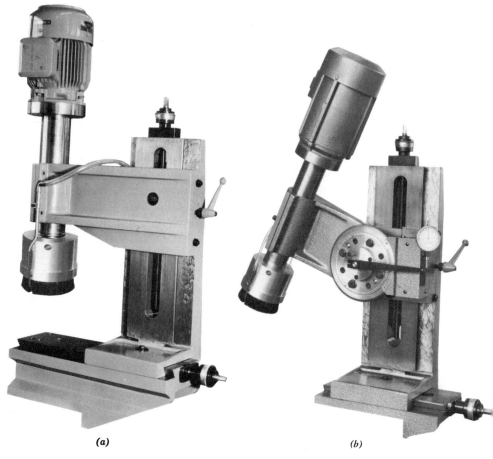

(a)   (b)

Fig. 2.14. Two LOH attachments used on the LOH universal milling machine (Fig. 2.12) for milling the glass angular surfaces on prisms and optical components.

7. GENERATING LARGE-DIAMETER RADIUS SURFACES WITH SMALL CUTTERS

Small optical shops rarely offer diamond cutters larger than 6 in. (15 cm). Smaller cutters cannot generate large diameter work in one setting. To change the cutter to a new angle and overlap distance, as in Figure 2.16, is a simple matter. By resetting the tilt of the cutter to a new angle large diameter work can be generated by the "step-step" method.

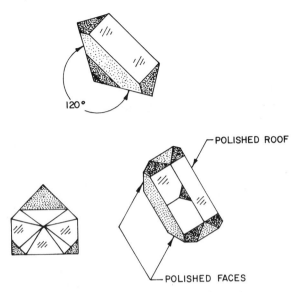

Fig. 2.15. Precise glass reflecting angles and small facets are required on Amici and similar prisms. These surfaces require the special fixture shown in Fig. 2.13.

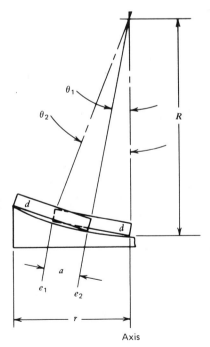

Fig. 2.16. The geometry of the generator's grinding wheel and a smaller diameter wheel for large diameter work. Angle O_1, first tilt angle; angle O_2, second tilt angle; R, radius of curvature; r, semidiameter of blank; d, diameter of cutter; a, distance between e_1 and e_2.

21

The geometry of the "step-step" setup is illustrated in Fig. 2.16. Two basic formulas are involved. The first tilt angle of the small cutter is calculated with

$$\sin A_1 = d/2R \tag{2}$$

where A_1 is the first tilt angle of the diamond cutter, d is the diameter of the cutting edge (for concave-surface generating the cutting edge is slightly inside the peripheral edge; for convex generating it is the inner edge), and R is the radius of curvature.

The second formula for the new tilt angle of the same diameter cutter by the step-step method is

$$\sin A_2 = [(R^2/a^2 - 1)^{1/2} + (4R^2/d^2 - 1)^{1/2}]\,(ad/2R^2) \tag{3}$$

where A_2 is the new angular tilt of the cutter, a is the overlap distance of the two cuts, R is the radius of curvature, and d is the diameter of the cutting edge.

To generate a concave radius surface on a large diameter mirror or lens by the step-step method, the cutter is set to the angular tilt according to the first formula. The full depth of the concave surface generated in the glass blank is

$$h = R - (R^2 - y^2)^{1/2} \tag{4}$$

where h is the sagittal depth, R is the desired radius of curvature, and y is the semidiameter of the glass blank. After the central area is generated to the full calculated depth the distance can be measured. (See Fig. 2.16.) The distance is one-half the glass land.

It is *not* good practice to change the tilt angle to allow the overlapping distance between the new cuts to cover the maximum diameter of the cutter. It should also be realized that the mass of glass being removed is proportionally larger by its square.

The new tilt angle is calculated by using the distance a in conjunction with the $\sin A_2$ formula.

Hint. The new angle tilt is always a little larger than the first calculated tilt and is not additive to the preceding angular tilt $\sin A_2$.

The diamond cutter is advanced vertically into the glass until the cutter reaches the generated surface. It *is* good practice to provide a precut plastic template of the desired radius to check the generated surface.

8. CENTERING AND EDGING

Centering and edging of glass elements may be done on a cylinder-type edger. (See Fig. 2.17). To edge a glass blank it is pitch-mounted on a

Fig. 2.17. The Strasbaugh Model H-6 edger with interchangeable quill holders. Glass disks as large as 16 in. (40 cm) can be edged if a 0.7 zonal flat holder is used in its mounting.

machined brass holder and then screw-mounted to a demountable quill which can be rotated. A number of quills is available. while one glass blank is being edged another can be centered on an auxiliary stand. (See Fig. 2.18.) The pitch mixture by weight is three parts beeswax, three parts water-white rosin, and two parts Montan wax (Brazilian Carnauba wax can be substituted for Montan wax).

To edge a blank of glass it must be slowly fed in by hand until the wheel just touches the outer periphery of the grinding wheel. Check the eccentric throw of the grinding wheel; *never* permit the wheel to go beyond the glass

Fig. 2.18. A quill-holder stand with a precision dial indicator for centering lenses in a safe horizontal position.

blank in either direction. A good rule is always to have one-half the wheel on the blank.

Direct the coolant to jet between the blank and the rotating wheel, set the micrometer close to the desired tolerance for the diameter, and turn on the automatic feed. A soft pencil is used to mark the periphery to indicate that it has been edged. Now remeasure the diameter of the disk.

The difference between these two readings is the amount remaining to be edged off.

Centering is the process by which finished glass or crystal elements are edged so that the centers of each radial surface are concentric. Before starting, however, the brass holders must be made to run true on their peripheral

(a)

(b)

Fig. 2.19. Chuck holders on a LOH portable grinder. Note the setup for grinding their outer and inner edges.

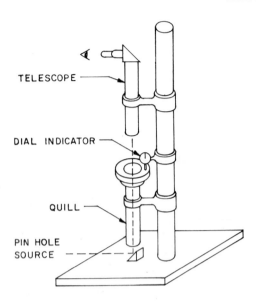

TELESCOPE

DIAL INDICATOR

QUILL

PIN HOLE
SOURCE

Fig. 2.20. A special fixture for centering a large objective lens that permits simultaneous checks to be performed. It is obvious that the fixture shown in Fig. 2.18 could also be converted.

edges. They may be checked while the brass holder is on the quill of the grinder or auxiliary stand. (See Fig. 2.18.) Four or more brass holders should be trued if a large number of elements is to be centered. Truing is, in general, a machine-shop operation, but occasionally it must be done by an optician. If a small lathe is available, the brass holder mounted on the quill can be used to true the edge. An alternative is a LOH WG grinder. (See Figs. 2.19a and b.)

In another method the edger itself is used by mounting the holder on the quill and then touching the edge of the holder slightly against the edge of the rotating diamond wheel. Only very little need be removed—perhaps 0.1 in. (0.025 mm). The steel holder may then be rotated and the small lands and bevels removed with a fine mill file and scraper. Finally, the edges are cleaned with 600 carborundum paper and the sharp parts are rounded with Croceus cloth and a soft wooden stick. The spindle should rotate at its highest rate during this operation. Because the outer edge of the diamond wheel does the truing, no harm results from loading one edge; this edge can be removed by using a carborundum stick.

Several methods are available for centering optical elements. In the most common the small lens elements are heated on a plate on white paper toweling to protect the polished elements from scratching or overheating. If the paper becomes scorched, the temperature is too high. Next mount a brass holder on the screw of a hand holder and heat the holder in the open flame of a Bunsen burner. Soft pitch (pitch formula, p. 37) is then smeared generously around the rim of the holder and inside the cup. The lens element is placed on a towel and the holder is centered on the element by using a machinist ruler and measuring around the lens and holder. After the steel holder has been mounted, heat gently around the steel holder with the open flame of the Bunsen burner to allow the pitch to melt down to the surface of

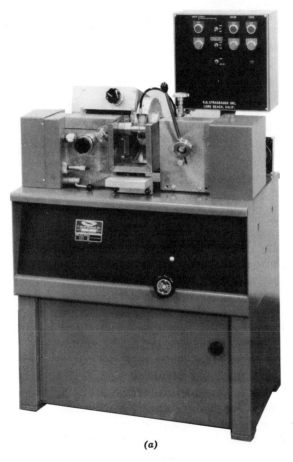

(a)

Fig. 2.21. The Strasbaugh Model 70 bell-chuck centering machine for high production includes these features: bell centering, automatic cycling, quick change, hollow transfer spindles [it also automatically] edges and bevels lenses up to 6 in. (150 mm), and built-in trueing cut of bell chucks without removing them from the machine. The LOH (Wetzler) bell chuck edger has many of the same features.

(b)

Fig. 2.21b (*Continued*)

the lens. Sometimes it is necessary to drop some pitch between the holder and lens.

The final centering is done when the assembly is mounted on the quill. To center the lens a small light source is placed about 9 ft (3 m) away. The operator looks directly into the lens and compares the outer with the inner image of the rotating lens. If both reflective images are superimposed and not wobbling, the lens is ready to be edged to its given diameter. Generally, the images are wobbling and the lens must be translated on the holder. Heat the brass holder very slightly while the lens is rotating at 5 rpm. Do not overheat or the lens may fall on the metal plate of the machine (place a rubber corrugated mat under the lens assembly). Place a small wooden dowel on the steady rest, gently move the stick into the lens, and watch the two reflective images move slowly together.

A more critical method of centering medium and large lenses requires an auxiliary device to hold the demountable quill. (See Fig. 2.20.) With this device the lens is held in a horizontal position and centered with a dial indicator reading in tenths of a thousand at the periphery. The lens is first mounted on the brass holder and then carefully screwed onto the demountable quill. It can be slightly heated and moved laterally until a zero reading is obtained on the dial indicator. The operator must always move the lens one-half the difference of the total reading of the dial indicator.

In another method often used to center lenses with this auxiliary device a light source is reflected from a microscope lamp with an adjustable diaphram which is projected through a hole drilled in the quill. A low-power telescope views the projected image that rotates in the field of the eyepiece. The lens is moved laterally until the image is stationary on the cross hairs. With this method lens elements as large as 8 in. (20.32 cm) can be centered, provided that the brass holder measures about 0.7 of the diameter of the lens.

Another method of centering involves a clamping bell-type center. Bell-type edgers and bevelers are made in Germany by Strasbaugh and LOH (Wetzlar). (See Figs. 2.21a and b.) These machines center lenses automatically by clamping their surfaces securely. Lenses with steeper curves are more precisely centered than those with flatter surfaces.

9. PLASTER AND PITCH BLOCKING

Prisms are often blocked in plaster to work the flat angular surfaces. Lens elements are pitch-blocked in a recess holder or circular assembly, depending on the radius of curvature.

We now consider the fabrication of a single large prism in a circular glass block. The glass ring that holds the prism must be sufficiently thick and core-drilled to a diameter slightly larger than the base of the prism. In core drilling the circular-edged glass blank should be placed on a section of plywood to which two thin strips of wood are nailed to forming a right angle. The glass disk can be centered with a machinist's vee square on a ruler marked with a diamond scribe. Next, a "wobbler" is placed in the drill chuck and the wobble is removed while the drill press is running. The glass disk is placed in the vee channel (90° wood strips) and the true-running wobbler locates the scribed center mark. The table with the glass disk must be clamped securely before the core drill with a running coolant is placed in the drill chuck.

The glass disk must be held tightly against the wooden strips. The core drill is turned on (150 rpm) and slowly advanced into the glass with a light, even pressure. The cut is made well past one-half the thickness before being turned over and drilled from the other side (make sure the glass disk is against the vee strips). The reason for drilling through on one setting is that as the drill approaches the bottom it will break through and chip the surface.

After core drilling the glass ring must be strongly beveled on both sides. Several channel slots are grooved into the inner circle with a three-cornered file and 120 carborundum. For even grinding of the whole assembly (prism plastered into the glass ring) a number of shallow cuts should be made across the bottom. The prism and ring assembly are illustrated in Fig. 2.22.

To make up the assembly first brush some light oil on a flat grinding tool and seat the prism firmly against the surface of the tool. Place the glass ring over the prism and wring it firmly onto the tool. Pour in the plaster mix. Before placing the prism on the flat tool, it must be shellacked or painted with soft pitch (the prisms must be slightly heated). The areas covered depend on the angular surfaces to be fine-ground and polished. Figure 2.22 shows one 45° face of a right prism being polished. The sides of the prism to be shellacked or coated are one-third of the hypotenuse surface and the entrance face. The parallel rays of the autocollimator are marked by arrows in the figure. All shellac and pitch must be removed from the surface of the prism being worked.

Next mix a batch of Hydrocal and pour it into the recess between the glass ring and prism. The mix consists of seven parts of plaster and four parts of water (by weight). Stir the plaster for several minutes and break up any lumps with the fingers. Pour the thick mixture around the prism and use a small tongue depressor to force it down to bring up bubbles of air. Hard water will cause a delay in the hardening of the plaster. Too much water will stiffen it and make it difficult to remove the prism from the glass ring.

To remove the plaster brush the mixture out with a brass-wire brush to a depth of about 0.020 in. (0.5 mm). Paint the ring and prism with shellac or melt some pitch into the recess. Clean all glass surfaces above the ring and the bottom surface of the assembly before testing. In the autocollimation check all glass angles contribute. (See Fig. 7.2.)

In multiple plaster blocking (three or perhaps as many as 38 prisms) a

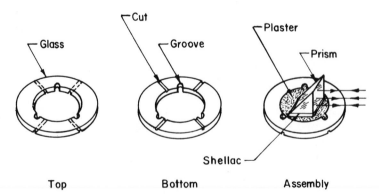

Fig. 2.22. Plastering a prism in a precut circular glass ring for correcting the second 45° glass reflecting angle complementary to the first 45° face.

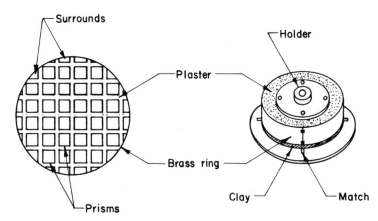

Fig. 2.23. Makeup of a plaster block of prisms used during fine-grinding and polishing. Note that no prism is placed in dead center.

similar technique is applied. Pregenerated prisms with slightly concave surfaces are cleaned and beveled with 145 grit at all corners and edges. Some optical shops shellac the warmed prisms or dip them in thin hot pitch and allow them to dry on a wire screen.

In plaster blocking a brass ring is required to enclose the prisms. It must be high enough to be at least 1 in. (2.54 cm) above their apex. A flat cast-iron tool larger than the plaster block to which the prisms are wrung, an aluminum plate equipped with a threaded boss at the bottom or top to serve as a driving and holding tool during fine grinding and polishing, and a number of plastic vessels for mixing the plaster are also needed.

In the procedure for plaster blocking the flat tool is screwed to the vertical post and the prism surfaces are cleaned with wood alcohol or xylene. A shop razor may be used to remove excess shellac or pitch. The cleaned prisms are wrung down firmly on the oiled surface of the flat plate. Fill any vacant area with glass surrounds. The brass ring is placed on top of the flat tool encircling the prisms and raised slightly above it with match-sticks. Long strips of modeling clay are placed on the periphery of the brass ring to prevent the liquid plaster from creeping out. Sawdust is pured around the prisms about ¼ in. (6.35 mm) down. The dental plaster or Hydrocal is thoroughly mixed in a large plastic vessel with a large paddle and all lumps are broken with the gloved fingers. Masses of plaster are now laid in the recesses and the semiliquid plaster is poured to a depth of at least 0.5 in. (1.27 cm) above the apex of the prisms. The aluminum driving plate is lowered onto the assembly. Do not allow it to sink down to the prisms. The use of heavy masking tape strapped across the top will prevent it. The finished assembly is shown in Fig. 2.23.

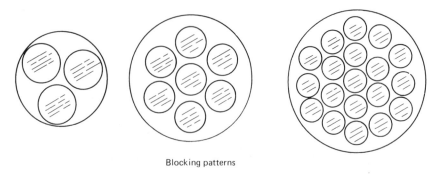

Blocking patterns

Fig. 2.24. Blocking configurations for mounting a multiple number of lens elements on a radius tool holder with pitch mallets.

Pitch blocking is used for fine grinding and polishing. Three or more elements may be assembled on one block as required. Figure 2.24 shows the usual blocking configurations. In this procedure pitch mallets, cast in special aluminum holders, perhaps 50 or more at a time, are inserted between the elements and the metallic holding tool. The mallets are released from the recesses of the mold by deep freezing the assemblies at 10°C and are placed on the surfaces of the elements which have been warmed to 100°C (warm enough to semimelt the pitch in the pitch mallets). The surfaces of the elements to be fabricated are wet-wrung to the grinding tool used to grind them the first time. A heated aluminum radius holder is lowered onto the pitch and glass elements. (See Fig. 2.25.)

The blocking tool which holds the assemblies is generally made of aluminum. Its diameter must be slightly larger than the encircled configuration of the blocked elements.

A small arbor press is used to press the pitch onto the elements while they are still warm. When the pitch has spread slightly and is more than one-

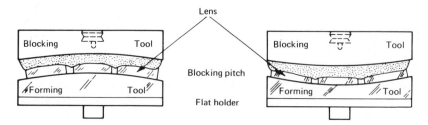

Fig. 2.25. Pressing the lens elements into the semiliquid pitch of a radius forming holder. An arbor press is often used. Pitch must be well above the middle half of the thickness of the elements.

third the thickness of the elements, the warm aluminum holder is cooled with sponged water. Figure 2.25 shows assemblies for convex and concave surfaces.

10. GRINDING AND POLISHING MACHINES

Fine grinding and polishing are done on machines whose moving parts are a bottom rotating spindle for holding the glass element or grinding tool, as the case may be, and a translating eccentric arm that carries a driving pin across the bottom spindle. A translating pin lever attached to the eccentric

Fig. 2.26. This four-spindle Strasbaugh polisher and grinder Model P6Y shows a setup for grinding.

arm can be set at any desired throw and rpm. The bottom spindle can also be adjusted to any desired rotation.

In principle when the glass elements are worked on top of the grinding tool the work becomes concave; conversely, when the grinding tool is worked on top of the glass elements, the work becomes convex. Polishing an optical surface can be considered as ultrafine grinding.

This effect can be offset to some extent by changing the rotation of the bottom spindle and slowing the rate of the transversing arm that carries the driving pin. Another technique changes the work (glass elements or components) from the bottom to the top. Figure 2.26 illustrates the rotating bottom spindle and the driving pin on the eccentric translating arm which can also be moved in and out. Note the two possible eccentric arm settings. Figure 2.27 shows the Strasbaugh Model R6Y-1 polisher with a random motion setting at the bottom of the translating arm (this offsets any repetitive stroke). Note also the air pressure holder which prevents the driving pin from moving upward away from the work during polishing. The more

Fig. 2.27. The Strasbaugh Model R6-Y machine with a random motion under the translating arm. Note the possible controls for adjusting the rotating spindle and translating arm and the eccentric adjustment of the length of the stroke, this model also has an air hold-down cylinder.

Fig. 2.28. A mass production Strasbaugh 10-spindle Model 6 AA machine. A slurry of polishing compound is pumped for each mounted element.

adjustment variables available on a machine, the more control the optician will have during optical figuring. The optician will often engage in hand polishing with only the slow, bottom rotating spindle during optical figuring of lens elements or components. Hand strokes are not repetitive.

The mass production of many similar elements requires a multiple-spindle operation. Figure 2.28 shows a 10 spindle machine for polishing concave or convex surfaces. The tubes direct the slurry polishing compound onto each spindle. A number of variable adjustments can be obtained individually to change the work from the bottom to the top spindle, and vice versa, or the driving pin can be moved forward off-center. Of course, any change in the rotation of the spindles or eccentric arm and its throw affect all polishing efforts will materially.

For very large work, such as astronomical mirrors, a large grinding and polishing machine is required. (See Fig. 2.29.) This machine is patterned on the design of a famous scientist and writer, Henry Draper (ca. 1860). Ritchey, another famous optician who built the 100-in. (2.5-m) mirror at Mt.

Fig. 2.29. The Strasbaugh (Draper-type) Model 6-AC with a tilting table for testing in a vertical position. Note the driving spindle that can be rotated counterclockwise to the rotating table that carries the large mirror. The table measures 60 in. (153 cm) and can be set to rotate at 2 to 6 rpm. Many extras are available for edging and beveling.

Wilson, added the tilting table for testing. A similar machine was used by H. H. Brown on the 200-in (5-m) mirror at Palomar near San Diego. This Strasbaugh machine can handle 60-in. (152.4-cm) work and many added features are available. It edges, bevels, core drills, and with cam follower and cam support brackets, generates curves. A quill traverse mechanism is used during the final figuring of the aspheric mirror because smaller polishers can be counter-rotated against the rotating mirror, thus keeping astigmatism to a minimum. This machine has a height of 7 ft 10 in., a depth 7 ft, and a length 10 ft 6 in. (the overarm protrudes a maximum of 3 ft).

LOW (Wetzler) and a number of other firms have developed high-speed diamond lapping (50-diamond) machine for mass production of lens elements. The fine diamond-lapped surfaces are polished on a similar machine with plastic-type polishers that can withstand high-speed and pressure action. This LOH (Welzlar) Model PLM 100 M lapping machine polishes fine (50-grit) diamond-generated surfaces 5 to 15 min, depend on the diameter of the work, with polyurethane plastic polishers. See H. F. Horne, *Optical Production Technology*, Crane, Russak & Co., Inc., New York, 1972.

PITCHES, CEMENTS, AND SHELLAC: FORMULAS AND SOURCES

Cementing pitch

1. By weight: three parts water white rosin, three parts bees wax, and two parts Montan wax.
2. By weight: six parts shellac, three parts pine tar.
3. By weight: three parts water white rosin, three parts bees wax.

Blocking pitch or shellac

1. By weight: 5 lb water white rosin, 5 oz pine tar. Add an equal volume of Agroshell (ground nut shells) or yellow ochre.
2. Hold-Tite Blocking Cement: Ernst Zobel Co., 521 Third Ave., Brooklyn, 11215, NY. Melts at 275°F.
3. Universal Shellac & Supply Co., 540 Irving Ave., Brooklyn, NY 11227. Three grades: No. 1 (year round), No. 8 (winter grade), and No. 16 (summer grade).

Polishing pitch

Gugolz brand: A Swiss pitch, Nos. 55, 64, 73, 82, and 91; No. 91 is extremely hard and No. 55, extremely soft; No. 72 is a medium soft pitch used for hand polishing and most machine polishing; No. 64 is used for polishing aspherics and No. 82, for steep radius work with heat generation. Source: Adolf Meller Co., 120 Corliss St., P. O. Box 6001, Providence, RI 02904.

3

Hand Polishing

Hand polishing is a skill that is acquired with time and experience. A few fundamentals in hand polishing are discussed in this chapter.

Pitch polishing optical surfaces is called lapping. The pitch polisher, or lap, is a mixture of compounds which has a slight flow at normal room temperature. Two types of pitch polisher are used for polishing optical surfaces. One is the facet type with smaller embossed facets impressed on the larger facets. The second is the solid type with scratched lines.

1. PITCH FORMULAS

Pitch formulas are often trade secrets. Each optician's favorite pitch resembles a witch's brew. There is no limit to what an optician will mix in his cauldron. The supervisor and the firm usually maintain a policy that requires each new optician to use standard pitch mixtures. Some typical polishing pitch mixtures are listed:

In Pounds

1. Optical D, 1 lb and 1 oz of paraffin.
2. Optical D, 1 lb and 2 oz of paraffin (for figuring).
3. Optical D, 1.5 lb, ¾ oz of ozhinite, and 5 oz of pine tar.
4. Optical D, 2 lb and 2 lb Zorphalac.

Pitch mixtures (2) and (3) were developed for polishing crown glasses, pyrex, window glass, and quartz. Pitch formula (4) is used for polishing rare earth glass and flints.

Pitch formulas divised by the late Don Hendrix followed a simple rule: mix similar compounds; for example, a pitch mixture of rosin from pine trees and castor or linseed oil from plants combines vegetable with vegetable. Optical D from oil refineries mixed with a soft asphalt compound is a mineral/mineral mixture.

Hendrix developed a number of pitch mixtures:

In Pounds

1. Water-white rosin, 5 lb, castor oil, 6 oz.
2. Optical D, 2 lb, linseed oil 6 oz.
3. Optical D, 2 lb, pine tar 4 oz, beeswax 1.5 lb.

Add 1 to 3 teaspoons of turpentine.

I use a pitch formula that consists of 5 lb of water-white rosin and 6 oz (liquid) pine tar. Pitch mixtures can be purchased ready for use. Zobal brands are soft and hard pitch. The mixture is transparent and the harder pitch can be softened by adding some of the softer. Semihard polishing pitch No 450 offered by Universal Shellac and Supply Co. (Brooklyn, NY) is widely used. This company also provides a number of burgundy pitches of various hardnesses and ranges of melting points. Wm. Dixon and Co. (Newark, NJ) is another supplier of good optical pitches. Some imported pitches are also very good and are quality controlled. Although this is not a complete list of the many formulas, it is apparent that pitch mixtures can be developed to specific requirements.

2. MEASURING THE HARDNESS OF PITCH

There are several methods of measuring the hardness of pitch. A penetration instrument which consists of a 6-mm inch ball bearing mounted on a sliding 0.5-kg brass weight is commonly used. A dial indicator in thousands of an inch is mounted on cylinder. The penetration of the ball bearing into the pitch is read after 5 min. (See Fig. 3.1.)

Another method is patterned after the one used by many earlier opticians who made their measurements by pressing down with the thumbnail at a pressure of approximately 2 to 3 lb, generally for 1 to 2 sec. This is a crude procedure, but a similar one can be devised by fastening a sharp wheel to a hinged bar with a sliding, 8-oz brass weight set at 10 cm from the fulcrum of the bar. This instrument measures the penetration of the cutter into the pitch after 60 sec. The width and depth of the cutter wheel's penetration into the pitch is measured with a machinist's ruler. (See Fig. 3.2.)

Although the gravity weight penetration method is more precise, it does suffer from the fact that the dial indicator, an expensive item, slowly becomes inoperative because of the pitch fumes that condense on its stem.

The hardness of pitch mixtures used with machine or hand polishers can be estimated. The room temperature should between 70 and 72°F. For

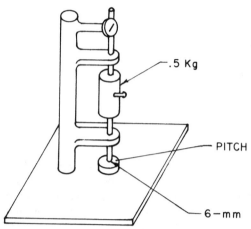

Fig. 3.1. A typical pitch penetration apparatus that controls the hardness of the polishing pitch.

polishing pyrex, quartz, crown glasses and Cer-Vit, a dial-reading instrument, the measured penetration should be 0.012 to 0.016 in. This is also an excellent penetration for hand polishing prisms and small flats. Pitch laps of 0.018 to 0.025 in hardness are used to figure aspheric surfaces. For flints the hardness is 0.025 to 0.030 in. and for rare earth, for example, EDF3, the hardness ranges from 0.030 to 0.040 in. These are only representative values for 40-hr milled Barnsite and polishing compounds.

All pitch laps become progressively harder as the essential tempering mixtures such as pine tar or linseed oil slowly evaporate and the lap becomes charged with the polishing compound. Because most pitch laps with precut facets have a "marquisette" pattern imprinted on them, the pitch lap can be treated by scraping the facets with a razor blade to renew the pattern.

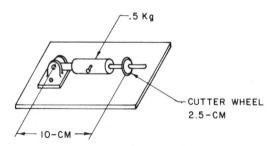

Fig. 3.2. Another design similar to the thumbnail for controlling the hardness of pitch.

3. FACET PITCH POLISHERS

The pitch lap used in hand polishing prisms is made as follows: 5 lb of water-white rosin is slowly melted and 6 liquid oz of pine tar is stirred thoroughly into the melted rosin. Sufficient pine tar to soften, or rosin to harden, is then added to the pitch to approximately 0.014-in. penetration.

Two 10-in. diameter polishers should be made at the same time. The aluminum holder should be at least 1 in. thick and the hub diameter, 1.5 in. or larger to prevent the lap from rocking on its mounting screw during polishing. One of the two holders is warmed under a faucet until it is hot to the touch. The holder is then screwed down on the post with very light pressure; if it is forced down too tightly while the tool is hot, the large lap will be hard to remove. Encircle the periphery of the holder with ¾-in. masking tape to a thickness of ½ in. above its surface. The semicooled pitch is poured slowing onto the warmed holder, starting at the center. The thickness of the pitch can be ¹⁄₁₆ to ³⁄₈ in. The lap is allowed to cool slowly to room temperature.

The second pitch lap is made in a similar manner, but the pitch mixture now contains ground nutshells; 1.5 cups of the shells are added slowly to a quart of the melted pitch.

Each polishing lap is laid out in ¾-in. squares with ¾-in. strips of brass. Each facet is separated by a channel ⅛ wide at a depth of approximately ³⁄₁₆ in. The pitch lap is cut with a single-edge razor blade. It is good practice to cover the lap with detergent water after the initial layout and then cut one side of all facets to a shallow depth. The channel cuts are made deeper with each pass. A right-handed person should always cut the channel on the right side (vice versa is left handed) by turning the lap 180° for alternate cuttings. Greater pressure can be applied to the razor blade after several cuts. The pitch chips are removed from the recessed channels with a 4-in. watchmaker's brush after flushing with detergent water.

The precut polishers must be made smooth with a 12-in. cast-iron tool with a flat surface. This tool is heated in running hot water until it is hot to the touch. The pitch lap is covered generously with thick polishing compound in a watery solution. The hot tool is placed on the pitch lap and several circular strokes are made across the lap. It is easy to determine whether the edges of the lap are compressed evenly all around the periphery in relation to the center. After this operation the facets must be recut and further embossed with smaller facets. For this procedure a "marquisette netting" or piece of plastic door screen is used. The netting is hard over the pitch lap and generously wetted with polishing compound. The heated cast-iron flat tool is placed on top. Careful note must be made of the impressions made in the lap. They should always be less than one-half the diameter of

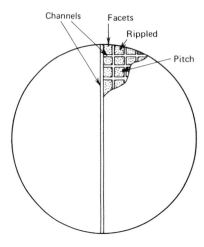

Fig. 3.3. A facet-cut pitch polisher with smaller embossed facets for polishing by hand or machine. The contour of the polisher is controlled by pressing plates.

one cord. If the impressions are too deep, it will be found when the netting is removed from the pitch lap that these embossed facets will break away from the main body of the lap. The pitch lap must be recut to remove pitch flow into the precut channels. Figure 3.3 illustrates a pitch lap.

The polishers should be allowed to cool for several hours. Most opticians like to charge their polishers on a machine by using a smooth flat tool with a surface quality of four waves or less. A 10-min run with a ⅓ diameter stroke at 30 rpm and an overarm rate of 20 rpm.

The flat surface contours on the polisher are controlled by Pyrex pressing plates. Three 10-in. Pyrex blanks are needed to compress the lap into the desired contour. If both sides of each Pyrex blank are used, a series of six pressing flats can be made. One blank may have a flat-surface configuration of two waves convex on one side and two waves concave on the other. The second Pyrex blank may be flat with a convex surface departure of three or four waves on one side and concave on the other. The third Pyrex blank can have flat surfaces of six to eight waves convex on one side and concave on the other. Each Pyrex blank should be inscribed with the wave departure to indicate whether it is concave or convex. To compress the pitch lap to a convex contour the concave pressing flat must be used; for a concave surface contour on the pitch lap the convex pressing blank is used.

4. SOLID POLISHERS

The pitch used in this type of polisher, called medium hard, is slightly soft. The hardness measured by the penetration apparatus ranges from 0.025 to

FOR CONCAVE SURFACE   FOR CONVEX SURFACE

Fig. 3.4. Solid-type polishers for polishing small flat surfaces by hand. The series of radial lines is made by razor-blade scratches of small depth. The contour of the polisher is controlled by pressing plates.

0.030 in. for a 5-min period. Some typical pitch mixtures are listed:

In Pounds

1. No. 835 Medium Burgundy Pitch, 2 lb.
   Optical Quality Rosin, 1 lb.
2. Zobal soft pitch, 2 lb.
   Zobal hard, 0.5 lb.
3. No. 850 Soft Polishing Burgundy Pitch, 2 lb.
   Optical Quality Rosin, 2 lb.

Many variations of pitch mixtures can be prepared. The Swiss soft pitch (Grezoly), Nos. 55 and 64 is one.

Solid polishing laps are generally small in diameter: typically, for 2-in. diameter work, a 4-in. diameter polisher; for 3-in. diameter work, a 5-in. diameter polisher; and for 4-in. diameter work, a 6-in. diameter polisher. The thickness of pitch on the toolholder is approximately ¼-in. These polishers are randomly grooved with a razor blade. The light grooves are ¼-in. wide across the polisher in two directions and ¹⁄₃₂-in. deep. Some opticians like to drill a ¼-in. hole in the middle of the polisher to allow the pitch to flow inwardly toward the center as the lap is polished. (See Fig. 3.4 for pattern contours scrapped in solid polishers.

5. HAND POLISHING USING FACET POLISHERS

For flat work blocked in plaster of paris during polishing a polisher without nutshells is used. For fine grinding and polishing the polishing lap with the added nutshells is used. In the latter case 5 to 6 lb of pressure must be applied across this lap as the finely ground surfaces are polished. The harder pitch lap does not lose its contour so readily as the softer pitch lap.

The technique of hand polishing is based on a few fundamentals that are easily acquired. The new polishers are charged with polishing compound by running a smooth cast-iron plate over them. This causes the lap to assume a convex surface contour. It is good practice to stir the polishing compound with a slightly moistened watchmaker's brush before applying the pressing plates to the two polishers. The length of time that the pitch lap must be compressed depends on the weight of the pressing plates. A pressing plate can be placed on a polisher overnight if several precautions are taken. First wet the polisher slightly with polishing compound. Second, place the pressing plate on the polisher; care must be taken to see that all areas of the pitch are covered. Third, encircle the pressing plate and pitch lap with 12.5-mm masking tape. This keeps the water from evaporating and the pitch from spreading beyond the periphery.

The concave pressing plate is removed from the polisher and its surface is now convex. Wet the top of the facets with water with a watchmakers' brush and stir the settled polishing compound. The prism is placed in the center of the polisher and grasped firmly with both hands. If the optic is finely ground and all grinding pits have been polished to a surface quality of one-eighth fringe, 5 to 6 lb of pressure is applied and a zigzag stroke across the polisher in both directions is used. Polishing should be done in a 45° direction to the precut channels in the pitch. The number of strokes averages 30 to 40 per minute. This seems slow but the time to polish the optic surface is proportional to the amount of presure applied. Little is gained by speeding up the number of strokes per minute because the optic surface then begins to skip across the pitch lap, thus acquiring a turned down edge. After a 5-min period of polishing the surface quality of the optic should be checked.

If the test flat shows that the optic surface is convex, it is not necessary to wait for the latent heat to leave the optic because an optic surface generally changes from a concave to a flatter surface. Continue to polish for three or four wets and hopefully the surface will progress toward the concave. Care must be taken when the optic is picked up that the same side is not always the same leading edge at the onset of a new wet. If the surface continues to turn toward the convex side after two or more wets, it has not been properly fine ground.

Perhaps the optic surface revealed that it was concave after the first wet. Immediately reduce the pressure to 2 to 3 lb while polishing and cut down the rate of the number of zigzag strokes across the polisher. If the optic surface does not become less concave after several wets, change the contour on the pitch polisher by using the two-wave convex pressing plate which will change the contour to concave surface. It will probably take 15 to 20 min to change the polisher but the process can be speeded up adding a 10-lb lead weight to the pressing plate with a piece of cardboard on top to prevent scratching. It is important to stress that all the optics that must be polished from ground surfaces must also be given a preliminary polish before the number of convex and concave surfaces can be figured.

If a number of surfaces can be polished, continue to use the polisher as it is. It must be kept in mind that the polisher changes slowly from a convex contour to one less convex as the polishing progresses. The opposite is true if the polisher is concave; it works slowly toward more concavity.

If the optical surfaces are free of pits, digs, sleeks, and scratches, release the pressure on the optic while polishing, perhaps to 2 lb or less if it is concave. It is good practice for beginners to use the polisher without the nutshells if some difficulty is experienced with sleeks. I like to finish optical surfaces with the softer lap pressed by the two-wave convex pressing plate, using the same zig-zag stroke at 20 strokes per minute, depending on the surface quality of the optic.

The use of flats to test the surface quality of polishing surfaces is described in Chapter 4. It is important to mention here that each polished surface must be carefully analyzed before work is done on it. Never treat a piece of optic under the assumption that it will improve with polishing.

Often in the polishing of rectangle surfaces of prisms and other optics, the surface will be astigmatic. One surface, perhaps in the long plane of the optic, is concave and the narrow plane is convex. This is called a "saddle back" curve and may be corrected by turning the prism's surface through a 45° angle and applying pressure at the highest corners. Another common surface fault is the "flying bird pattern" with belly up or down in the parlance of the optician. This means that if the center is low (concave), and intermediate area will be high (convex) and the edges turned down. If the center is high, the intermediate area will be low and the edges turned up.

To correct the surface on which the intermediate and center areas are high apply pressure at these areas and polish the optic slightly off the edge of the polisher. (See Figs. 3.5a and b.) This must be done carefully in short runs. When these humps become less pronounced on the prism, smooth the surface out with a slower rate on the polisher by exerting most of the pressure with both hands around the convex surface of the prism. The second or opposite surface pattern is corrected by applying the pressure alternately at the center and edges with an increased rate and polishing in shorter zigzag strokes.

As the surface improves on the prism or optic it is often found that small hills (called corns) develop near the corners. These corns are frequently stubborn and cannot be removed by pressure alone. Many experienced

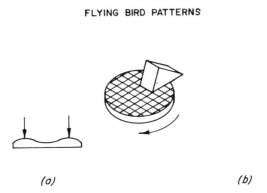

FLYING BIRD PATTERNS

(a)   *(b)*

Fig. 3.5. Typical slope departures polished in a prism's flat surface. The lower part shows a method of correcting humps.

opticians resort to finger localizing. The thumb or middle finger is used with much care not to over do it. After a 10-sec period at each corner the prism must be set aside for 5 min to allow the latent heat absorption to normalize. These localizations of the prism cause a number of irregularities on the surface and must be removed by polishing with a slow stroke and light pressure on the softer lap, which has a two-wave contour compressed on it.

Hints

1.   Because the many faceted polishers are initially file-like and can cause demarcations in the optical surfaces, an optical surface should never be finished until the small facets have developed appreciable lands.

2.   Glass surfaces that are worked in plaster blocks and must be retouched to a higher surface quality should be polished on the pitch lap free of the nutshell additive, although some opticians use only the pitch lap with the nutshells.

3.   When polishing, never start the wet at an edge area. Start at the center of the polisher or in a new area each time.

4.   Using a zigzag stroke, make at least two passes across the lap and finish at its center. Do not rotate the prism with pressure at the center of lap when changing to a newer position around the periphery of the polisher. This often causes the surface to remain concave.

5.   When polishing across the pitch lap at approximately 45° to the large facets of the lap, there are only four positions in which to begin the stroke.

6. The addition of fresh polishing compound should be broken down with a 3-in. glass bruiser which must be at least three-wave flat. This is used with a rotary motion across the lap. Stir up the renewed polishing compound in the recessed channels with a slightly wet watchmaker's brush.

7. Thin, watery polishing compounds on a pitch lap tend to keep the surface of the prism or optic closer to the pitch and may cause a concave surface to develop. Heavier polishing compounds have a tendency to polish the optic toward a convex surface.

8. The concavity of an optical surface can be reduced by polishing the optic or prism on a concave pitch polisher formed by a convex pressing plate. Likewise, the convexity of an optical surface on a prism or optic can be reduced by polishing it on a convex contour polisher which is formed by a concave pressing plate.

9. From time to time the channels and peripheral edges of the polisher should be retrimmed and the edges resharpened. A solid polisher's edges are retrimmed by rotating it at high speed and using a single-edged razor blade. Much care must be exercised to hold the blade stationary with both hands and not to feed too deeply into the pitch.

10. Opticians who use solid polishers should have a number of "bruisers." A set of three should suffice. They should be 1 to 3 in. in diameter and have small triangular ⅓-in. stub handles of glass epoxy.

11. When an optician has difficulty polishing a concave surface he will often place one of his bruisers in the middle of the lap and drill a recess area in the pitch lap. Lowering the pitch to flow inward into the recessed hole in the middle of the lap causes the lap to become concave in this area and less polishing action will take place. The diameter of the bruiser is determined by the nature of the optic to be polished.

12. Rippling of the facet polishers is renewed only when the smaller facets are merged.

13. The amount of land on the small facets rippled on the pitch lap can be increased by running the lap on the polishing machine and using a cast-iron flat tool. This can also be done by leaving the pressing plates on the polisher overnight.

6. HAND POLISHING WITH SOLID POLISHERS

Solid polishers are generally used to touch up or figure optical surfaces which have had preliminary polishing by blocking in plaster of paris or some other blocking method. These optical surfaces are astigmatic and the surface quality is at best only one or two waves.

The diameter of the optics worked is generally no larger than 4 in. Optics of larger diameters are better figured on facet-type polishers. This does not mean that these larger optical surfaces cannot be figured on solid polishers if certain precautions are taken. Larger surfaces are hard to work because tension holds them tightly to the lap.

The general procedure shapes an even contour across a solid polisher by using contour pressing plates. This is best accomplished by convex-type pressing plates of one to four wave contours which press a concave contour onto the solid polisher. Before placing the pressing plate on the polisher, it is necessary to rub a pinch of dry polishing compound gently over the full diameter of the polisher and then blow off the excess.

If the pressing plate is weighted down with a lead weight, it acquires its contour in a short time, perhaps within 5 min. When the pressing plate is removed examine the polished surface and determine its surface configuration. If it is concave, take a razor plate and dig out a small hole in the center and scratch some 10 or more radial lines from the hole to the edge. Then wave a small brusier about one-quarter the diameter of the solid polisher over the wetted polishing compound until a clear sheen appears. The optic is then placed in the center of the lap and grasped with one hand if the diameter of work is small or with both hands if large. Here procedures differ from one optician to another. Some opticians like to work around a stationary post. Others prefer a small spindle on a stationary post which can be rotated with one hand while the optic is worked with the other. This method is popular because the optician can be seated while working. An alternate method features a slow rotating spindle (e.g., 1 rpm) on a polishing machine which allows the optician to work with both hands on the optic. If the surface of the optic is concave, the optic is moved in a zigzag stroke at a rate of three strokes per second across one-half the diameter of the polisher and back again. The pressure on the optic varies. Occasionally, if the surface is two waves, it is good practice to lift the optic from the surface of the polisher. This keeps the adhesion and cohesion of both surfaces working against the optician during polishing.

Working convex surfaces requires different techniques. The pressing plate contour is generally only one-wave convex, which impresses a concave contour on the polisher. After pressing, remove the pressing plate and scratch a series of squares in both directions across a precut hole in the center of the lap. A small bruiser is worked across the wet polishing compound, but this tends to make the polisher convex because it is worked mainly around the peripheral. Caution: examine this glass bruiser and keep it reasonably flat. Do not allow it to become too astigmatic or convex; it should be ground on a two-wave convex tool.

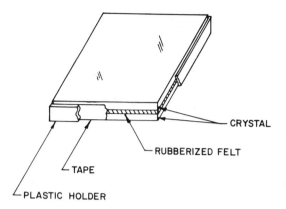

CRYSTAL

RUBBERIZED FELT

TAPE

PLASTIC HOLDER

Fig. 3.6. A sandwich holder for polishing thin optical flat materials to minimize the latent heat from hand polishing.

When the polishing compound is broken down and the polisher acquires a soft sheen, the optic is worked across the polisher with perhaps 2 lb of pressure and five zigzag strokes per second across the aperture.

Some opticians like to remove all the loose, broken down polishing compound with a sponge and then sprinkle several drops of water on the polisher before polishing. This tends to keep the edges from turning down. If the edges are turned up, the loose compound is allowed to remain on the polisher.

As pointed out earlier an astigmatic curve (in the optician's words, astigic) is often polished into the optical surfaces. This is called saddle back because in one plane it is convex and in the other, concave. The direction of the highest edges should be found and the greatest pressure placed at those areas that require a slow zigzag stroke. Four corners often have corns near the edges that are difficult to remove. Treat them with care by localizing them with thumb or middle finger.

It is essential that the optican make a careful survey of optical surface faults before reworking the polisher. It will be found that two or more surface faults are present when the optical surface is one-quarter wave or better. If the surface is two-to-three waves convex, it will soon become obvious that it does not pay to wait for the latent heat absorbed by the optic to dissipate because an optical surface becomes flatter from a concave than a convex surface.

The special holders that glasses or crystals need because they are subject to larger surface contour changes due to polishing heat have not been discussed. Special holders are essential to prevent the optic from absorbing latent heat. Rubberized felt must be inserted between the plastic holder and optical material in thin sectional areas. (See Fig. 3.6.) Prisms and odd-sectional pieces of crystal and glass are often handled with gloved hands to keep to a minimum the heat absorbed from the hands.

These hand-polishing methods have been practiced for many years. Learning how to figure a flat surface to one-tenth wave or better is like learning how to paint. Some say the art cannot be taught. The late B. Schmidt, inventor of the Schmidt camera, was polishing a lens while some of his friends were visiting him. When he left the room for a short time someone moved his polisher several strokes. Schmidt asked who had moved his polisher. This is not unusual because each worker possesses a polishing technique that, like his fingerprints, cannot be duplicated by another.

4

Testing Flats and Radius Test Plates

Perhaps interference color bands were observed first by ancient man in thin films of oil floating on water. Opticians in the United States generally call the interferometric bands "fringes," whereas in England they are "Newton rings" or "rings."

Fringes are produced when the two optical surfaces of a radius test plate are placed on a lens surface. The interference fringe is spread across the lens surface like oil on water. In this chapter we discuss flat surface testing using interference bands.*

These tests are difficult because of the wide variety of zonal faults that can be hand polished into a glass surface. It is important that the optician analyze the work carefully and know what is present thereon. In general two faults are evident in an optical surface; look for the poorest zone and correct the highest zone first.

1. EQUIPMENT

Many types of reflecting viewer are used in optical shops. The most common is the direct viewer which is a reflecting plate glass in a painted black box. The monochromatic light source consists of uncoated fluorescent tubes on a sheet of vellum or mylar (tracing paper). (See Fig. 4.1.) The work can be tested at a 10 or more diameter distance. For large work 7 to 10 in. in diameter the correct viewing distance should be 15 or more diameters from the fringes.

Many optical shops now have plano-interferometers for testing optical flat work. These instruments pay for themselves quickly. The optical polished surfaces are unmarred by digs, scratches, or sleeks because no optical flat is placed in direct contact with them. Also, there is no contact point to be located; simply press the eyepiece to produce a shorter light between the two plane flats. Because fringe bands are like water, which

---

* A "Band Pattern Guide" is available for $2.00 from Van Keuren Co., 176 Waltham Street, Watertown, Mass. 01272.

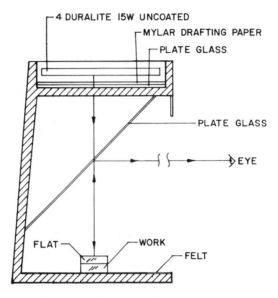

Fig. 4.1.   Direct viewer for an optical shop.

always run downhill, the runoff point of a convex surface and the collection point of a concave surface can be easily located.

Let us examine some of the aspects of interferometric methods of testing optical flat surfaces. Shop flats may be controlled by checking periodically against a master optical flat. There must be a number of flats of various sizes and weights for large and small optical flat work. It is not good practice to use a large flat on small areas of prisms or other work.

2. TESTING PATTERNS

Several fundamental rules must be followed to determine whether a flat surface is convex or concave. When a checking flat is placed on another flat, contact points and air wedges are created: for example, a flat resting on the middle point of a convex surface forms one contact point; a flat resting on a cylinder or saddle-back glass surface forms two contact points. The thickness of the air wedge separating the surfaces determines their slope. (See Fig. 4.2.) By applying pressure on the test plate with a finger the air wedge can be changed to another position but the contact points remain the same.

This leads to an important point in determining whether a surface is concave or convex. If the radius of curvature of the lines moves over to the point at which pressure is applied to the glass surface, it is high or convex.

Likewise the glass surface is concave or low if the radius of curvature of the lines moves away from the point of pressure. Work may be tested at any pressure point; when the pressure is applied on the test flat, the fringes will return or right themselves to their original points of contact.

A number of fringe patterns present when slight convex, concave, or other surface irregularities appear. (See Figs. 4.3A-I.) The apparent zone or zones of the glass surface are shown in Figs. 4.4A–F. It is apparent that the zonal areas have any diameter and these illustrations which contain some odd-ball fringe patterns are merely representative.

The placement of test plates on an optical surface requires some care. The surface is wiped with a soft, smooth towel and then cleaned with a

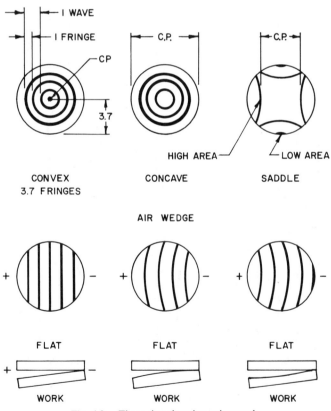

Fig. 4.2. Three situations in testing work.

static or camel brush to removed any large particles of dust or towel material. The test plate is laid on the optical surface; beginners should support both arms while grasping the test plate before lowering it to the optical surface to prevent what is called "playing castinets."

Other types of fringe pattern are illustrated in Figs. 4.4A–C. One is the spheroid, which is an astigmatic surface, both concave and convex. This surface occurs most often in correcting angles of prisms when pressure is applied in one area while polishing on a convex polisher.

Another type of astigmatic surface is the saddle back which occurs in thin optical flat or radius work. This refers to two and four points of contact or high areas. In the case of two points one area is concave, the other is convex

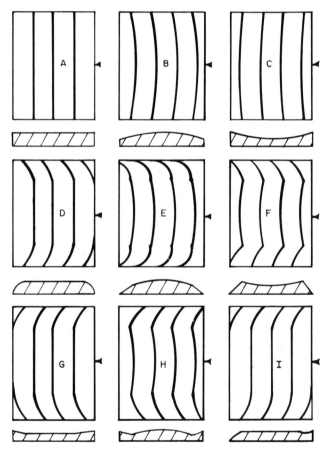

Fig. 4.3. Fringe patterns of surface irregularities.

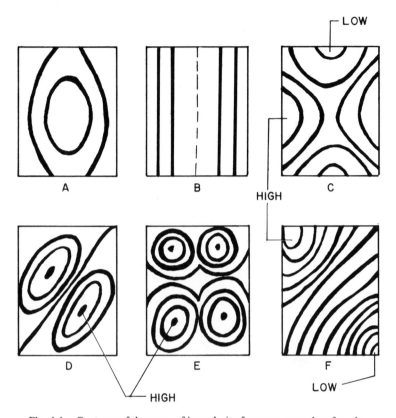

Fig. 4.4. Contours of the zones of irregularity for some examples of work.

(this occurs in oblong work). If the two contact or high areas occur at the edges, the patterns in Fig. 4.4C and F are observed. Two and four contacts located inside the corners of the work are shown in Fig. 4.4E).

3. USING THE PLANOINTERFEROMETER

The planointerferometer is an optical instrument with which the flatness of materials such as polishing crystal, glass, or lapped metal surfaces is measured. It can also be used to check the parallelism of planoparallels. There is no physical contact between the master reference flat and the surface of the work; this eliminates the use and replacement of costly optical flats. The separation also keeps digs and scratches from appearing on the work. The light source in the planointerferometer is a reed-type mercury

Fig. 4.5. The Davidson Model D-305 interferometer.

lamp filtered for 5460 A. The Davidson Model D-305 interferometer is shown in Fig. 4.5. Interferometers range in size from 2 to 12 in. in diameter. Figure 4.6 illustrates a simple and cheaply made planointerferometer. The planoconvex lens is made of preselected Cer-Vit or quartz which is reasonably free of striae. Small striae can be ignored if they fall into view under observation. The lens should be about f/6.0, and the light path is folded to a compact system with one or two folding optical flats.

4. A MOIRE FRINGER FOR TEACHING SURFACE TESTING

Beginning optical technicians have some difficulty learning how to interpret surface quality with interferometric fringes. It is well known that Newton's rings or interference fringes are formed by a monochromatic light source with two flat or two close-fitting curve surfaces. An optical setup can be built with nonoptical elements and a simple Moire fringe.

The simple arrangement shown in Fig. 4.7 consists of a Ronchi grating of low frequency placed on various slope areas of an open book. The Ronchi grating rests on a blank page free of printed matter which forms a biquadratic slope with convex and concave departures from a plane. An obvious convex slope area exists near the bound edges of the pages and a concave slope area near the free edges (positions 1 and 2, respectively, in Fig. 4.7).

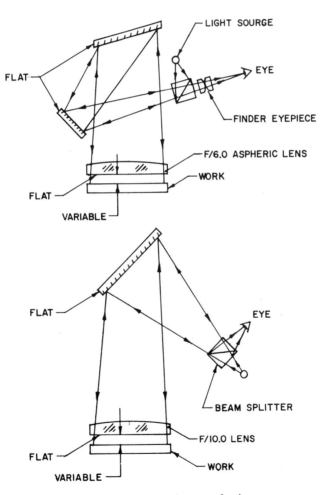

Fig. 4.6. A home-made interferometer for shop use.

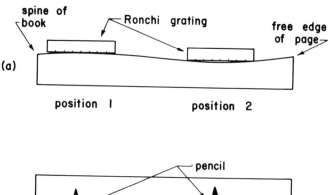

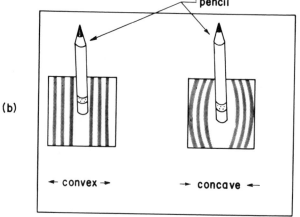

Fig. 4.7.   The Moire fringe principle using a Ronchi ruling.

The use of Moire fringes with noncontacting optics provides an experiment in the physics laboratory for demonstrating to the student some aspects of the interference of light.

The Moire fringes observed in this setup will necessarily originate from cylinder slope departures with straight-line Ronchi gratings. Circular Moire fringes similar to Newtonian rings are formed by circular gratings with some astigmatic contours.

Consider position 1 in which the Ronchi grating is placed on the obvious convex surface near the spine of the book. Use a finger on the eraser of a pencil to push against the middle of the grating and observe, as pressure on the grating is increased, the fringes running toward the edges of the grating like water off a hill. Consider position 2 in which the Ronchi grating is placed at the obvious concave slope near the free edges. We can observe the concave fringes running toward the center as pressure is applied to the center of the Ronchi grating.

Another method can be used to determine whether the observed fringes from flat or curved test plates develop on a concave surface. Here the observer bobs his head up or down. For a concave surface contour the fringes move inward, that is, toward the center of the grating, if the head is bobbed down, and outward if the head is bobbed up. The converse is true for a convex surface. Here the viewpoint is actually being changed by decreasing or increasing the length of the light path. Under white light it is difficult to observe the movement of Moire fringes, although this movement is easily .seen under monochromatic light which produces thinner fringes. Perhaps this is the first experimental correlation of Moire fringes and Newtonian rings or fringes observed under a monochromatic light and the changing point of view of the observer.

EXERCISES

1. Make a Pyrex flat with a 2-in. or 50.4-mm diameter and $\frac{1}{6}$ in. thickness. One side polished is to one fringe or better, the second to $\frac{1}{10}$ wave or better. The irregularity should be less than $\frac{1}{20}$ wave.

2. Make a rectangular flat with a length twice its width and a thickness of $\frac{1}{10}$ the width. One surface is one fringe or better and the second, $\frac{1}{10}$ fringe or better and $\frac{1}{20}$ or less irregularity.

3. Name four advantages of a planointerferometer over a master test plate.

4. What happens to observed fringes on large optical flats when viewed at less than five diameters?

BIBLIOGRAPHY

Biddle, B. J. "A Non-Contacting Interferometer for Testing Steeply Curved Surfaces," *Opt. Acta.*, **16,** 139–157 (1969).

Cooke, F. "Aid to Viewing Test Plate Interference Fringes," *Appl. Opt.*, **10,** 2216–17 (1971). *See also* Bibliography in Chapter 15.

De Vany, A. S. "Compensator Curves of Test Plates," *J. Opt. Soc. Am.* **49,** 1131 (1959).

De Vany, A. S. "Testing the Sphericity of Short Radii Convex Surfaces," *Appl. Opt.*, **13,** 229–230 (1974).

Deve, C. *Optical Workshop Principles*, (translated by T. L. Tippell), Hilger and Watts, London, 1954.

Johnson, B. K. *Optics and Optical Instruments*, Dover, New York, 1960, pp. 183–187.

McQuire, D. "Test Plates", *Scientific American*, January and February 1945.

Short, C. "The Use of Fringes in the Determination of Spherical Test Plates," *Nortronics Division Memo*, Northrop Corp. (1955).

Smith, W. J. *Modern Optical Engineering*, McGraw-Hill, New York, 1966, pp. 411–418.

5

Testing the Homogeneity of Glass

Testing the homogeneity of glass for optical elements is essential because large nonhomogeneous areas cannot be removed by optical figuring. It is necessary for the optician to "know his glass." Glass that meets the Mil 174-A specification for homogeneity is often not good enough. Therefore a simple method should be used for testing purposes. In this chapter tests of homogeneity are divided into three areas: the Williams interferometer, a laser interferometer, and a Babinet compensator interferometer.

To test glass, planoparallels with some degree of accuracy must be available. Accuracry is called for because the transmission fringes must be observed with the reflection fringes, which means that elements must be within 20 sec or better. Three types of interferometer are discussed. The Williams interferometer makes use of two (nearly equal radii) spherical mirrors whose radius of curvature must meet Murty's standards for various thicknesses. In general spherical mirrors of R/D 10 will suffice; R is the radius of curvature and D, the diameter. A Murty laser interferometer makes use of a f/12 spherical or f/8 parabolical mirror. Another interferometer, the Babinet compensator, has as its primary use the measurement of strain in glass.

1. THE WILLIAMS INTERFEROMETER

Since the advent of bubble chambers wind tunnels, and large orbiting telescopes there has been a need for large pieces of thick glass with a high degree of homogeneity. Some of these glass pieces have been 152 cm in diameter and 16 cm thick. The Williams interferometer offers a low-cost method of checking the homogeneity of glass. This instrument can also be used to check the parallelism of planoparallels and provides an alternate method of making divider plates. The Williams interferometer is a simple optical system that can be built with elements found in most optical shops or laboratories. The optical setup is shown in Fig. 5.1. The two mirrors are $1/32$ wave spheroids and have equal radii of curvature and diameter. One of the two mirrors, which are set at 90° to each other, must be fastened to a

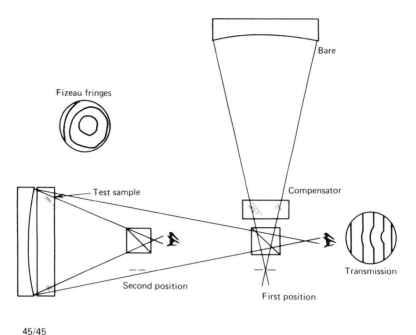

Fig. 5.1. The Williams interferometer setup with two spheroids of equiradius and meeting Murty's R/D number for testing the homogeneity of optical materials. Note that the Fizeau reflecting fringes can be observed in the second position.

sliding stand so that it can be adjusted to the stationary mirror. For a typical optical shop setup the spherical mirrors could have diameters of 30.7 cm and radii of 310 cm. The radii of the mirrors can be unequal by 50 mm. This setup would be suitable for glass thicknesses ranging from 5 to 200 mm. For glass thicker than 200 mm an R/D of 14 is required.*

A small beam splitter is positioned on three axial holders and placed slightly inside the radius of each mirror. Because this is an equal-path interferometer, a compensator of near equal thickness of glass must be placed close to the beam splitter. Now because the light path cone is small in diameter, it cannot materially effect the results. A pinhole of 0.2 mm diameter with a mercury reed source completes the setup.

One of the mirrors must be aluminated with a partial 45/45 coating; the second mirror can be uncoated. Ajustment of the interferometer is critical. Before the test sample (a planoparallel element) and compensator are

---

* This is described by M. K. Murty, *Appl. Opt.* **2,** 1337, (1963).

placed the interferometer must be zeroed. The two mirrors are adjusted until the reflective images are observed in the beam splitter and then superimposed on each other. First small circular fringes, which will become enlarged as the sliding mirror is adjusted, are observed. When there are straight and equidistant parallel fringes of approximate 1 in. (2.5 cm), the test sample and the compensator are placed in their respective positions.

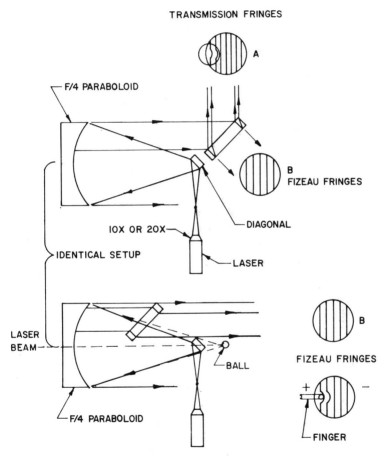

Fig. 5.2. The upper part of the figure shows Murty's lateral-shearing interferometer in a Newtonian setup. Here the transmission and Fizeau fringes can be observed on a frosted screen. The lower part of the figure shows the Fizeau fringes observed from the test sample placement in the parallel rays. Two setups appear: the Newtonian system and the illuminated ball source with a laser source passing through the mirror.

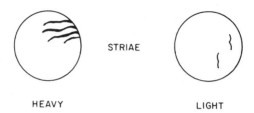

STRIAE

HEAVY   LIGHT

Fig. 5.3.   Two types of stria observed in samples during the interferometric testing of the glass disks.

Some further adjustment is usually required because of unequal glass thicknesses or wedge angles in the test sample or compensator.

Like all long-path interferometers, moving air currents affect the Williams interferometer. They can be blocked by cardboard concrete-type tubing used in construction work or the instrument can be shortened proportionally by the use of smaller R/D numbers for the mirrors; for example, if the glass thickness were 25 mm or less, the R/D number would be 7.

The semipolished planoparallels of the sample glass being tested for homogeneity must be a 20-sec wedge or less with a surface quality of one wave over an area 150 mm in diameter. All glass samples should be placed close to the coated mirror because the Fizeau reflection fringes of the test samples are also checked. This subject is discussed more fully in the next section. The glass compensator is placed close to the beam splitter and should be of comparable thickness to the glass samples. The sliding mirror holder, as well as the beam splitter, may need further adjustment, as well to permit superimposition of the two returning images from the two mirrors. When circular fringes are observed the sliding mirror is adjusted along the axial plane until the fringes increase in size—the proper direction of adjustment. Move the sliding mirror until the fringes are approximately 1 in. (2.5 cm) apart. If the test samples are free of inhomogeneities, the fringes will be

OFF CENTER FRINGE PATTERN

Fig. 5.4.   The concentric pattern of a typical crystal is off-axis because of the growth pattern. This pattern can be observed in the knife-edge or interferometric testing setups. The sample can be used if it is edged concentric with the pattern; if used without edging, astigmatic images are the result.

equidistant and parallel. These transmission fringes will probably wiggle in the air flows that cross the intermediate light paths.

Now make a quick, qualitative examination of the homogeneity of the glass. Watch for boxed in areas where the 1 in. (2.5 cm) width lines are parallel to one other (one pattern is shown in Fig. 5.2). Mark these areas with a grease pencil. Recheck again by causing the fringes to run in a horizontal plane, thus boxing the nonhomogeneous area. Striae will appear as water running out of an open faucet. A number of these glass faults are illustrated in Figs. 5.3 and 5.4. Although some faulty glass would not do for prism making and would have to be discarded, it may serve well in other optical elements.

2. THE MURTY INTERFEROMETER

The Murty is one of the least expensive interferometers for large systems. Figure 5.2 illustrates a 30-cm paraboloid of f/4.0 ratio in a Newtonian collimator. A low-power visible laser source (about 0.5 mW) and a 10 or $20\times$ microscope objective complete the collimator. A 0.25 polished ball bearing can be substituted for the microscope objectives. As shown by the dashed lines in Fig. 5.2 (see also Chapter 20), when a parallel plate with or without a slight wedge is placed in the parallel or collimated beam, it will project a lateral shear interferogram of the correction of the spherical aberration of the $f/4.0$ paraboloid. This is a versatile interferometer because planoparallels, finished lenses, and so on, can be tested interferometrically by a slight modification and rearrangement of components in the collimated beam. Because of the interest in testing the homogeneity of optical glasses and crystals, this interferometer is widely used.

2.1. Striae

A stria is a vein or streak with an index of refraction different from the surrounding glass area. Heavy striae will appear in the lateral shear interferograms which are projected on a screen or wall for safe viewing. See Fig. 5.3 for a number of typical patterns. If the test glass samples are not polished, two finished planoparallels are used. These flats are attached to finely Blanchard-ground surfaces of the blank by a matching oil. Matching oil of a large range of indices can be purchased or made to meet the requirements. The planoparallels must be Schlieren quality with a parallelism of 10 sec or better and are polished to one-quarter fringe or better or both sides. Heavy striae cause serious flares in image-forming systems, and many light and veil striae may produce scattering of light and a general blurring of the image. These heavy striae are cut away for prism making.

The edge-on checks of test samples are carried out in a similar manner. Large windows can be edge tested to determine whether the sample veinlike streaks are running horizontal to the diameter of the glass blank. If vertical loops or jagged streaks are present, the glass or crystal should be rejected.

2.2. Bubbles

Unless bubbles are large or numerous enough to scatter a considerable portion of light in an optical system, they are generally ignored. Occasionally, however, large bubbles or masses of smaller ones require the rejection of portions of the glass.

2.3. Stones

Stones are present in glass as small, undissolved particles. They may be part of the original mixture but are usually small pieces of the pot or stirring rod which have broken off during melting. Besides being objectionalble, they sometimes leave trails of comet like striae due to their partial dissolution in the glass. Bubbles and stones can be observed by rubbing matching oil on both sides of the Blanchard-ground test samples and then locating them with a microscope illuminator lamp placed under the glass.

2.4. Strain

Strain is not easily detected by the Murty interferometer. It is better detected by a cross polaroid sheet placed about the test samples on a lighted frosted glass sheet. An even and gradual cut off of the light as the polarized sheet is rotated reveals whether the glass or crystal is free of strain. Badly strained glass or crystal forms a Maltese cross.

Strain is often set up when lens elements are joined with resin cements. I recall that in a certain missle program a small optical guidance system was difficult to make. The secondary system of the Cassegrain telescope was a cemented doublet with a high amplification ratio around 8. After a period of rechecking test plates it was found that the resin cementing the two elements was distorting the spherical surfaces. The strain pattern was a Maltese cross and the surfaces were astigmatic by four fringes. Before cementing they had been one-eighth fringe!

In Fig. 5.2 note the formation of two sets of fringe patterns; the lateral shear transmission fringes projected on the wall and the Fizeau fringes on the back of the test samples. Photograph or draw in full scale the contour of the fringes and the location of stones, bubbles, and striae. Be sure to count the number of Fizeau fringes, (one fringe per 1 in. equals 1.5 sec).

2.5. Other Nonhomogeneities

Sometimes samples will exhibit small irregularities in the fringe patterns which do not move when the test samples are translated in the collimated laser beam. These irregular patterns are not from the uncorrected zones of spherical aberration in the paraboloid. They come from small nonhomogeneous areas in the test samples, which always remain fixed in their respective areas when the optical materials are translated.

Most optical crystals are grown around a seed and the crystal materials collect in circular wavelets like ripples on a pond when a small rock is thrown into the water. Some crystal optical materials, such as quartz or calcium fluoride, float off center from the seed and thus cause the nonhomogeneous circular wavelets to be off center. Figure 5.4 is a drawing based on a photograph of an optical quartz window on which the knife-edge test and an R/D 15 spherical mirror were used. If this ring system is in dead center and shows few rings (each ring in quartz is a plus or minus 1×10^{-6} refractive index variation), there will be little if any imagery deterioration in the lenses or corrector plates. Some spherical surfaces on large aperature lenses may require localizing.

A final note of interest on the Murty interferometer is that it can test a window 1.4 times the diameter of the interferometer because of the foreshortening effect of its 45° placement in the collimating beam.

3. THE BABINET COMPENSATOR INTERFEROMETER

A Babinet compensator instrument may be used to measure strain by polarized light. It may be converted into a polarization interferometer of the Lenouvel type and has the additional capability of determining the axial chromatic abberation (Δ chr) for each wavelength by interposing narrow band filters and changing the light source. (The Δ change is the repositioning of the focus from a preselected wavelength.) This instrument is made to fit easily over any Babinet strain measuring device.

The assembled unit is illustrated in Fig. 5.5. It consists of two glass polaroids and a small right-angle prism with a pinhole in a reed mercury light source. One 12-mm square polaroid has as 5-mm slot to hold the 5-mm right-angle prism. A second circular glass polaroid is cemented over the entrance face of the prism after the unit is assembled. A small mercury reed source with a 0.1-mm pinhole must be mounted 2 mm from the cemented prism to minimize the heat from the reed lamp.

The sliding crystal wedges in the Babinet compensator are similar to a Wollaston prism in the Lenovel polarizing interferometer. With the Babinet

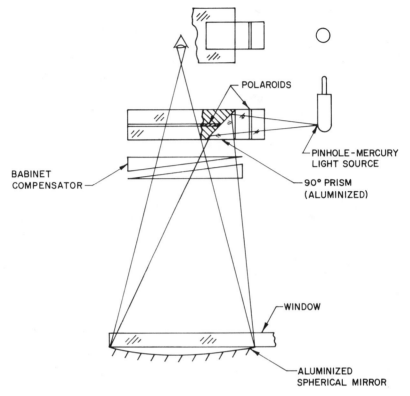

Fig. 5.5. The makeup of the Babinet compensator interferometer which forms lateral-sheared interferograms. This setup is used for testing the homogeneity of the glass disk.

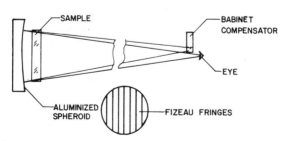

Fig. 5.6. Observation of Fizeau fringes at one-half the radius of curvature of the reflecting spheroid.

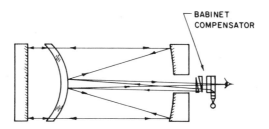

Fig. 5.7. Testing a Maksutov lens system or another lens system in an autocollimating setup. This produces laterally sheared interferograms of spherochromatism.

compensator it is possible to measure qualitatively or quantitatively the homogeneity of glass or crystalline materials by using the Twyman-Perry method. This measurement relies on the double optical path difference observed in the fringe patterns. With Murty's method to measure the variations of the order (10^{-7}) in the refractive index, $\frac{1}{20}$ of a wave $(5.5 \times 10^{-4}$ mm) should be detectable in a spherical mirror. The spherical mirror must be $\frac{1}{32}$ wave or better with an R/D of 10 for a glass thickness of 62 mm or thinner and 14 for a glass thickness of 200 mm or thicker. The 14 R/D spherical mirror may be used for the thinner test samples. If the test sample is small in diameter, its obstruction ratio will fulfill the R/D requirement.

It is possible to test the homogeneity of Blanchard-generated samples by attaching Schlieren-quality planoparallels to the test samples with a matching oil. If these high quality plates are not available, plate-glass covers attached to the test samples with a matching oil can be substituted. (It is good practice to test the plate-glass covers in the interferometer drawing patterns.)

Fizeau fringes can be viewed with this instrument by moving the test unit at one-half the radius of the spheroid. (See Fig. 5.6.) Note that the fringes are observed off-axis.

It is of interest that the Babinet interferometer gives lateral-sheared interferograms and can be used to test autocollimated lens systems. As already pointed out, this instrument can determine the axial chromatic aberration. Figure 5.7. shows a Maksutov telescope system being tested while the corrector lens or primary mirror is being optically figured.

EXERCISES

1. Name three types of interferometer discussed in this chapter. Which one is most easily setup?

2. Which fringes observed in an interferometer are caused by striae in the optical transmission material?
3. In a divider plate for an interferometer what fringes must be straightened by optical figuring?
4. How can Blanchard-generated glass blanks be tested for homogeneity?
5. How many seconds of wedge angle are there in a 4-in. (115-mm) diameter of BK 7 glass blank showing four Fizeau fringes?

BIBLIOGRAPHY

De Vany, A. S. "A Note Testing Homogeneity of Glass," *Appl. Opt.*, **3**, 645–646 (1964).

De Vany, A. S. "Supplement to A Note on Testing the Homogeneity of Glass," *Appl. Opt.*, **3**, 901–902 (1964).

De Vany, A. S. "An Instrument for Measuring Strain and Homogeneity in Glass," *Appl. Opt.*, **4**, 513–514 (1965).

De Vany, A. S. "Supplement to An Instrument for Measuring Strain and Homogeneity in Glass," *Appl. Opt.*, **5**, 756–757 (1966).

De Vany, A. S. "Using a Murty Interferometer for Testing the Homogeneity of Test Samples of Optical Materials," *Appl. Opt.*, **10**, 1459–1461 (1971).

Deve, C. *Optical Workshop Principles*, translated by T. L. Tippell, Hilger and Watts, London, 1954.

Johnson, B. K. *Optics and Optical Instruments*, Dover, New York, 1960.

Murty, M. V. R. K. "A Note on Testing Homogeneity of Large Aperture Parallel Plates of Glass," *Appl. Opt.*, **2**, 13337–1340 (1963).

Murty, M. V. R. K. "Homogeneity Measurements with, Fizeau Interferometer," *Optical Shop Testing*, Wiley, New York, 1978, pp. 29–30.

Francon, M. "Polarization Interferometers Applied to the Study of Isotropic Objects," National Physical Laboratory, *Symposium No. 11, Interferometry*, Her Majesty's Stationary Office, London, 1959, pp 129–136.

Post, D. Characteristics of the Series Interferometer, *J. Opt. Soc. Am.*, **44**, 243–244 (1963).

Saunders, J. B. "In-Line Interferometer," *J. Opt. Soc. Am.*, **44**, 241–242 (1963).

Twyman, F. *Prism and Lens Making*, 2nd ed. Hilger and Watts, London, 1952, pp. 411–413.

Part II

Fabrication of Prisms and Lenses

6

Autocollimation of Prisms

The autocollimator is similar to the telescope. There are two types: the off-axis and on-axis. The off-axis is used more frequently because it offers stronger illumination than the on-axis. The off-axis autocollimator projects a lighted, etched, prismatic reticle onto the objective lens. A planoflat projects the light rays onto an eyepiece for observation. This instrument will often have a bifilar with high-power eyepieces.

A refracting off-axis autocollimator is shown in Fig. 6.1. A bright light source with a small condenser system is projected onto a prismatic reticle. One half of the prismatic reticle is aluminized; the other half is not. The reticle assembly is slightly off-axis. The on-axis autocollimator is shown in Fig. 6.2. Here the illuminated reticle is projected on-axis by a thin membrane pellicle set at a 45° angle. The light source for the on-axis auto-collimator must necessarily be brighter than the light source for an off-axis instrument.

One of the most commonly used autocollimators is the Angle-Dekkor made by Hilger & Watts of England. (See Fig. 6.3.) The object lens measures 1.2 in. or 3.05 cm and has a focal length of 17 in. or 43.2 cm. The TA-121 Model has a micrometer drum reading to 0.5 sec. but is not really practical because the autocollimated image from the prism is too difficult to observe. The TA-130 instrument is equipped with a bright field to overcome this difficulty. A collimator with dark lines on a bright field, however, is impractical for internal autocollimation of prisms.

1. MAGNIFICATION OF IMAGES BY THE AUTOCOLLIMATOR

Because the autocollimator magnifies images in the same manner as a telescope, it is important to understand magnification and the principles involved. This section discusses magnification briefly. Magnification, or the increase in size of the image of an object by a lens, occurs with a change in the angle at which the light rays from the object enter the eye. If an object is brought nearer the eye, it apparently increases in size; if it is moved away from the eye, it appears to be smaller. If the same object is viewed through

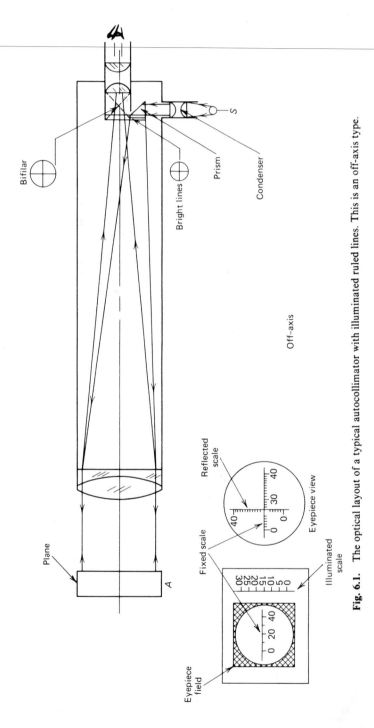

Fig. 6.1. The optical layout of a typical autocollimator with illuminated ruled lines. This is an off-axis type.

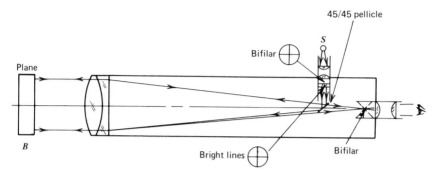

Fig. 6.2. The design of an on-axis autocollimator with a partial aluminized pellicle. This is a bright field.

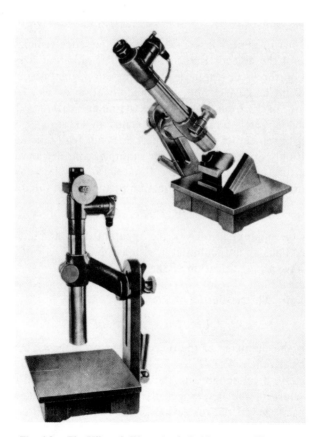

Fig. 6.3. The Hilger & Watts Angle Dekkor autocollimator.

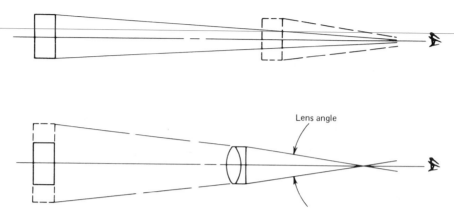

Fig. 6.4. The normal eye viewing distance of an object 10 in. (25.4 cm) away for 1×: dashed lines are for an object at 2× [5 in. (12.7 cm)]. The lower part of the figure shows a lens subtended angle of viewing distance with increase (dashed lines) of object.

a telescope, it will appear larger because the telescope reduces the angle of convergence of the object's rays at the eye. The apparent size of the object will depend on the focal length or magnification of the telescope. An object placed at 10 in. or 25.4 cm has unit magnification or 1×. If the object is placed at 5 in. or 12.7 cm from the eye, its magnification is 2×. Therefore magnification is merely a matter of subtended angle ratios. (See Fig. 6.4.) It is important to remember that the object does not really increase with magnification; it only appears to do so; for example, consider an autocollimator with an aperture of 1.2 in. or 3.05 cm and a 15-in or 38.1 cm focal length. Neglecting the eyepiece, the magnification by the lens of the telescope is

$$M = A_f/T_f \tag{1}$$

where M is the magnification, A_f, the apparent field, and T_f, the true field of telescope. Thus an autocollimator with an objective lens of 38.1-cm focal length will have a magnification of $38.1/25.4 = 1.5\times$.

The practical limit of resolution of a lens is 4.56 sec of arc, which is called the Dawes limit. The resolution of a lens is

$$4.56/Do = \text{sec of arc} \tag{2}$$

where Do is the diameter of the objective lens in inches. Then for a 1.2-in diameter objective lens $4.56/1.2 = 3.8$ sec of arc is the limit of resolution.

Theoretically, a telescope with these specifications should resolve to 3.8 sec, but because of the inhomogeneity of the glass and the small area of prism's surface viewed the resolution is nearly 5 sec for direct comparison collimation. If, however, the amplification ratio involved, for example, in a planoparallel plate amplification is 3× and that of a 90° glass reflecting angle of a prism, 6×, about 1 sec of arc can be resolved.

2. ERROR AMPLIFICATION BY AUTOCOLLIMATORS

The average optician does not often engage in making prisms. Consequently it is easy to forget some of the important details of autocollimation and the interpretation of the imagery patterns formed. The amplification caused by internal reflections of the glass-reflecting angles raises particular difficulties. The amplification factor for testing a planoparallel plate by the autocollimator is 3×. This is obtained with Snell's law; the reflective value of 2 × 1.5 (the refractive index of glass) equals 3. It is possible to make use of both external (air angles or paths) and internal (glass reflective angles) autocollimation setups. A description of these setups is given here.

From external autocollimation setups we can obtain amplification of 2× or 4×, two from a single surface reflection, four from two reflections. Because of the many surfaces involved in the internal autocollimation of prisms, there is 3× to 18× amplification.

In the following, autocollimation procedures and their observed errors in relation to the actual glass-angle errors are described.

In the first example reflection is taken from a polished glass plane as shown in Fig. 6.5. In this case the size of the reflective image equals the angle of incidence image. Because this is an external autocollimation setup, the observed error is 2× times the actual error.

Consider a second planoparallel element, shown in Fig. 6.6, for which the autocollimator is set so that the line of sight at normal incidence on the surface will be reflected along its original path to form one image. Another ray will enter the glass at normal incidence, strike the surface, and be reflected at an angle. Refraction at the surface will then increase the deviation to

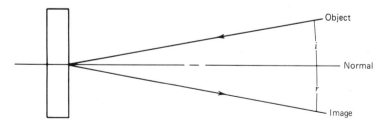

Fig. 6.5. Illustration of reflection. The angle *i* equals angle *r*.

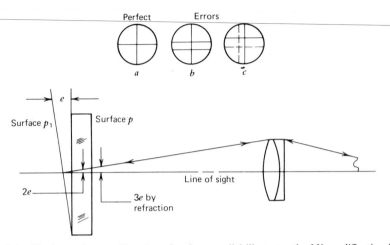

Fig. 6.6. The internal autocollimation of a planoparallel illustrates the $3X$ amplification factor for a $2e$ error.

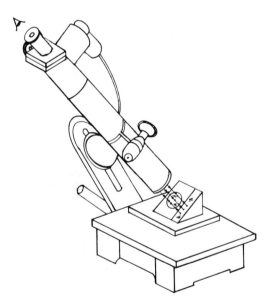

Fig. 6.7. The direct comparison external autocollimation setup of a prism's glass angle compared with a master gauge block.

three times the actual error. In the image pattern *abc* the separated lines in *b* and *c* reveal glass-angle errors. Pattern *c* can be rotated into *b*.

3. EXTERNAL AUTOCOLLIMATION METHOD

External autocollimation can be approached in various procedures; for example, direct comparison by (see Fig. 6.7) with a master angle while fine-grinding prisms or by external autocollimation between two polished faces, one on the prism (generally the reference side), the other on an aluminized flat. The first gives an amplification factor of 2×, the second, 4×. If the surfaces of the prism are only fine-ground, they can be made reflective with aluminized flats wet-wrung on its surfaces.

The direct comparison method makes use of an Angle-Dekkor TA 121 Model autocollimator manufactured by Hilger-Watts. (See Fig. 6.7 or 6.3.) It has magnification power of about 14× and an arm length that makes it comfortable to place the prism in a stable position.

4. TESTING THE COMPOUND REFLECTING ANGLES OF PRISMS

It is the purpose of this chapter to describe how one autocollimator can be used to test any compound reflecting glass angles of prisms. In the production of a penta or Pechan prism, which has a one-half-degree glass angle as a fractional part, it is not apparent that one autocollimator would suffice. An examination of Fig. 6.14 will show that a penta or Pechan prism with a 22.5° angle is used (in a penta prism this is a supplementary angle) the 90° pyramidal angle can be checked by simply extracting the prism from the waxed nylon balls and presenting its polished reference surface to the autocollimator. There is a wide range of angular settings on the autocollimator—from 87° in the vertical plane through 4° in the horizontal; therefore the 90° pyramidal angle can be externally autocollimated.

How does one determine where to press the surface of the prism while fine grinding to correct the error in the glass angles? Press the top of the prism while using the autocollimator and note where the horizontal image tends to close up. (See Fig. 6.8.) Mark the appropriate corner with a pencil and then set up the prism as shown in Fig. 6.8*b*. Press the prism on the top or near the bottom and again observe where the separated images tend to approach. Mark this area with a pencil. Of course, it is possible to find no separation of the imagery in Fig. 6.8*a* but Fig. 6.8*b* may show error or (*b*) may show no error, yet (*a*) will.

Note in the illustration that the prism is resting on three nylon balls which are tightly waxed (lab wax) to an aluminized flat. The flat rests on a

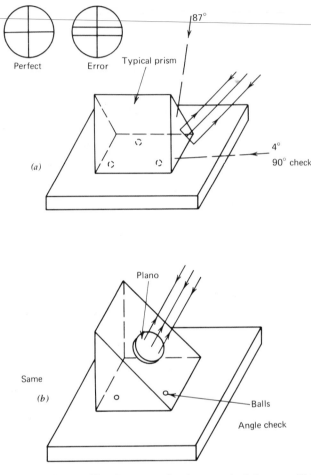

Fig. 6.8. Two external autocollimation setups for the control of the pyramidal angle; each glass angle of the prism is rotated in turn.

felt pad to protect the polished metal plate of the autocollimator. The reflecting surface of the prism to be fine-ground must be made reflective by using a small precisely aluminized planoparallel flat, wet-wrung to the ground surface. It is assumed here that one semipolished reference surface with a surface quality of one-half wave is maintained (preferably two that are parallel for very precise prisms and internally autocollimated for pyramidal control). The autocollimator is preset to the prescribed reflecting angle by placing a master angle block on a precise planoparallel flat. The master angle has an elongated base which rests on waxed nylon balls. Some care must be exercised to make sure that the width of the three nylon balls encompasses a stable area for the base of the prism.

There are alternate methods of grinding the surfaces of the glass angles of prisms. Several choices are available. For the right-angle or 60° prism it is good practice to grind two sides alternately with a fine grit of one size. *First grind the hypotenuse face of right-angle or one face of a 60° prism using the autocollimation in Fig. 6.8b only.* In many prisms, penta and deviation roof prisms, for example, one surface is fine-ground and checked for a pyramidal angle. (See Fig. 6.8a.) When the second surface of the glass angle is fine-ground, the reflecting planoparallel flat is wet-wrung to the finished fine-ground surface and the second ground surface of the prism is set down on the nylon balls. Both arrangements for autocollimation of the prisms are then used. The method of grinding a first surface reference is used by opticians with considerable experience. Beginners should learn to use both set ups in Figs. 6.8a and b.

Preliminary checks must be made before nylon balls or ball bearings are used. Care must be taken not to scratch or cause digs in the finely ground surfaces. The procedure is as follows:

1. Check, with a differential micrometer, for three balls of nearly equal diameter.
2. Wax the balls tightly to an aluminized optical flat, taking the wax slightly above one-half the diameter. The waxed balls should encompass a stable area of the base of the prism at 120° intervals.
3. Place the whole assembly on a felt pad to protect the polished metal surface of the autocollimator.
4. Place a small planoparallel flat, which has an error of 5 sec or less, on the three balls. Nullify any error in the flat by placing a sheet of lens paper on the waxed balls and rotate the flat until the two images are left-to-right of the horizontal line of the bifilar setting. This becomes a near zero wedge in the planoparallel flat. Place a piece of masking tape on the flat at this point for future reference as this zero wedge is positioned at each ball.
5. Remove the planoparallel flat and the paper and compare the reflected image from the aluminized flat. Any drift from the bifilar setting is now due to the error in the diameter of the balls. One or two balls must be honed with a white emery stone or a finely ground glass microscope slide. Little honing will be required if the balls were carefully selected. These balls should be checked from time to time to determine whether they are fastened tightly to the flat and that no uneven wear is caused by rotating the prism during checking. The prism should always be lifted from the balls for repositioning.

I have come to the conclusion that if one or two waxed balls have an error of 0.0001 in. or 0.0025 mm the pyramidal glass angle will be nullified by two factors. First, the master angle block is always set by the planoparallel zero-wedge direction. Second, the drift arc set by the uneven diameters of the balls is always in the same direction *if all* the reflecting glass angles are fine-ground to the first established ball setting.

6.   Set the master angle block so that its elongated base runs parallel to the zero direction of the planoparallel flat, as marked by the masking tape.

5.   OTHER TEST METHODS

Several other methods can be used for checking the direction of the glass-angle error. One defocuses the eyepiece inward toward the prism. Move the eyepiece approximately 1 mm while observing through the eyepiece. If the images draw together, the glass angle is minus or less than 90°. If the images move apart, the glass is plus or greater than 90°. To make use of another simple method touch the prism lightly with a finger while observing the separated image in the eyepiece. The location of one pressure point or area must be found in *both* autocollimated setups. First determine where the separated images tend to *approach* each other when the prism is touched. Do this for each external autocollimation setup, as shown in Figs. 6.8*a* and *b* and mark this point or area on the prism. If two marks are made, the point midway between them is the one pressure point that will correct for both errors simultaneously. Only one mark will indicate that only one glass angle is in error.

6.   DEVIATION PRISMS: CONTROLLING THE 90° ROOF ANGLE RIDGE

Deviation prisms with 90° roof angles may have entrance and exit faces with angles of 45 to 120°. This class of prism includes the Schmidt (45°), Frankford No. 2 (60°), Amici (90°), Frankford (115°), 120°, and the penta roof prism.

In the fabrication of these prisms the control of the 90° roof line, which must be 90° to the entrance and faces, is difficult. It is the purpose here to describe how one autocollimator can be used to control the angle and ridge of the 90° roof angle. Consider a Schmidt prism.

If this prism is placed on one of its 90° roof surfaces, the vector angle will be 74° 18′ 4″ from the perpendicular plane or 15° 41′ 56″ from the horizontal plane. It is impossible to make up an angle of 15° 41′ 56″ from a commercial set of master angle blocks. Moreover, odd angles for many prisms would result.

To overcome this difficulty place a typical deviation prism with one of its 90° roof surfaces resting on three nylon balls as shown in Fig. 6.9*a*. If the

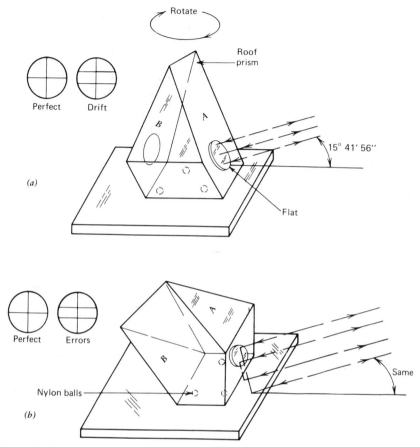

Fig. 6.9. Two external autocollimation setups for the control of the compound slant in the 90° roof ridge of the deviation prism.

entrance and exit faces are not semipolished, they can be made reflective with a small aluminized planoparallel mirror, wet-wrung to each face. The autocollimator is set to the reflective image and its bifilar setting is placed on the horizontal line. In the next step the prism is picked up (same side) and its second face is rotated. Note whether the new reflective image is below or above the preceding bifilar setting. If the 90° roof ridge is perpendicular to both faces, there will be no drifting of the image above or below the first setting of the bifilar. Some autocollimators lack a bifilar, in which case fine-grind toward the midpoint of the horizontal drift. In any case a null point must be achieved.

The second side of the 90° glass angle is also fineground as already described. At this point of fabrication the 90° roof angle is subjected to the test described in the preceding section in which the prism is picked up and repositioned on the three waxed nylon balls. The 90° glass-angle is presented to the autocollimator setting shown in Fig. 6.9*b*.

7. INTERNAL AUTOCOLLIMATION

The difference between external and internal autocollimation is that the former deals with "air angles" between the reflective surfaces of prisms and the latter with the reflective internal glass surfaces. The internal auto-collimation of a prism's compound reflecting angle will produce an amplification. To find the amplification factor divide the glass angle of the prism into 360° and multiply by 1.5; for example, a 45° glass angle will give an amplification factor of 12×. If error is present in the reflecting glass angle, the autocollimator will show separated images in the field of the eyepiece. This section illustrates internal autocollimation setups and shows how assembled prisms can be tested.

First examine the intensity of the light of the internal autocollimated images from a prism being tested. Figure 6.10 illustrates a conventional setup for checking the internal 90° glass angle of a right-angle prism. Disregard the rotating image on surface (a). In consideration of the relative brightness of both reflective images, the glass-to-air reflection will be approximately 4.5% of the intitial light from the autocollimator. Therefore 4.5% will be reflected from the glass surface (a) to the autocollimator to form the image (A). Roughly 95% of the light will enter the surface (a) and form images (B) and (C) by transmission and reflection. Because of the loss of light by reflection, refraction, and absorption, however, the intensities of images (B) and (C) will contain only 38.13% of the original intensity. Thus the internal images are brighter than the external image.

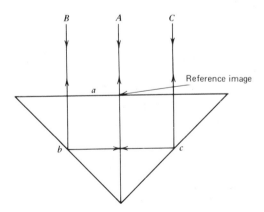

Fig. 6.10. The intensity of the external and internal image values in a prism.

It is rather difficult to remember where the fainter image is for the many possible internal autocollimation setups that can be used for each prism. One simple test is called the breath test for want of a better name. In this test the optician breathes heavily on the prism's surface away from the front surface nearest the collimator. He then watches for the internally autocollimated image to reappear as the moisture evaporates. The external autocollimated image always remains fixed in the eyepiece; it does not disappear and reappear like the internal image. The optician should mark the location of this fixed external image, for it will be used to correct the glass reflective angles of the prism. (See Section 7.2.)

8. INTERNAL AUTOCOLLIMATION SETUPS

Collimation testing of compound reflecting glass angles is done by internal autocollimation. Study some of the typical image patterns in internal setups. Figure 6.11 shows a 90° glass reflective angle being checked by the collimator. Note the multiple-image pattern and the amplication factor $6\times$ of the glass angle (x) and pyramidal angle (p). Some image patterns rotate

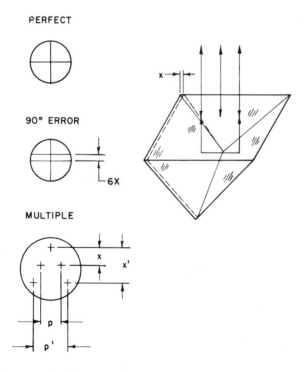

Fig. 6.11. The external and internal images observed in a collimator of a right-angle prism with the parallel light testing the 90° glass angle.

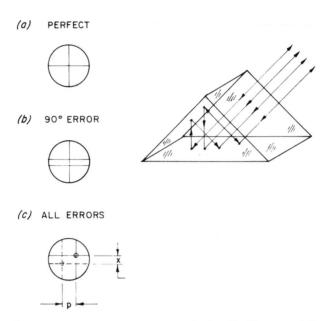

Fig. 6.12. The internal autocollimation of a typical prism (in this case a right-angle prism) which gives a 6× amplification factor of the sum total of all angular errors.

if the prism is rotated. The pattern with five images reveals several minutes or more of glass-angular and pyramidal errors. If there are no errors, the two sets of images are exactly superimposed. Any horizontal separation of the images is an error in the reflecting glass angle. Any separation in the vertical plane is an error in the pyramidal angle or slant of the apex of the prism. In the image pattern (b) will remain stationary, but (a) and (c) are subject to rotation. Image (c) is actually quadruply reflected from the 90° right-angle faces of the right-angle prism. Do not make use of this complex internal setup during fabrication of prisms. Its description is for the purpose of information only. (See Fig. 6.11.)

An internal autocollimating device often in use is illustrated in Fig. 6.12. This optical setup gives a 6× amplification of the glass reflective angles of a finished prism. If the prism is placed on one of its 45° faces, the setup will be identical. The images are not subject to vibrational rotation if they are placed on a felt pad for testing.

9. SINGLE REFLECTING ANGLES

Twyman has shown that a single reflecting angle which can be evenly divided into 360° will if perfect, project one superimposed image. Errors in the angle will autocollimate into two separate horizontal images. Pyramidal or slant error does not autocollimate.

Consider the single reflective glass angle of 45° on prism shown in Figs. 6.13a and b. Assume that the pyramidal control of surface (h), the hypotenuse face, is to be checked by the internal autocollimation setup in (b).

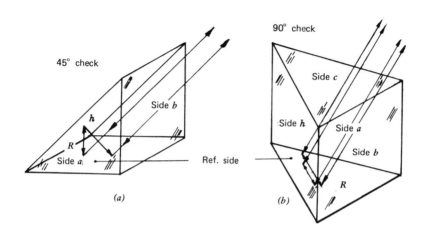

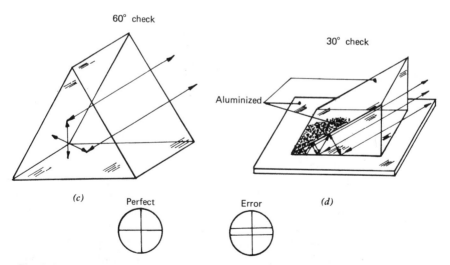

Fig. 6.13. Internal autocollimation of a single glass reflective angle of various glass angles.

Three polished glass surfaces h, R, and C are required for the test. Two sides of the prism are semipolished; the chosen reference surface R and the hypotenuse surface (h) are being corrected for an perfect 90° angle. The chosen reference side R (best surface quality) is always placed on a piece of felt and the surface being polished faces the collimator. The reference side of the prism (down) and hypotenuse side (h) form a 90° angle. The top surface of the prism's side serves as a window for the 90° check. No appreciable error results if this side (C) is not parallel to the reference side R. The hypotenuse side is completed when it is one-eighth fringe and the corrected 90° angle shows superimposed images.

The next side to be corrected is a 45° reflective glass angle. Check the angle with the setup in Fig. 6.13a. The autocollimator looks directly into the 45° angle. Surface (a) is the glass angle side being polished and surface (b) is the window for the parallel rays from the autocollimator.

In summary, work on only one compound reflecting angle has actually been done, but each single angle has been taken in turn. First, the 90° glass angle formed by the chosen reference side and the hypotenuse surface was checked. Then the 45° reflecting glass angle formed by the hypotenuse surface (h) and side (a) was checked. To finish the prism, side (b) can be polished and corrected in like manner by using the internal autocollimation setup shown in Fig. 6.13b.

Other angles used in prisms are 60 and 30°; these angles can be internally autocollimated. (See Figs. 6.13c and d.) Here it is assumed that the hypotenuse side of the prism has been worked and its 90° side angle checked by (b) in Fig. 6.12. The 60° autocollimation check is easily done but the 30° check is handicapped by a weak imagery return. The finished side of the prism (top) is aluminized and is the surface of the flat corrected by polishing. For the 30° check there is an 18× amplification factor; for the 60°, a 9× amplification.

The imagery patterns illustrated at the bottom of the figure are identical to those of the other reflective glass angles discussed.

10. OTHER INTERNAL AUTOCOLLIMATION SETUPS

Other internal autocollimation setups may be used, some of which are shown in Fig. 6.14; others have been shown before. Of special interest are the double-image patterns for an autocollimator that looks directly at two surfaces of the prism from one surface. It is also possible to look through two surfaces and check the 90° glass angle with two surfaces of right-angle, penta, or Amici prisms. These setups are seldom used during the fabrication of prisms. The first five figures illustrate setups have extensive application.

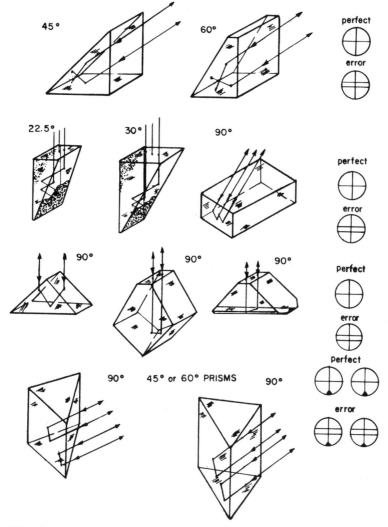

Fig. 6.14. A number of typical single- and double-image patterns for autocollimated prisms being tested internally.

EXERCISES

1. Describe the difference between an on-axis and off-axis autocollimator.

2. Describe the autocollimation of a prism's compound glass reflecting angles with one collimating setup.

3. What is the "breath test" and what purpose does it suggest in making prisms?
4. What is the difference between external and internal autocollimation of prisms under working conditions?
5. What is the internal autocollimation amplification factor of a right-angle prism, 60° reflecting angle and a 30° glass reflecting angle?
6. Is is possible to autocollimate internally a glass reflecting angle of 22.5°?
7. List the four reflecting glass angles that can be internally autocollimated.
8. Why are wax nylon balls used on an aluminized flat during the autocollimation of a prism?
9. Can two reflecting glass angles be corrected simultaneously by using one pressure point while grinding or polishing? How is this one pressure point determined?
10. What is the amplification factor while testing an air angle of a 90° angle of a right-angle prism?

BIBLIOGRAPHY

De Vany, A. S. "Testing a Prism's Compound Reflecting Angles," *Appl. Opt.*, **14**, 2791–2792 (1975).

De Vany, A. S. "Testing Glass Reflecting-Angles of Prisms," *Appl. Opt.*, **17**, 1661–1662 (1978).

Ferson, F. B. "Prisms, Flats, Mirrors," in *Amateur Telescope Making—Advance*, A. G. Ingalls, Ed., *Sci. Am.* 75–93 (1959).

Habell, K. J. and Cox, A. *Engineering Optics*, Pitman, London, 1956, pp. 218–282.

Johnson, B. K. *Optics and Optical Instruments*, Dover, 1960, pp. 190–198.

Moffitt, G. W. "Autocollimation of Prisms," *J. Opt. Soc. Am; Rev. Sci. Instrum.*, **7**, 41–51 (1923).

Ryland, F. and Chalmers, S. D., "Prisms," Optical Society, London, 36–39 (1904 and 1905).

Wagner, A. F. "Prisms," *Experimental Optics*, Wiley, New York, 1929, pp. 109–120.

Wright, F. E. "Prism Making," *J. Opt. Soc. Am.*, **5**, 148–152, 1921.

Wright, F. E. *The Manufacturing of Optical Glass and Optical Systems*, Bureau of Ordnance, Washington, D.C., 1921.

7

Finishing Prisms

In this chapter the correction of compound glass angles of prisms is discussed. Chapter 6 covered the internal autocollimation of glass reflecting angles but did not tell how to correct the angles.

1. PRESSURE POINTS

When one surface of a prism is worked, only.one pressure point is applied during polishing to correct the 90° glass angle to that surface. There is also one pressure point that corrects two glass-angle errors in the compound angle simultaneously. In this section the pressure point is located for the first surface of the reflecting glass angle of the prism which is 90° to the surface. Table 7.1 can be used to locate the pressure point for correcting the 90° to the surface. Table 7.1 can be used to locate the pressure point for correcting the 90° side angle in relation to the first surface polished. The reference surface of the prism is shown in Fig. 7.1*a* and can be on the left or right side. Once a side is chosen it must be marked by a diamond scribe for future identification.

2. FINISHING THE 90° SIDE ANGLE

The largest surface of the prism is generally finished first and its 90° side-angle is corrected. See Fig. 7.1*a*. The techniques to be described can be used for any type of prism, optical square, or rhomboid, but are illustrated for a right-angle prism. The largest surface is the hypotenuse which is to be corrected with its 90° side angle. The 90° side angle is internally autocollimated. (See Fig. 7.1*a*.) The reference surface (premarked *R*) is always placed downward on a piece of felt. The collimator is set approximately to an angle of 45° (this can vary from 70 to 40°). The parallel rays from the collimator are returned by the 90° side angle formed by the prism's hypotenuse and the reference surface (*R*). The image patterns can generally be of two types; one superimposed image, indicating a correct angle, the other, two separated horizontal lines, indicating glass-angle error. The glass-angle

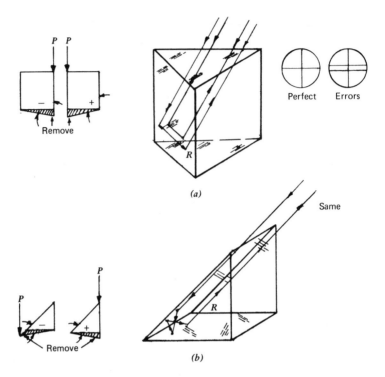

Fig. 7.1. (*a*) Internal autocollimation of a prism's 90° side angle (pyramidal control) with a surface of 45° reflecting angle and the reference side of the prism. The reference side is resting on a felt pad. (*b*) Internal autocollimation of a prism's 45° reflective angle. See Figs. 6.13 and 6.14 for other glass angles that can be internally autocollimated. The small figures on the left show the location of the wedge of glass that must be removed by applying pressure during polishing.

error is plus (obtuse) or minus (acute). A single pattern reveals a perfect 90° glass angle.

There are several methods of determining whether the glass angle is acute or obtuse:

Rule I. Defocus the eyepiece from the best focus. If the two horizontal images close up, the glass angle is acute (minus).

Rule II. Defocus the eyepiece as above. If the two horizontal images tend to separate, the glass angle is large (obtuse). (See also p. 577.)

Another method, called the knife-edge test, will verify the glass angle. Shade the bottom of the eyepiece with a machinist's rule. If the top of the horizontal image disappears, the glass angle is obtuse or greater than 90°. If the bottom of the horizontal image disappears, the glass angle is acute or smaller than 90°. Figures 7.1a–b shows that pressure is always applied during fine-grinding or polishing on the minus angle (acute). The surface is finished if the 90° side angle is corrected and one-tenth fringe or better.

3. FINISHING OTHER ANGLES

Consider the first 45° glass angle on the right-angle prism. The hyptotenuse surface is finished but there are two 45° reflecting glass angles. Correct one 45° angle as carefully as possible and let the second 45° angle balance out the residual errors. The surface of the hypotenuse is finished but the 45° glass angle shares another surface with it. The result is a compound glass angle of 45° with a 90° side angle. For testing the internal autocollimation setups shown in Figs. 7.2 must be used. The first 45° glass angle to be polished and corrected is checked by an internal autocollimation. The reflected parallel rays are returned to the eyepiece of the autocollimator. The same procedure is used for checking the 90° side angle. Rules (1) and (2) or the knife-edge test are applied to determine the accuracy of the glass angles. Corrections are made by polishing as pressure is applied at a point that corrects two glass-angle errors simultaneously. If the first 45–90° compound glass angle is being fabricated, columns 2, 3, and 5 in Table 7.1 are of primary interest. (check the premarked reference side). Table 7.1 lists the 45° glass angle and its 90° side angle (for pyramidal control). The glass angles are identified as acute (−), obtuse (+), and perfect (0). The pressure point or area is indicated by 1 to 8. These numbers correspond to pressure

TABLE 7.1. CORRECTING THE GLASS ANGLES OF PRISMS WITH ERRORS: FOR ANY TYPICAL PRISM
(Right-Angle Prism Illustrated)

				Pressure point	
Row	90° side angle	Pyramidal direction	45° reflecting angle	Reference/left (R/L)	Reference/right (R/R)
1	Perfect (0)	Zero	Acute (−)	7	7
2	Perfect (0)	Zero	Obtuse (+)	3	3
3	Obtuse (+)	Right	Perfect (0)	5	1
4	Acute (−)	Left	Perfect (0)	1	5
5	Acute (−)	Left	Obtuse (+)	2	4
6	Acute (−)	Left	Acute (−)	8	6
7	Obtuse (+)	Right	Acute (−)	6	8
8	Obtuse (+)	Right	Obtuse (+)	4	2

Note. For one compound glass angle use columns 2, 4, and 6. For all compound glass angles use columns 3, 4, and 5. If the reference side is on the right (R/R), the numbers are different (see Fig. 7.2 (1–8)).

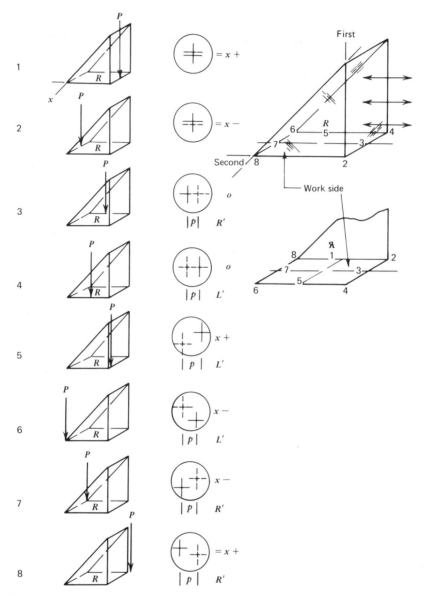

Fig. 7.2 (1–8)   The location of the pressure point applied during polishing or fine-grinding to correct simultaneously any of the eight sets of image patterns. R is the polished reference side and is generally on the left. R' and L' next to the image patterns are the directions of the pyramidal angle (see text). If the reference side is on the right of the prism (reversed R), all the pressure points p are renumbered anew as shown; p and x are, respectively, pyramidal and glass reflective errors. This is an internal autocollimation of the prism's combined glass angles in all parts of the figure (1–8). The working order of the prism's compound glass angles is first and second.

points shown in Figure 7.2 (1–8) for a prism surface corrected in a downward position during the collimation check.

4. FINISHING THE LAST SURFACE

The last surface of the right-angle prism to be finished is the second 45° face. The combined effect of the two glass angles and the last compound glass angle being worked is checked by a single internal autocollimation shown in Fig. 7.2. The residual errors in the first 45° face plus the errors in the second 45° face produce two internally autocollimated images that can be located in any quadrant sector, separated or together in the field of the eye-piece. It is difficult to tell which of these images comes from the 45° reflective face nearest to the autocollimator and which derives from the internal autocollimation. A simple procedure identifies these images. Breathe heavily on the surface of the hypotenuse and watch for the reappearance of the image as the moisture evaporates. We call this the breath test. Mark the side of the prism on which the image disappeared. This is important because this side is always one of the two possible pressure points during polishing.

In Fig. 7.2 (1–8) the dashed lines are the internally autocollimated images and the solid lines are the reflected images from the 45° face nearest the collimator. The vertical separation of the lines denotes pyramidal error. The horizontal separation of the images is the total of the residual glass-angle errors in the prism. In Fig. 7.2 (1–8) (s) and (x) and (p) are the glass-angle error and pyramidal direction respectively. Note that R is on the left. If the reference side is on the right, the pressure points differ from those shown. In Table 7.1 we can locate the one point at which pressure is to be applied during polishing; for example, suppose the 45° reflecting angle is acute (−) and the breath test reveals that the pyramidal direction is on the same side as the reference surface. Row 7 in Table 7.1 applies and the pressure point is 8.

Glass angles and pyramidal direction change rapidly if the images are close together and must be continuously checked. This means that the pressure point as well will change continuously. Pressure is applied on the minus glass angle, which is always opposite a plus angle.

Although we have used the right-angle prism to describe the correction of glass-angle errors, these techniques can be applied to most typical prisms.

5. COUNTER REFLECTIVE IMAGES

Three images will often be observed in autocollimation setups even for corrected prisms. There will be one image surrounded by two counterreflective

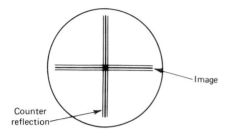

Fig. 7.3. The counterreflections of images observed in internal autocollimated prisms. Ignore the ghost images.

ghost images. In a corrected prism these images are in near coincidence, but blemishes and/or inhomogeneity of the glass cause light scattering that creates ghost images. The optician must learn to concentrate on the two separated brighter images (prism with glass-angular error) and ignore the counterreflective images. A typical image pattern with counterreflective ghosts is shown in Fig. 7.3.

5. MIRRORLIKE SURFACES ON PRISMS

Mirrorlike surfaces that exist on prisms sometimes have rainbow tints surrounding observed objects; others are true mirror surfaces without color. True-mirror surfaces are reflective surfaces, whereas surfaces with images that show rainbow tints transmit light that is broken down into spectral colors. See Figs. 7.4a–c.

Interest is shown primarily in the true-mirror surfaces because they can be used to autocollimate a prism internally while the glass-angle errors are checked; for example, in a right-angle prism a mirrorlike surface can be observed while looking into one of the 45° faces of the prism. If a pencil is placed on the prism's second 45° face or hypotenuse surface, its movement can be observed. This information is useful in using the collimator. In another application of the mirror surfaces as follows, the reflection of the eye is observed in the hypotenuse face of the prism. The 90° glass angle (mirrorlike surfaces) returns the image of the eye. If the 90° glass angle is minutes of arc in error, the eye will be distorted. Observation of the mirrorlike surface provides a 90° check of the two 45° complementary glass angles. This check can have several aspects. In Fig. 7.4c the image of the eye is eggshaped, which indicates that the 90° glass angle is less than 90°. In this case one 45° face could be perfect and the second could have several minutes of error. In another both 45° faces could have off-setting errors that would give the net result of a zero 90° error. What if the image of the eye were true? Two situations can develop:

Case 1. The two 45° faces could be corrected without glass-angle error.

Case 2. The 90° angle could consist of two opposite but equal glass angle errors that would nullify each other; for example, 6 min + in one face and 6 min − in the other.

The 45° roof deviation prisms are unique. Many mirrorlike facets can be observed in a face of a Schmidt prism resting on a piece of felt. Care must be taken that the observations made are always 90° to the surface and not off-axis or a color tinted image will appear.

True mirrorlike surfaces can be autocollimated internally for checking individual reflecting glass angles or compound glass angles. An important point to keep in mind in the internal autocollimation of deviation prisms is that all facets of the prism must be generated because misleading image patterns are often observed.

One use for these mirrorlike surfaces is in making ocular prisms for range finders. Ocular prisms have metallized surfaces and a complex number of

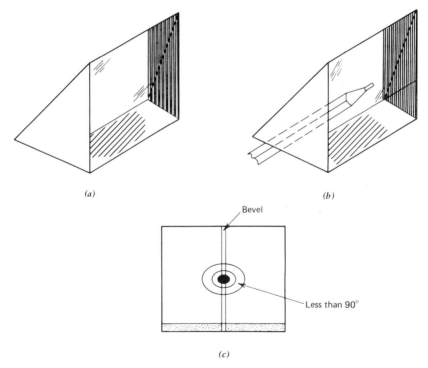

(a)   *(b)*

Bevel

Less than 90°

(c)

Fig. 7.4. Mirrorlike images observed in different sections of a prism. These sections often betray possible autocollimation setups.

individual prisms of various types cemented together. They are costly to make but lend themselves to autocollimation.

6. THE THREE GROUPS OF PRISMS

Prisms can be divided into three groups. (See Figs. 7.5a–x.) Group I contains individual prisms, Group II, two prism assemblies, and Group III, assemblies of three or more prisms. One of the better methods of internal collimation is illustrated.

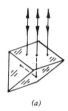

(a)

Fig. 7.5 (a).   Right angle.

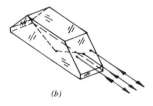

(b)

Fig. 7.5 (b).   Harting-Dove.

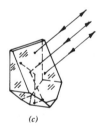

(c)

Fig. 7.5 (c).   Schmidt.

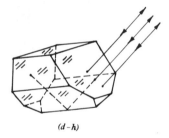

(d–h)

Fig. 7.5 (d–h).   Five types.

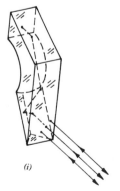

(i)

Fig. 7.5 (i).   Leman.

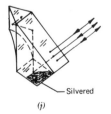

Silvered

(j)

Fig. 7.5 (j).   Frankford No. 4.

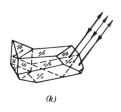

(k)

Fig. 7.5 (k).   Frankford No. 6.

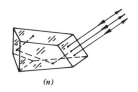

(l)

Fig. 7.5 (l).   Frankford No. 7.

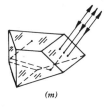

(m)

Fig. 7.5 (m). Frankford No. 5.

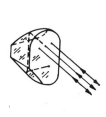

(n)

Fig. 7.5 (n).   Frankford No. 3.

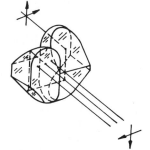

(o)

Fig. 7.5 (o).   Porroprism assembly.

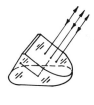

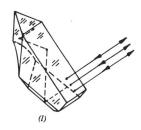

(p)

Fig. 7.5 (p).   Abbe assembly.

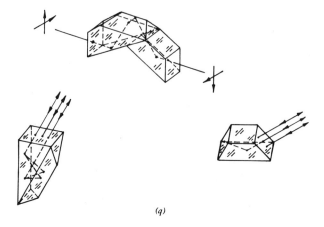

Fig. 7.5 (q). Abbe prism, Type A.

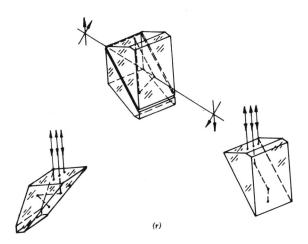

Fig. 7.5 (r). Pechan assembly.

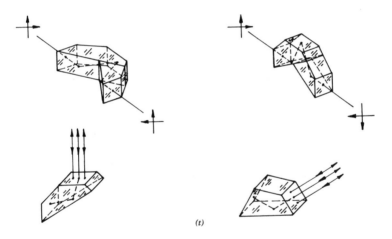

Lenses

(s)

Fig. 7.5 *(s)*. Two doves.

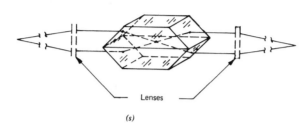

(t)

Fig. 7.5 *(t)*. Reversion.

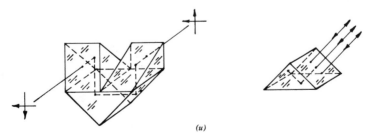

(u)

Fig. 7.5 *(u)*. Erecting assembly.

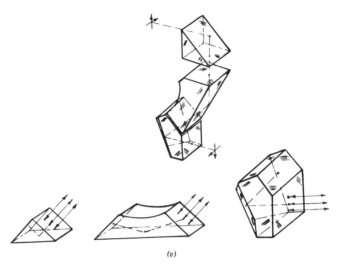

Fig. 7.5 (v).   C. P. Goetz prism system.

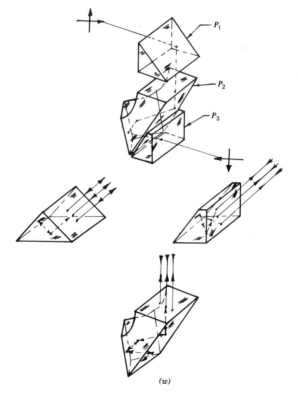

Fig. 7.5 (w)   Carl Zeiss prism system.

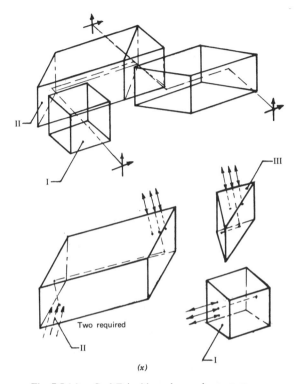

Fig. 7.5 (*x*).   Carl Zeiss binocular-ocular system.

To check even multiple reflecting glass angles refer to Fig. 7.5*x* for internal collimation. For example, the Pechan prism, which consists of two prisms, has two silvered, slightly separated surfaces. (See Fig. 7.5*r*.) One consists of a even 45° glass angle; its other angles are 67.5° (odd). The second prism has one angle that is 22.5°, one that is 45°, and the third, 112.5°. Angles 22.5 and 45° are even, whereas 112.5° is not. It is important that the observation of multiple glass angles in any prism be thoroughly understood because clues to their construction are given. This subject is discussed at length in the chapters on typical prisms.

7. TYPICAL PRISMS

7.1. Individual Prisms

Figure 7.5*a* is a right-angle prism that will deviate the line of sight through an angle of 90°. The image will be inverted in this position. If it is placed on one of its sides, the image will be reverted. Illustrated is the best normal internal autocollimation optical setup.

Figure 7.5*b*, which is a Harting-Dove prism patterned on a large right-angle prism, has different base lengths and heights when different glass types are used. The apex of the right-angle prism is generated to form the dove prism if the right prisms are small. It is a direct-vision prism and can be used only in parallel light. The image will be inverted when the prism rests on its base. This prism is often used in scanning because it will rotate the image at twice the normal rate. The best internal autocollimation setup is shown.

Figure 7.5*c* is a 45° roof deviation prism (a Schmidt) with entrance and exit faces that form an included angle of 45°. This prism will invert and revert the image and at the same time deviate the light of sight through an angle equal to the included glass angle of the entrance and exit faces. The best internal autocollimation for the roof of the prism is to rest it in a 90° slot in a wooden block.

Figures 7.5*d–h* shows a prism (Frankford Arsenal Prism No. 2) representative of the group of five which also includes 80, 90 (Amici), 115 (Frankford Arsenal Prism No. 1), and 120° prisms. Any intermediate included glass angle can be designed to meet the deviation of the line of sight of the instrument's requirements. The best internal autocollimation is described in Fig. 7.5*c*. See also the discussion of 90° roof prisms.

Figure 7.5*i* is a Leman prism with a 90° roof that will revert and invert the image. The line of sight will be displaced laterally by an amount equal to three times the diameter of the entrance pupil. The prism can be placed on its side for internal autocollimation setup.

Figure 7.5*j* is a Frankford Arsenal Prism No. 4 that does not have a 90° roof. The line of sight is deviated through an angle of 90° in the horizontal plane and at the same time through an angle of 45° in the vertical plane. The image will be reverted to an observer standing at a right angle to the line of sight. The prism can be placed on its side for autocollimation. Note the parallel rays that enter the face of the 45° glass angle of the prism.

Figure 7.5*k* is a Frankford Arsenal Prism No. 6 that deviates the line of sight through an angle of 90° in the horizontal plane and at the same time through an angle of 60° in the vertical plane. The image is inverted by the prism. For internal autocollimation the prism's 90° roof angle is placed in a 90° slot in a wooden block and the parallel light enters the vertical face of the prism.

Figure 7.5*l* is a Frankford Arsenal Prism No. 7. It has a 90° roof and will deviate the line of sight through an angle of 90° in the vertical plane and at the same time through an angle of 45° in the vertical plane. The image from the prism is neither inverted nor reverted to an observer standing at a right angle to the line of sight in the prism. The prism can be placed on its side for its internal autocollimation.

Figure 7.5*m* is a Frankford Arsenal Prism No. 5. The line of sight is deviated through an angle of 90° in the horizontal plane and at the same time through an angle of 60° in the vertical plane. The image is inverted and reverted to an observer standing at a right angle to the line of sight. The prism can be placed on its base and the parallel light, received by its side face.

Figure 7.5*n* is a Frankford Arsenal Prism No. 3 that will deviate the line of sight through an angle of 60° in the vertical plane and through an angle of 45° in the veritcal plane. An observer, standing at a right angle to the line of sight, will observer the image reverted and inverted. The internal auto-collimation position is similar to that in Fig. 7.5*l*.

7.2. Two Prism Assemblies

Figure 7.5*o* is a Porro prism assembly which consists of two, usually identical, right-angle prisms place at right angles to each other. The internal autocollimation is similar to that described for the right-angle prism (see Fig. 7.5*a*). One prism is checked at a time.

Figure 7.5*p* is Abbe's modification of the Porro assembly. It consists of two identical prisms cemented together. The image is inverted and reverted. Like the Porro prism assembly it is a direct-vision system. The line of sight will be offset by a distance equal to the thickness of one of the prisms in the same plane. Each prism is internally autocollimated by directing the parallel light through its polished side.

Figure 7.5*q* is an Abbe prism, Type A. It will not deviate the line of sight and is called a direct-vision prism. This prism assembly inverts and reverts the image, and each prism is internally autocollimated as illustrated. For the autocollimated light to be retrodirected from the first prism a small aluminized flat on a gimlet can be used. The second prism, which is actually a 60° equiangular prism with one 60° angle cut off, is autocollimated by directing the parallel light through the two polished sides. There will be two identical images observed in the eyepiece of the autocollimator.

Figure 7.5*r* is a Pechan prism assembly with two prisms. This assembly is similar to that of the Harting-Dove but can be positioned in both convergent and divergent light, an added advantage. It will invert or revert the image, depending on its lateral orientation. One surface of each prism is silvered and the unsilvered reflecting surfaces have airspace of approximately 0.002 in. Each prism, independently autocollimated, is placed on its side with parallel light rays enters the entrance and exit faces.

Figure 7.5*s* is a double dove prism which in essence, is a pair of Harting-Dove prisms. The reflecting bases are silvered, assembled, and cemented together. This assembly cuts the length of the single Harting-Dove prism in half. It must be used in parallel light only. The autocollimation is set up by lightly silvering the base (see Appendix 9), setting the prism on its base and directing the parallel rays into one of its faces.

Figure 7.5t is a reversion prism which consists of two elements cemented together and is a modification of the Abbe prism, type A. (See Fig. 7.5p.) Like the Pechan prism, it may be placed in parallel, converging, or diverging light. It may be used to revert the image if its top reflecting surface is perpendicular to the horizontal plane and will invert the image if its top reflecting surface is parallel to the horizontal plane. The prism must be decentered vertically because it will decenter the line of sight by twice the centering error, although the line of sight is not deviated. Autocollimation setups for checking each prism are shown.

7.3. Three or More Prism Assemblies

Figure 7.5u is an erecting prism system. It consists of one large right-angle prism with two smaller right-angle prisms cemented to the larger prism in proper orientation to accept a converting light beam. Each of the right-angle prisms can be autocollimated as shown in Fig. 7.5a.

Figure 7.5v, a C.P. Gorez prism system, consists of three single prisms. The light is received on an objective placed in front of or behind the right-angle prism. This prism system will invert or revert the image. The original direction of the line of sight will not be deviated, but it will be displaced by an amount that will depend on the separation between the prisms P_1 and P_2. Autocollimation setups for checking each prism are shown. Note that prism P_3 has a 90° roof.

Figure 7.5w is a Carl Zeiss prism assembly with three prisms. The objective is usually placed between the right-angle prism P_1 and the 90° roof prism P_2. It can be placed in front of the right-angle prism P_1. This assembly of prisms will invert and revert the image but will not deviate the line of sight. Like the Gorez prism system, the line of sight will be displaced by the amount of separation between prisms P_1 and P_2. Internal autocollimation for each prism is shown. Prism P_3 is actually a 60° wedge that can be checked by using the autocollimation setup shown in Fig. 7.5d.

Figure 7.5w is a Carl Zeiss binocular-ocular prism assembly. It consists of four prisms, two rhomboid prisms, R_1 and R_2, a right-angle prism, P_2 with a 45/45 aluminized coating on its hypotenuse side cemented to the rhomboid prism R_1, and a glass block with two polished parallel surfaces. The autocollimation of each element is divided into three parts. The glass block B is checked as a planoparallel (see Fig. 6.6). The rhomboid prisms are checked in a manner similar to the checking of the planoparallel, but the collimator is directed into the hypotenuse face of one 45° face.

8. CONCLUDING OBSERVATION

A prism is like a planoparallel plate in that it is an external autocollimation device and has a vector wedge caused by glass angular errors in both optical elements. The mystery that surrounds prisms and their making vanishes as an awareness of this fact increases.

EXERCISES

1. What causes the counterreflections in the observed superimposed imagery during the collimation of prisms?
2. Of what use are the mirrorlike reflections in the autocollimation of prisms?
3. Name three multiple assemblies of prisms used in optical sights.
4. Why does 22.5° internal collimation check require coated surfaces?
5. What is the angular range for collimator's setting during internal auto-collimation of a 90° glass angle?
6. What is meant by amplification factor of the glass angle errors during internal autocollimation?
7. How many degrees are there in a planoparallel plate?
8. How many typical dove prisms are there?
9. Name six deviation prisms with a 90° roof?
10. Make a small aluminized planoparallel plate with 10-sec parallelism and a surface quality of one-eighth fringe.

BIBLIOGRAPHY

De Vany, A. S. "Correcting Glass Angles of a Prism," *Appl. Opt.*, **16**, 2019–2021 (1977).

De Vany, A. S. "Testing Glass Reflecting-Angles of Prisms," *Appl. Opt.*, **17**, 1661–1672 (1978).

Military Standardization Handbook—Optical Design-(MIL-HDBK-141,5th October 1962).13 Mirror and Prism Systems 13–1 to 13–52.

See bibliography in Chapter 6.

8

Right-Angle and Dove Prisms

In this chapter it is assumed that the following phases or tests in the construction of prisms have been completed.

1. The glass disks have semipolished surfaces of two fringes or less and around 10 sec or less of wedge. The glass is tested by the Murty method for homogeneity.

2. The prisms are in finished form and at top dimensional tolerances.

3. All reflecting faces have the same reference side for checking the 90° side angles.

1. THE HYPOTENUSE FACE

The general practice in making a small number of prisms (10 or fewer) is to cement small glass runners of the same type of glass to the prism for easier handling during grinding and polishing. This is more fully described in Appendix 3. Briefly, the small, thin, rectangular glass slabs are pitched to the sides of the prisms as shown in Fig. 8.1a. Three or four prism assemblies should be made up at one time. To make up the assemblies, the prisms and glass runners are heated in an oven at 85°C. Pitch is lightly smeared on the prism and lens paper is placed between it and the glass runners. Lens paper must always be used between all polished glass surfaces to eliminate the Twyman effect (the twisting of a finished polished surface when another adjacent surface is ground and polished). For prisms with semipolished surfaces sawn from the tested semipolished planoparallels choose the side of the prism with the best surface quality, make it the reference side, and mark it with a diamond scribe. This reference side will be used throughout the prism's fabrication.

The prisms and glass runners are placed on an asbestos pad preheated to 85°C. The prisms are taken out singly and two of the glass runners are cemented to each of them with a small sheet of lens paper between. The assembled prisms are placed on a flat tool which is covered with a sheet of paper. It is essential that the assembly be carefully checked around all the bevels. The lens paper should be slightly above both surfaces, for if it is allowed to creep below the two surfaces the assembly will be misaligned. The whole assembly is pressed together firmly on top of the runners and prism. A wooden block can be used to square and set the proper extension of the glass runners around the prism. It is good practice to assemble a series of four prisms in this manner before working the hypotenuse side.

After the assemblies have cooled they are washed in xylene. The pitch and paper should be removed around the bevels and glass-runner surfaces.

During grinding (with three grits—225, 145, and 95 aloxide) check the prism in an external autocollimation setup as shown in Figs. 8.1(a) and (b). The prism, and the glass runners, rests on three nylon balls that are tightly waxed to an aluminized optical flat. The light from the collimator strikes the polished surface of the prism and is reflected onto the aluminized surface of the flat and finally back to the collimator. If the 45° reflecting face, the 90° air angle formed by the prism, or the surface of the flat is in error, two separate horizontal images will be observed in the collimator.

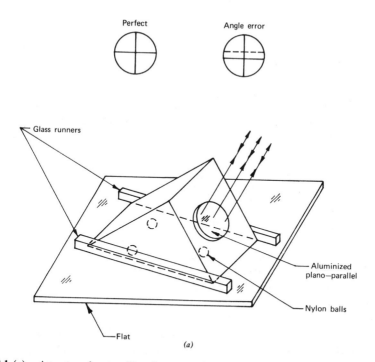

Fig. 8.1 (*a*).   An external autocollimating setup used during fine-grinding of the hypotenuse surface of a right-angle prism. The autocollimator is preset by a master 45° angle on a precise planoparallel that is resting on three nylon balls. A small aluminized working planoparallel is wet-wrung to the ground surface to make it reflective.

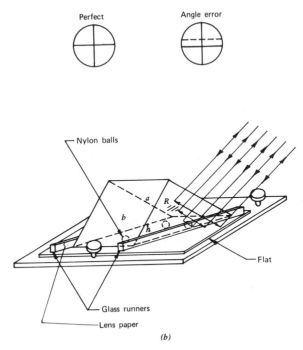

Fig. 8.1 (b). The external collimation of the 90° side glass angle (pyramidal control) to the prism's hypotenuse side. This is a 90° air angle formed by the aluminized flat and the prism's reference side which is generally polished; if it is not polished the small planoparallel can be wet-wrung to this 90° side angle.

How is the direction of the error located on the prism? While observing in the eyepiece of the autocollimator, place a finger on an area of the prism in which the horizontally separated images tend to come together. This area is used as the pressure point when fine-grinding the hypotenuse side. Work for a superimposed image, especially when fine-grinding with 95 grit. This collimation setup produces a 4× amplification factor and the changes are rapid during grinding. After fine-grinding the prism, check with a loupe for any digs or scratches around the glass runners. Remove any loose grit in the bevels. If any bevels have been lost during grinding, they can be renewed by using a brass shim and 145 grit.

The prism assembly is polished on a faceted pitch polisher. (See Chapter 3.) After several wets of polishing the prism, the 90° side angle (pyramidal control) is checked in the autocollimation setup. Figure 8.2 illustrates the autocollimation of the prism's 90° glass angle which is formed by the reference side and the hypotenuse side being polished. Note that the reference side is resting on a piece of felt under which is a small wooden block to make it level. The autocollimator is set at approximate 45° and then anchored securely. The three polished surfaces are used to return the parallel light through the collimator to the viewer. This internal collimation amplifies the errors in the 90° glass angle by 6×.

A good optician should observe angular errors and treat them as such. The glass-angle errors in prisms are shown by horizontal separation. (See Fig. 8.2.) Any doubling of the vertical line is due to the poor quality of the surface on the top side of the prism, where the parallel rays from the auto-collimator enter and return to the autocollimator. The surface quality polished on the hypotenuse face should be one fringe or better before checking the 90° glass-angle errors.

If the glass angle is in error, an acute or obtuse angle will be made apparent by the amount of separation of the horizontal images. How is it known whether the glass angle is obtuse or acute?

Rule 1. Defocus the eyepiece inward from the best obtainable focus. If the two horizontal images of the cross arms approach each other, the angle is acute.

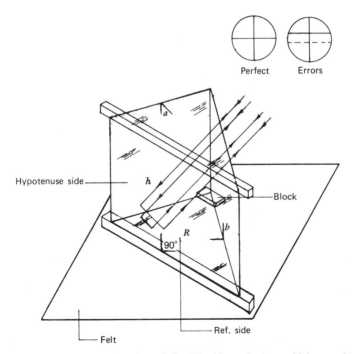

Fig. 8.2. The internal autocollimation of the 90° side angle (pyramidal control) with the polished reference side on a piece of felt.

Rule II. Defocus the eyepiece inward from the best obtainable focus. If the two horizontal lines separate, the angle is obtuse.

Widely separated images are difficult to check by this method. Use a steel machinist's rule as a knife edge and shade the eyepiece at the bottom. If the top image of the separated images disappears, the glass angle is obtuse. If the bottom of the separate horizontal images disappears, the glass angle is acute.

To correct an acute angle on the 90° glass angle apply pressure while polishing the hypotenuse side near the *reference* (marked) surface of the prism. To correct for an obtuse glass angle apply pressure at the *opposite* side of the prism's reference surface. The pitch polisher should always be compressed into a convex contour with concave pressing plates. In this way a concave area is polished in the face of the prism. The surface quality is then slowly increased as the glass angle is corrected.

The hypotenuse face is finished when the surface is completely polished and superimposed images can be seen in the autocollimator. The surface quality should be free of astigmatism, without digs, scratches, or sleeks, and one-eighth fringe or better.

2. THE 45° REFLECTING FACES

The glass runners are removed from the polished hypotenuse face by placing them in a controlled oven heated to 85°C. New glass runners, strongly beveled, are cemented around both 45° faces (The technique is similar to the one described in Appendix 3.) The prism assembly with its glass runners is illustrated in Fig. 8.3. Note that the glass runners extend past the 45° faces and butt together at the vertex of the prism.

Recheck the 45° setting of the autocollimator by placing the plano-parallel on the three nylon balls with the 45° master angle block on it. The autocollimator can be readjusted as required. Check the bifilar, set it to zero, and secure the autocollimator.

In this procedure grind the two 45° faces alternately. Here a compound angle of the prism must be checked; that is the 45° angle formed by the hypotenuse face, one face of the prism, and the 90° (pyramidal angle) side angle. (See Fig. 8.3 *a* and *b*.)

When grinding either of the two 45° faces, make use of the polished hypotenuse face with the preground 45° faces resting on the three nylon balls. Each glass angle is rotated alternately by picking up the prism and replacing it in a stable position on the three nylon balls. During any fine grinding the two glass angles must be corrected simultaneously.

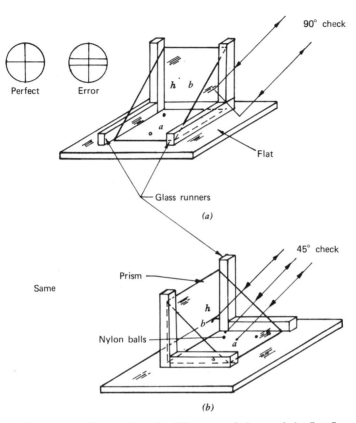

Fig. 8.3 (a) The alternate fine-grinding of a 45° compound glass angle by first fine-grinding the 45° face and then checking the 90° side air angle and rotating the 45° face to the collimator. (b) The external collimation of a 45° glass angle by a preset 45° comparison master angle.

As stated before, there are eight possible conditions for glass-angle orientation. (Review Section 7.4, Chapter 7.) The technique for correcting glass angles is easily learned if it is recognized that there is always one pressure point on the prism. One way to find this pressure point (during grinding) is to touch the prism lightly to locate an area in which the horizontal images approach other. This must be done for each alternate auto-collimation of the two glass angles. It is good practice for beginners to mark this point with a felt pen or crayon. Use this marked side while externally autocollimating the 90° side angle.

The location of the pressure point, which will correct simultaneously for any two glass-angle errors is midway between the two marked areas of the prism. The correction of glass angles can be overshot on fine grinding, especially when 225 or 145 grits are used. Great care must be taken to observe changes in the angular errors or to mark their absence.

When the 95 grind of the two 45° faces is completed and the 45° glass angles and their 90° side angles are corrected, polish the surfaces for several wets. Make sure that each face has a surface quality of two fringes or better before checking.

Check the 45° reflecting angles by placing the prism on a felt pad: this is an internal autocollimation of the 45° glass angle. Note that only the glass angle is involved. (See Fig. 8.4.) The internal angular errors are revealed by the horizontal imagery in the field of the autocollimator's eyepiece. If the images are superimposed, the 45° glass angle is correct. In the event of error there would be a horizontal separation which would show an obtuse or acute glass angle.

The amplification factor of the glass-angle errors is 12×. By the use of Rules I and II it can be determined whether the glass angle is large or small. This also applies to the 90° internal autocollimation used to check the 90° side angle of the hypotenuse side. The reference side of the prism must be placed downward on the felt pad and the proper 90° side angles autocollimated. (See Fig. 8.2.)

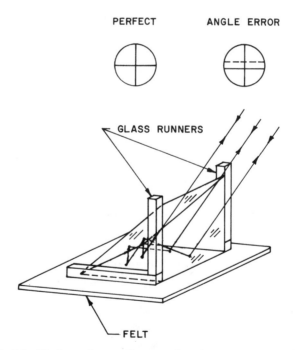

Fig. 8.4. The internal autocollimation of a 45° glass reflecting angle.

3. REFLECTING TOOL PRISM OR DEVIATION PRISM

Fabrication of this prism differs according to the intended use. Tooling prisms often must sometimes have aluminized reflecting angles with errors of 2 sec or less. Deviation prisms, which are used in sighting instruments, are classified as precision prisms if their total deviation is 5 sec or less.

Consider the reflecting tool prism whose required angular errors are 2 sec or less. Recheck the 45° glass angle and mark *A* on one side, *B* on the other. If there is little choice between the 45° faces take one that has the better 90° glass angle. The correction of the 45° glass angle moves at twice the rate of the 90° side angle during polishing. The amplification factor of the 45° glass angle is 12×; that the 90° glass angle is 6×. The rate of correction is due to the amplification factor and the fact that it is hard to maintain even pressure on the angular sides of prisms during polishing.

The 45° face (*a* or *b*) chosen must be worked free of observable angular error when checked by autocollimation (see Fig. 8.4.) and its 90° side angle should be checked. (See Fig. 8.2.) It must then be completely polished and free of any astigmatic surface.

Persistent astigmatic surfaces are often caused by humps (called corns by opticians) at the corners of the prism. These faults must be localized by the use of the thumb or middle finger, but care must be exercised or a hole may be formed in its surface. Allow sufficient time for the latent heat introduced by the hands to leave the prism. The surface is retouched on the polisher after refiguring and is completed if the compound reflecting 45° glass angle is without observable error and has one-eighth wave surface quality. The prism should be free of any digs, scratches, and sleeks.

The second 45° reflecting face, which is worked by a different technique, is placed in a new internal autocollimation setup. (See Fig. 8.5.) Here, for the first time, the pyramidal and sum total of the angular errors is observed. We observe the pyramidal error (now due to this last 45° face) by the extent of the separation of two *vertical* lines. The extent of the separation of the two *horizontal* lines is the glass-angular error of the 45° face being polished. Note this double-cross imagery in Fig. 8.5. The error can be in any quadrant (the dashed line is the pyramidal direction) and can be one of eight possible sets. (Review Section 7.4.)

One of the best methods of locating the error is to find the counterreflection image retrodirected by the prism. Breathe hard (moisture condensation) on the 45° face, away from the autocollimator. Watch in the autocollimator's eyepiece for the reappearance of the retrodirective image as the moisture evaporates from the surface. One image is always reflected from the front semipolished surface nearest the autocollimator. Locate the direction of the pyramidal angle and mark it with a felt pen. This will *always* be

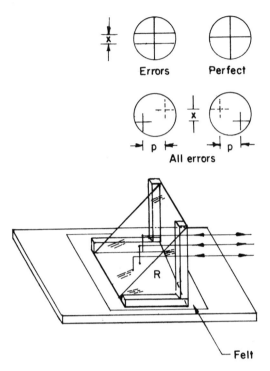

Fig. 8.5. The final internal collimation of a right-angle prism showing all angular glass angles.

one of pressure points used to correct the angular errors in the prism. Again, the separation of the horizontal imagery denotes angular errors in this unfinished 45° face.

The polishing phases for the prism's 45° face are exactly like those for the first 45° face. Pyramidal error and 45° glass-angle error are shown in Fig. 8.5 by (p) and (x), respectively. First, determine whether (x) is obtuse or acute by using Rule I or II (the direction of the pyramidal error has already been determined). Suppose, for example, that the 45° reflecting glass angle is acute (small) and the 90° pyramidal error is opposite the prism's reference surface which has been used throughout while working the prism. The one pressure point that will correct simultaneously for both errors is located at the "toe" (the toe is the end of the 45° glass angle and the heel is the 90° glass-angle corner) and on the opposite side angle of the prism's reference side. By exerting pressure on the toe of the 45° face opposite the prism's reference side this prism's angular errors can be corrected. This example considers only one of the eight possible errors that can occur. The fully reflecting prism is finished when all autocollimated imagery is one on one and its surface quality is one-eighth fringe or better.

4. TWO-SECOND DEVIATION PRISMS

One fabrication method for two-second deviation prisms makes use of precision fixtures that hold the cemented prisms at the prescribed glass angles while generating the surfaces. Another method makes use of plaster blocking during grinding and polishing. Plaster blocks that contain many prisms and large diameters (37 cm) can be made up.

Right-angle prisms made by these methods of fabrication are used in cheaper binoculars. Various assembly procedures are applied in collimation. These prisms have rather large glass-angle departures, as high as 10 min of arc. In mass producing prisms by these methods two sides can be worked by either of the two methods (holding fixtures or plaster blocking). The two surfaces of the prisms that are fabricated in plaster of paris blocks are the hypotenuse and the 45° face. The third side is altered by alternately fine grinding and semipolishing until superimposed imagery is observed in the autocollimator. (Use Fig. 8.5; see also Chapter 2, Fig. 2.22.)

It is good practice to make use of glass runners because most glass-angle errors are generally widely separated in the field of the autocollimator's eyepiece. The assembly of prism and glass runners is given a hasty 95 grind and semipolished for a period of 5 min before checking.

The direction of the pyramidal error (p) is the distance between the two vertical lines; its slant angle can be in any quadrant. The separation of the horizontal imagery (x) is the total of all the glass-angular errors. For wide separation of the two sets of imagery the set from the 45° face must first be determined. The breath test is used. With the autocollimator setup illustrated in Fig. 8.5 one set of images is derived from the side being worked and the second set is the internal autocollimation. The autocollimated set will disappear and reappear as the moisture evaporates from the hypotenuse. Mark on the side of the prism the direction of the *vertical* line of the *second* set. For the time being ignore horizontal separation.

For widely separated images it is necessary to use the machinist scale to determine whether the horizontally separated images denote plus or minus glass errors.

All the necessary information for determining the one pressure point for grinding the prism in the correct direction is now at hand; for example (one of the eight possible conditions), the pyramidal *vertical* line's direction was marked on the reference side and the glass angle is obtuse or plus. See Rules I and II or the machinist scale method. Where is the one pressure point? It is on the heel at the reference side of the 45° face of the prism.

This process of regrinding and repolishing may have to be repeated until the technique has been acquired. No one should attempt by polishing to correct any imagery observed in the autocollimator (i.e., separated by three or more line widths) but should regrind for nearly superimposed imagery. This use of glass runners removes the introduction of gross overcorrection and undercorrection of the glass angles. Any attempt to free-hand the prisms without glass runners is time consuming.

Occasionally because of the large glass-angular errors the two sets of images are widely separated and only one set is located in the autocollimator's eyepiece. In this case the prism may be assumed to be excellent. To guard against this use the eye test. Look directly into the surface of the hypotenuse and observe the roundness of the pupil of the eye. If there is any contraction or expansion of the eye, the prism has large glass-angular errors.

5.   DOVE PRISMS

Right-angle prisms can be changed into 45° Dove prisms by generating and fine grinding a large facet at the apex. This is done, of course, during the shaping of the prisms. During fabrication the use of glass runners is essential. Testing the glass reflecting angles of Dove prisms in autocollimation setups is identical to the testing of right-angle prisms. (See Table 8.1.)

6.   OPTICAL FIGURING OF PRISMS

A precision prism is not finished even if its deviation is 2 sec or better and its surfaces are flat to one-tenth of a fringe. It is necessary to test most prisms for optical path difference (OPD) in an interferometer setup. The

TABLE 8.1.   HARTING-DOVE EFFECT ON THE
CONSTANTS OF PRISMS WHEN DIFFERENT
TYPES OF GLASS ARE USED

$n =$	1.5170	1.5725	1.6170	1.7200
$B =$	1.4849	1.4849	1.4849	1.4849
$L =$	4.6498	4.4303	4.2822	4.0072
$H_1 =$	4.5498	4.3303	4.1822	3.9072
$D =$	2.4498	2.2303	2.1822	1.9072
$d =$	3.7165	3.5071	3.3637	3.1084
$\dfrac{d}{n} =$	2.4499	2.2303	2.0801	1.8071

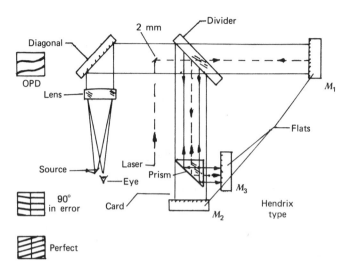

Fig. 8.6. A Hendrix interferometer setup with a third mirror (M_3) for maintaining a stable mode with mirrors M_1 and M_2; this makes placement of the prism easier during testing. Note the possible usage of a low-power laser for alignment.

OPD in prisms is caused by nonhomogeneity in the glass and for this reason the glass must be checked for homogeneity. Because optical glass is costly, especially in large blocks, the optician must deal with OPD introduced by nonhomogeneous glass.

Optical figuring is the technique of localizing one face of a prism to an optical-path difference (OPD) tolerance of one-quarter fringe or better. This quantitative measurement is a double-passage value of the transmission fringes when the prism is placed in the interferometer. Assume here that the transmitted fringes (OPD) have been divided by two and the localized face is the entrance face. The second face of the prism becomes the exit face and to prevent any mixup of surfaces is marked with a piece of tape. When set up in an interferometer some prisms have foreshortened areas and it is difficult to locate the interferogram. For better and faster replacement of the prism during optical figuring use a third retrodirective flat (M_3) near the prism. This permits stabilization of the interferometer and a reexamination of the vertical fringes in a normal setup. Another method uses a low-power laser (0.3 mW), directs the light through the beam splitter, the prism, and the third flat, and then superimposes each laser spot. Care must be taken to wear safety glasses and to turn off the laser after the setup has been made. (See Fig. 8.6.) The terms figuring, localizing, and retouching are synonymous.

The fringe patterns caused by OPD are generally astigmatic, as shown in

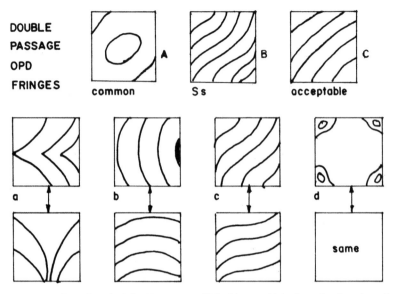

Fig. 8.7. A number of typical double-passâge OPD fringes which show the direction of the lack of homogeneity.

Figs. 8.7(a)–(d), but the astigmatic character of the polished surfaces contributes little.

It is important to localize only the face of the prism that becomes the entrance face. Do not try to balance the figuring between the entrance and exit faces. The reason for this is that large prisms, such as the penta, which has long light paths, sometimes may have two or more fringes of OPD. The exit face and changes made during retouching must remain as an auto-collimation device for rechecking the deviation. (See Chapter 9.) A rough calculation indicates that a four-fringe change of the prism's surface is required for one fringe of OPD (4:1). This is why beginners are startled by the amount of figuring.

Let us consider the figuring of a typical prism with OPD fringes, as shown in Figs. 8.7(a)–(d). Because the fringe pattern is astigmatic, it must be checked in two planes. Checking is done by adjusting the screws on the M_3 holder of the interferometer. (See Fig. 8.6.) Locate the poorest plane and determine whether it is convex or concave (high or low). Pull slightly back on M_3, thereby increasing the light-path distance between the prism and the flat. If the fringes run off the prism's edges, the surface is convex. Conversely, if the fringes run to the central area, the surface is concave. (See Fig. 8.8, which is similar to that described by Johnson.*)

---

* Johnson, B. K., *Optics and Optical Instruments*, Dover, New York, 1960, pp. 200–202.

By locating the poorest plane draw the main fringe pattern on the prism with drafting ink or rouge.

Polishers in several sizes are made of medium-hard pitch with 50% wood flour. Some eminent opticians, like the late Harold Hall and Don Hendrix, used polishers twice the diameter of the prism's largest surface. The laps had 0.5-in. (12.5-mm) square facets embossed with 0.020-in. (5-mm) facets. The pitch laps were contoured with concave or convex pressing plates and a surface quality ranging from two to six fringes. A convex contour lowers a high area and one that is concave brings up a low area. In any case, glass is always removed. Only computers can add glass to optical surfaces!

Consider Fig. 8.7(*b*). The surface OPD is one-half fringe convex across one plane and perhaps one-half fringe concave for the second. Choose a convex polisher with a two-fringe contour and polish across this *one* determined convex plane for several wets. Do not allow the prism to rotate in the hands either during the stroke or after finishing. Always begin a new series of strokes on a new area of the polisher (walk around it). Check the surface quality with an optical plane to determine whether progress is being made in the right direction. Find an astigmatic surface (one astigmatic surface actually is off-setting another during optical figuring). Recheck the interferometer when the surface is three or four fringes and allow 5 to 10 min for the latent heat to dissipate. The central area should become flatter, whereas the second plane will perhaps remain the same but will rotate to another quadrant. During the final figuring birdlike fringe patterns develop with wing tips turned up or down. An optician will often be seen using the thumb or middle finger to retouch the humps or corns of the fringe patterns. This can easily be overdone and the latent heat the prism has absorbed from the fingers must be considered.

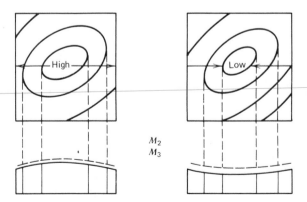

Fig. 8.8.   The movement of fringes on an optical surface for high or low areas.

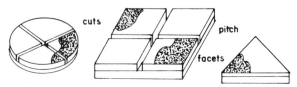

Fig. 8.9. Typical small polishers used to localize the OPD fringes.

Another method of localizing is the use of smaller polishers. (See Fig. 8.9.) The polishers are generally smaller than the surface of the prism. Likewise the polishers are contoured by placing them on top of the pressing plates with dry polishing compound to prevent them from sticking. In time they will anchor themselves and must be removed with a rawhide or plastic hammer. The concave area (low) on a prism's surface is raised by lowering the extreme edges. During this figuring the edges must be closely watched to prevent them from turning down.

Even if the surface were chopped up by bad localizing the surface could be reestablished to approximate the original surface. Perhaps making the surface one to two fringes convex before beginning the figuring on a new high area will work. It is impossible to describe in detail the figuring of all OPD fringe patterns. Figure 8.7(a) is a difficult case and others are of like nature. An extreme case is a penta prism with perhaps two to three fringes in one plane and a half-fringe in the second plane; then 8 to 12 fringes are required for figuring in the poorer plane. A loss of deviation previously worked for the prism during internal autocollimation is also apparent. Of course, to reestablish this deviation tolerance the entrance face must be reworked. *Do not* polish the corrected OPD face. Make use of tape to prevent a mixup.

EXERCISES

1. What is the difference between a right-angle prism and a 45° deviation Dove prism?

2. Which face of a right-angle must have the best surface quality and why?

3. Make a small right-angle prism from BSC-2 glass or its equalivalent with 45° glass angles ± 5 sec of arc and no pyramidal error. The glass surfaces to be one-quarter wave and have less than one-eighth irregularity.

4. Is a prism with the optical tolerance described in No. 3 required to meet the OPD tolerance? How is this checked and how is this one-quarter wave tolerance achieved?

5. Can a right-angle prism with one 45° glass angle 8 sec plus a second 45° glass angle, 8 sec minus, meet the deviation tolerance of 2 sec? Why?

BIBLIOGRAPHY

Chalmer, S. D. and Ryland, F. "Collimating Prisms," *Trans. Opt. Soc., London* (1904–1905).

De Vany, A. S. "Making Right-Angle and Dove Prisms," *Appl. Opt.*, **7**, 1085 (1968).

De Vany, A. S. "Testing a Prism's Compound Reflecting-Gass Angles," *Appl. Opt.*, **14**, 22791–2792 (1975).

De Vany, A. S. "Correcting Glass Angles of a Prism, *Appl. Opt.*, **16**, 2019–2021 (1977).

De Vany, A. S. "Testing Glass Reflecting-Angles of Prisms," Appl. Opt., **17**, 1661–1662 (1978).

Ferson, F. B. "Prisms, Flats, Mirrors," in *Amatuer Telescope Making—Advance*, A. G. Ingalls, Ed., *Sci. Am.* 75–93 (1959).

Glazerbrook, R. *Dictionary of Applied Physics*, Vol. 4. Macmillan, New York, 1923, pp. 117.

Guild, J. *Proc. Opt. Soc. (London)*, **22**, No. 3, 139 (1915).

Guild, J. *Proc. Phys. Soc. (London)*, **28**, 60 (1915).

Guild, J. *Trans. Opt. Soc. (London)*, **22**, 153 (1920–1921).

Johnson, B. K. *Optics and Optical Instruments*, Dover, New York, pp. 192–198.

Moffitt, G. W. "Collimation of Prisms," *J. Opt. Soc. Am. Rev. Sci. Instrum.*, **7**, 830 (1923).

Tyman, F. *Prism and Lens Making*. 2nd ed., Hilger and Watts, London, 1952. pp. 294–307.

Wagner, A. P. *Experimental Optics*, Wiley, New York, 1929, pp. 11–120.

Wright, F. E. *The Manufacturing of Optical Glass and Optical Instruments*, Bureau of Ordnance, Washington, D.C., 1921.

Wright, F. E. "Prism Making," *J. Opt. Soc. Am.*, **5**, 148–152 (1921).

9

Duplication of a Prism Angle with Master Angle Prisms and a Planointerferometer

Early use of Twyman-Green interferometer with which to compare glass angles of prisms was discussed by Twyman. The duplication of prism angles can be interferometrically controlled by using master angle prisms that are optically contacted on the polished reference surface of the prism. The planointerferometer is a modified Fizeau interferometer illuminated by a low power laser (0.3 mW) or a reed-type mercury source. Laser illumination makes it possible to project the interferograms on a frosted screen and the mercury source permits naked-eye observation of the interferometers. These bright sources make it possible to contact the master prisms optically several centimeters from the surface of the prism and the interferogram can still be observed. Another uses the reflecting angular surfaces in the grinding stage. Some general optical procedures for making a typical prism, such as penta prism, which has several large, varying glass angles, are discussed. Testing a typical prism by planointerferometer and duplicating the glass angles on a penta prism are compared interferometrically with the master angle prism. Other optical and test procedures are discussed.

1. PLANOINTERFEROMETER DESIGN

A typical planointerferometer design is one that uses a planoconvex lens in which the planosurface serves as an optical flat and the convex surface can be a spheroid for a laser-illuminated sources or a hyperbolic surface for the mercury source. The aspheric figuring is not difficult if the process is began with a rose-leaf polisher with a 0.707 lens diameter. Testing is done with a knife-edge or a Ronchi grating of five lines per millimeter and is figured null in conjunction with its optical back surface. This is, in essence, Norman's optical test for making hyperbolic secondary mirrors for a Cassegrain telescope. The light source for the pinhole must have the wavelength ·of the intended source for the planointerferometer. The lens material for a

4-in. (102-mm) or 6-in. (153-mm) aperture at f/6.0 can be ULE quartz or preselected CerVit. (See Fig. 9.1.)

One other design has an $F/6.0$, two-element objective that has been color corrected for the intended light source of the planointerferometer. The other element is a wedge whose first side serves as the master flat (one-tenth fringe); the second side has a wedge on it to reflect the internal light of the source out of the system.

In both designs chosen for the planointerferometer we fold the optical light path by incorporating a small reflecting flat to ensure a compact assembly. An eyepiece with a large field (Erfle) is used to help locate the two images that form the interferogram when they are superimposed. Of

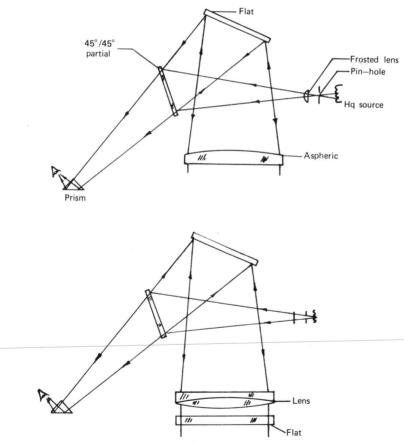

Fig. 9.1. Two planointerferometer designs used for testing flat optical surfaces.

course, the eyepiece is rotated away from the light path to observe the projected interferogram. Neutral filters of sufficient density must be used in laser-illuminated light sources and are located at the eyepiece opening. The tilting table for testing the prisms should be a separate unit. At times, although infrequently, the planointerferometer must be rotated from its conventional horizontal position; for example, when the optical contact has been broken during polishing and the master prism must be reestablished interferometrically on the penta prism's surface by optical contacting. This sounds unlikely, but it can be done.

A planointerferometer has many uses in an optical shop; for example, to test the parallelism of wedges, the surface quality of flat surfaces, and homogeneity of glass by the Twyman-Perry method, for duplication of prism angles, and for the height of master "Jo" gauge blocks. With this interferometer the polished surfaces are free of digs and scratches because they cannot be marred by an optical flat. Davidson of West Covina, CA, makes these planointerferometers in diameters of 4 to 12-in. (See Fig. 4.5.)

2. MASTER ANGLE PRISMS

Master angle prisms are made of optical glass, generally the BSC-2 or BK-7 type. Here the interest is primarily in glass angles of 22.5, 30, 45, 60, and 90° which can be tested interferometrically with three-angle prisms on a optical plane by following Deve's method of superimposition. One of the most commonly used master angle prisms is the right-angle prism, which has two 45° reflecting faces and one 90° face. These prisms are easy to make if the optical procedures outlined in Chapter 8 are followed. These masters are made with the corners and ends ground back and all edges have 1-mm bevels. These prisms can also have any base length, but for easier working 1-in. (25.4 mm) is perferable. The final shaping is done when the master prisms are finished. The strong bevels and ground-back corners make optical contacting and decontracting easier.

If these prisms are worked as right-angle prisms, there will be precise angles on all the reflecting surfaces of the masters. The 90° side angles to these surfaces must be polished and held to 5 sec. The glass angles can be improved by retouching them on a polisher and measuring the angles interferometrically by Deve's method. For ultraprecise measurements Deve should have expanded on one point: the fringes of the master prism resting on the optical plane must run with the horizontal plane of the prism angle and in turn the vertical plane must be perpendicular to the optical flat. The master prisms should not be aluminized because the contrast between the polished surfaces, one on the prism, the other on the master prism, is destroyed by this treatment.

3. MAKING THE PRISM

Now examine the way in which the master right-angle prism is used in the production of a penta prism. These techniques are applicable to many types of prism, such as Perchon, Porro, Dove, and Schmidt. Note that the 45° angle end of the penta prism is the most important because the light path is doubly reflected at this area. In this discussion it is assumed that some preliminary work has been done; for example, the penta prism is in final form and at top tolerances in all dimensions and the same reference surface has been used to grind in all the angles with a precision protractor and a precise 90° square to the reflective angles. The reference polished side must be protected in order to contact the master prism. This is done by using a cover glass and a sheet of lens paper cemented on the reference surface which help to prevent scratches. Any glass-angle error will be ground out later when the master right-angle prisms are used on the reference surface of the penta prism. The cover glass is removed by heating the prism in a 100°C (212°F) oven, cooling it to room temperature, and washing it in xylene or acetone and then in a detergent and water solution.

The clean prism is then placed on three prewaxed nylon balls as described in Chapter 6. Here small (2.5-mm), precise, aluminized planoparallel plates are used to make the ground surfaces reflective by placing them (wet-wrung) on the fine-ground surfaces of the prism. This is more thoroughly discusssed in Chapter 6.

Optical contacting is not covered in this section. (See Appendix 9.) It is sufficient to mention that cleanliness is essential and that all work must be done in a dust-free area. Several antistatic liquids are available on the market which are used on the prism after it has been cleaned with acetone and twice-washed linen. The master right-angle (the 45° side) is started at a corner and a contact is made. Watching for fringes to appear, move the prism slowly toward the central area. Then tap it into place until the images in the collimator are nearly superimposed. The optical contact is made firm by spraying with acetone; the bevels can then be lacquered all around the master prism with fingernail polish. Although difficult, this can be easily accomplished after some practice.

The contacted master right-angle prism with its 45° reflective angle running parallel to the penta prism's 45° face is shown in Fig. 9.2. Often only one face of the penta prism can be made nearly parallel as revealed in the image patterns. Small separations of several minutes of arc, shown as vertical lines in the collimator, will be corrected later. Here the master right-angle prism is contacted as closely as possible.

Decontacting is easy: first clean off the nail polish with acetone; then take an orange-stick wedge and pry up the right-angle master under one of the

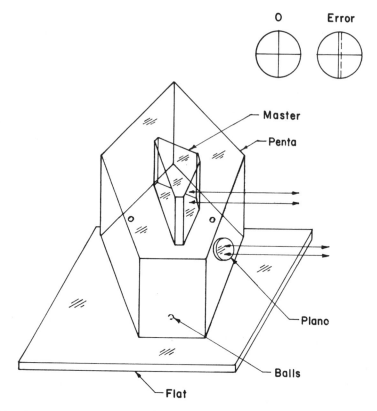

Fig. 9.2. Optical contacting of the master right-angle prism to the reference surface of a penta prism: the 45° angle of the master prism is made to coincide with the penta prism's 45° reflective faces with a collimator and a planoparallel.

ground-back corners or bevels and watch for the fringes to run away from the prism. Stubbornly contacted prisms are decontacted by using a small Bunsen burner and gently heating the master prism. A clear clicking sound will be heard when the prism is decontacted. Prying under the bevels with a shop razor blade is a technique oftern used.

4. GRINDING AND POLISHING THE 45° ANGLE

Use is made of the master right-angle prism in the grinding and polishing phases. Figure 9.3(*a*) and (*b*) shows the external autocollimation setups for checking the 45° compound-glass reflective angle with one collimator. This is done by rotating the prism alternately. Because the prism is resting on three prewaxed nylon balls on an old aluminized flat, this can be an accurate procedure. Before grinding the top surface wet-wring the aluminzed plano on the penta prism surface and compare the returned images observed in the autocollimator from the plano and 45° glass angle of the master prism. If any separation in noted, the penta prism must be fine-ground in the determined direction to observe the superimposed image

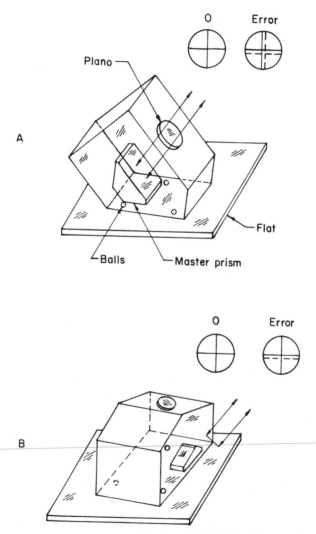

Fig. 9.3. Check the alternate external collimation of the 22.5° supplement glass-angle setup by first comparing the 22.5° glass angle; the 90° side air angle is compared by rotating its 90° side to the collimator.

in the collimator. Use the "touch method" in which the prism is touched with the finger to locate the pressure point for correcting the 45 or 90° glass angle that causes the separated images to draw together, at the same time observing in the eyepiece of the autocollimator. Make sure that the collimator is viewing the reference side of the prism and not the parallel side of the contacted right-angle master prism (it may not be parallel to the second side if separate images are observed). A cutout section of cardboard placed on the master prism is helpful. If only one image is observed when the cardboard is removed, both sides of the master prism are parallel. In any case, always use the reference side of the prism. It is often necessary to correct two glass angles on the penta prism simultaneously by locating one pressure point on the penta prism during fine grinding. Press in the desired direction on the penta prism until the images come together in the collimator. The midpoint between the two pressure points found on the prism is the area in which the pressure is applied during fine grinding. The penta prism is ready for polishing when the images in the two setups of the collimator are superimposed, as shown in Figs. 9.3. Note that the ground surface should be one to two fringes concave because it does not pay to polish two convex surfaces, one on the penta prism, the other on the convex contour (use of pressing plates) of a pitch polisher.

A fine-ground surface (95 grit) can be checked by placing a small optical flat (shop use) on the surface of the prism and observing the fringes under a monochromatic light source. The whole assembly must be held above eye level. The number of of fringes in this setup are multiplied several times.

The second side of the 45° surface of the penta prism can also by fine-ground by the same grinding and testing procedures. If the work has been carefully done in the fine-grind phase and the images in all the collimating setups are superimposed, and glass-angle errors can be corrected easily by polishing and the angles on both prisms can be interferometrically duplicated by the use of planointerferometer.

5. USING THE PLANOINTERFEROMETER

The first surface of the 45° surface is semipolished, placed in a felt-covered, precut wooden holder, and set on the tilting table of the interferometer. Lower the interferometer to 1 cm of the surface of the prism and in the eyepiece locate the two images that, when superimposed, form the interferogram. By rotating the eyepiece out of the way the interferogram can be observed directly. Adjust the tilting table and cause three fringes to spread over the 25-mm surface of the master right-angle prism. If equidistant, straight, and parallel fringes are observed on both right-angle master prism and penta prism, the two surfaces are duplicates. It is more likely that in the two planes checked one plane with parallel and straight-line fringes and the other with slant fringes in relation to the parallel fringes on the master right-angle prism will be observed. There can be a number of variations in the slant of the fringes in the two planes. The slant and number denote the glass-angle error which can be acute (minus) or obtuse (plus) on the penta prism. (See Fig. 9.4.) To determine the lowest area or runoff point press down lightly on the tilting table with the prism to cause a longer light path. The fringe runoff point on the penta prism is the lowest area provided that the referencing fringes on the right-angle master prism are properly set. A fellow worker always says of fringes: "Fringes like water, run downhill." The high area on the prism must be 180° from the runoff point on the penta prism. It is necessary at all times to check this point in two planes: one at right angles to the penta prism and the second perpendicular to the first plane on the master right-angle prism. (See Fig. 9.4.) In Fig. 9.4 the circle drawn with four quadrants illustrates the runoff area of the fringes on the penta prism. Observe it for both planes in relation to the master prism.

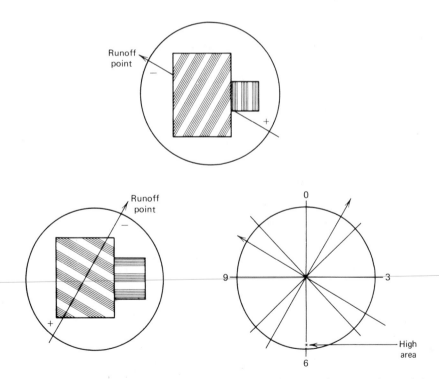

Fig. 9.4. Determination of the interferogram runoff in both planes of a penta prism and the location of the highest area.

Mark these low areas on the circle; the point midway between the two points is the lowest area on the penta prism. The point or area on the penta prism 180° or opposite this midpoint is the highest area. (See Fig. 9.4.) To locate the highest area of the penta prism apply pressure while polishing to correct the glass-angle error simultaneously in both planes of the prism. It is obvious that the planointerferometer will reveal surface faults, such as astigmatism, cylinder, and turned edges. The first surface is finished when it is one-eighth fringe or better and completely polished, with equiparallel and straight fringes on the penta-prism surface and the master right-angle prism.

Polish the second side in the same manner and again, to correct the glass angles, locate the highest area on the penta prism by applying pressure while polishing. Here perhaps the glass angles are duplicated by interferometric testing. Some experience must be acquired before it is realized how much faster penta prism and other precise prisms can be mass produced by using a number of master angle prisms. All the possible variations of the slanting patterns demonstrating glass-angle errors cannot be treated here. In many ways this system is analogous to the interferometric measurement of the height of an unknown against a known gauge block, the only difference being that on the gauge blocks the contact point must be found, and in this case the reference point is always the master angle prism on the reference surface of the prisms being made. A rough estimation of the error in a prism being worked is that one fringe over a 50-mm area of the prism surface is equal to 1 sec of arc.

6. GRINDING AND POLISHING THE 90° ENTRANCE AND EXIT FACES

Before proceeding with the next contacting of the master right-angle prism to the reference surface of the penta prism do some preliminary fine-grinding on one 90° surface. Consider the three 90° glass angles formed by the entrance and exit faces of the penta prism: the actual 90° glass angle formed by the entrance and exit faces and the two 90° side angles, one each at the entrance and exit faces. If the penta prism is set on one of its 90° faces in an autocollimation mode, it will form a 22.5° angle. A master 22.5° is built up from the set of master angular blocks and a precise planoparallel plate (5 sec or less) is placed on the three prewaxed nylon balls. Next the autocollimator is set to the reflective image from the surface of master 22.5° angle and for a better image relection the small planoparallel plate can be wet-wrung to its surface. The collimator must be securely tightened and the bifilar set to the reflective image. The planoparallel plate can be removed and the penta prism placed on the three balls. [See Fig. 9.5 (a) and (b).]

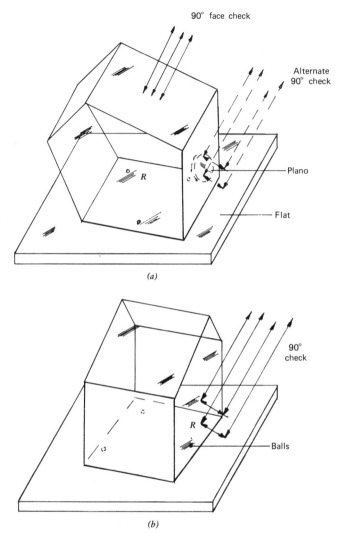

Fig. 9.5. The preliminary fine-grinding of the 90° entrance and exit faces of a penta prism before the optical contacting of the master 90° face of the right-angle prism. Note the alternate rotational check of each surface of the compound 90° glass angle in (*a*) and (*b*).

Now pick up the reflective image on the polished surface of the 45° glass angle of the penta prism. Any separation of the returned reflective image from the preset cross hairs in the collimator will show the error. In the next step the prism is rotated to check the 90° side angle shown in *b* by placing the small planoparallel plate on the surface of the prism facing the collimator. First check the 90° side angle, using the polished reference side of the penta prism and then rotate the penta prism to check its 90° side with the wet-placed planoparallel plate. Here is a 90° compound glass angle that must be fine-ground, as described in Section 4, and all reflective images must be referenced to the cross hairs of the bifilar of the collimator. After this preliminary grinding proceed to contact optically the master right-angle prism's 90° glass angle face which runs parallel to the ground face of the penta prism. Of course, no fine grinding has been done on the second face in conjunction with the contacted master right-angle but it will be in time. To contact the master right-angle prism follow procedures outlined in Appendix 9. After contacting the master prism according to Fig. 9.6 (a) and (b) fine-grind the entrance face and the exit face anew, but this time use the

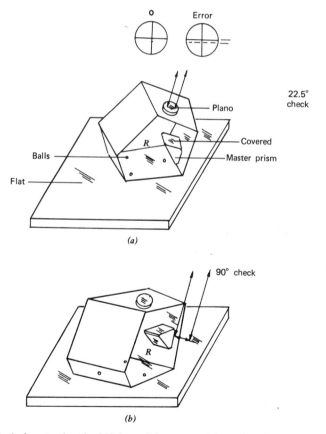

Fig. 9.6. Optical contacting the 90° face of the master right-angle prism parallel to the wet-wrung planoparallel on a penta prism: (a) 22.5° check; (b) 90° check.

90° angle of the master prism which has been optically contacted on the reference surface of the penta prism. Perhaps the slight error observed in the collimator is due to the errors introduced while optically contacting the master right-angle prism. These 90° faces contain small errors in relation to the critical 45° angle and no perceptible error in the performance of the penta prism will be noted. In any case, when the exit face is worked, all existing errors will be corrected. It must be stressed here that all glass angles are fine-ground to superimposed images in the cross hairs in the collimator test; that is three 90° angles—one formed by the entrance and exit faces and two side angles, one to the other. Again, care must be taken to check the 90° side-angles (pyramidal control) to make sure that the collimator is operating off the side of reference polished surface and not off the master right angle prism. Many master prisms may lack parallel sides. This can be checked by using a piece of cardboard to cover the polished surface of the master right-angle prism contacted on the penta prism.

The polishing and duplication of the 90° entrance and exit faces are described in Section 4. The entrance face is semipolished and a check is made with the planointerferometer. In review, check the penta prism's faces, referencing to three equiparallel and straight fringes on the master right-angle prism in the plane being checked. There will be two runoff points or areas if the interferograms on the polished surface of the penta prism have slant patterns. The duplication of both prisms must be one-eighth fringe and completely polished. This is true only if the prisms are homogeneous and need not be optically figured. (See Appendix 8.)

The last face of the penta prism, which is the exit face, is worked with the entrance face. Semipolish this surface according to the method used in completing the penta prism's entrance face, but stop short in this polishing phase when the duplicated glass angles are within a fringe or two of each other in both planes of the prisms and the surface quality is about one-quarter fringe. (Of course, all testing is done with the planointerferometer.)

To observe the internal reflective images in the collimator the penta prism must be aluminzed or coated with silver on two surfaces—one side of the 45° surface and the finished 90° entrance face. (See Appendix 7.) The coated prism with its semipolished surface is checked by placing it in front of the collimator. Here for the first time the pyramidal angle and all the angular errors are observed. Any pyramidal error is shown by two vertical lines and two horizontal lines indicate angular glass errors. These multiple image patterns are shown in Fig. 7.2 (1–8) and their correction is discussed in Chapter 7.

If the images in the collimator test are widely separated and polishing is not bringing them closer together, there can be a number of causes. First, duplication has not been faithfully carried out; second, the master right-angle prism is in error; and third, the master prism is too small in proportion to the size of the prisms. Regrind and repolish in the two right-angle directions and superimpose the images in the internal autocollimation setup. (See Chapter 7.)

Note contact was not broken with the master right-angle prism. Perhaps in checking the penta prism by autocollimation it became apparent that glass angles could be corrected by polishing. The polishing is continued and checked interferometrically with the planointerferometer and using autocollimation of the penta prism until the polishing is completed and the images in the reflective glass angles are superimposed. Of course, the autocollimation of superimposed images is the result and duplication of this glass angle of the penta prism is not required.

Surface quality at this point in the polishing can be one fringe or better and on the convex side. Because of the small variations in the refractive index, zonal refiguring necessary to best performance is begun in a high area. (See Appendix 8.) All prisms are tested in a Twyman-Green or Hendrix interferometer to determine optical path differences (OPD). At

Fig. 9.7. A Hilger-Watts spectrometer and goniometer which is used to verify the refractive index values of glass at various wavelengths. As a goniometer it is used to verify the glass angles on prisms and wedges. See Appendix 19.

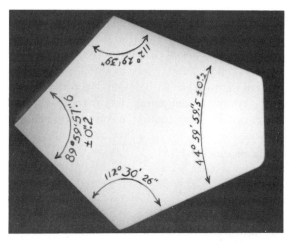

Fig. 9.8. A penta prism with the critical glass angle verified by the Hilger-Watts instrument. See Appendix 19.

times, which must be expected, too much retouching of a penta prism causes the deviation to change. The coated entrance face then must be reworked by polishing in the correction. The retouched exit face must be identified by a small pieces of masking tape to prevent the mistake of polishing the wrong face.

7.   CONCLUSION

These procedures have been used successfully in the mass production of precise prisms of many types in conjunction with master angle prisms optically contacted on their reference surfaces. The planointerferometer can control the duplication of the prism's glass reflective angles by comparison. With careful workmanship the autocollimated images observed in the collimator can be superimposed by proper methods of polishing the prisms for final correction. The acceptance of the prisms must be fixed by an optical path difference of one-quarter fringe (single passage in the interferometric test) or better, as determined by the interferometer check and its deviation angle. (See Figures 9.7 and 9.8.)

EXERCISES

1.   Name at least four uses of a planointerferometer for optical testing of prisms.

2. Name four advantages of its use on a contacted master prism while making a typical large prism.

3. Ray-trace a parallel light beam from an autocollimator through a penta prism.

4. Contact a small right-angle to a larger right-angle prism by optical contacting or by pitch cementing it to another prism. Place the assembly under a collimator and make each surface coincident with the other.

5. What are penta prisms used for?

BIBLIOGRAPHY

De Vany, A. S. "Reduplication of a Prism Angle Using Master Angle Prisms and a Plano-Interferometer," *Appl. Opt.*, **10**, 1371 (1970).

De Vany, A. S. "Testing a Prism's Compound Reflecting Angles," *Appl. Opt.*, **14**, 2791–2792 (1975).

Deve, C. *Optical Workshop Principles* (translated by T. L. Tippell), Hilger and Watts, London, 1954.

Johnson, B. K. *Optics and Optical Instruments*, Dover, New York, 1960, p. 4.

Smith, W. J. *Modern Optical Engineering*, McGraw-Hill, New York, 1966, pp. 93–94.

Twyman, F. *Prism and Lens Making*, 2nd ed., Hilger and Watts, Londong, 1952, pp. 294–313 and 406–413.

10

Rhomboid Prisms

Rhomboid prisms have limited use in optical instruments. They are found principally in binocular eyepieces for miscroscopes and in submarine periscopes, range finders, and artillery sights. Because these rhomboid prisms have long light paths (at least three to four times their apertures) the glass must be A-A quality. Never assume that it is free of striae, bubbles, or other nonhomogeneities even if it is the best grade. It should be emphasized that the glass blanks must be checked in three directions in tests of homogeneity. (See Chapter 5.)

Because rhomboid prisms have two parallel sides, use is made of these planoparallels. The test for homogeneity is as follows: the preselected blanks are made into planoparallel plates on a Blanchard generator by alternately generating each side in a vacuum holder. A minimum of glass is removed. Perhaps a thickness of 0.02 in. (0.5 mm) above the top tolerance of the finished prism will suffice for grinding and polishing. The glass blanks can be checked for larger bubbles, stone inclusions, and other defects by wetting them with kerosene or matching refractive index oil. If premade planoparallel blanks are available (5 sec or better, Schlieren quality), they can be tested for homogeneity by carefully placing them on the generated parallels. Plate glass can be substituted for these Schlieren elements, provided that it has been pretested. These plate glass disks may have a rainbow tint with an astigmatic appearance that will not materially effect the test (see Murty's test, Chapter 5).

After this preliminary test the glass blank must be checked through the edges in two directions. The edge windows must be ground and polished to a surface quality of one fringe or better and the parallelism, held to 0.003 in. (0.08 mm). This can be accomplished only by the use of glass runners cemented to the disks. Parallelism must be checked by a height gage. (See Fig. 10.1.) Cementing glass runners to optical elements is discussed in Appendix 3.

This method is time consuming, but it pays off because these elements have perhaps the longest light paths of any optical element. Also, the optical

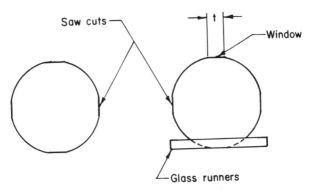

Fig. 10.1. Four small windows are saw cut in the glass blank to test for striae edgewise. If polished surfaces are required, glass runners are pitch-cemented to the glass blank.

figuring of the OPD tolerance is difficult because of the foreshortening effect caused by the 45° front reflecting surfaces.

1. WORKING THE PLANOPARALLEL ELEMENTS

The generated elements are made into plano parallel plates by the following methods: the elements are fine-ground with 250 grit by hand or by placing lead weights on the wedge areas. The wedge area is located by measuring with a differential micrometer. Loose-grit grinding by hand is done by applying pressure on the highest area while grinding on a flat grinding tool. The wedge angle moves to different areas as it becomes smaller. Only one surface is fine-ground at this time. When the wedge is within 0.004 in. (0.01 mm) or less, this side is worked by the polishing machine in a conventional manner. When the hand-grinding method is used, unequal grinding areas (called witness marks) develop.

Loose-grit grinding of planoparallel element to remove the wedge is discussed in Chapter 14.

Briefly, the second side of the element is fine-ground by placing a small lead weight on the wedge area. After several wets the glass disks are remeasured; six areas around the periphery must be used. On the opposite side mark the measured values with a red felt pen. Both surfaces are given an equal amount of grinding time with 250 aloxide grit.

Fine grinding with 145 and 95 grit is a similar procedure. Before leaving the 145 grit the elements must have less than 0.0002 in. (0.005 mm) of wedge. It is possible to do better if the optical shop has a three-point, circular checking fixture and an electronic Sheffield Accutron sensing instrument. Beginners have difficulty in learning how to use the differential micrometers because the surfaces are not free of the grit and lint which cause a false reading. The finished grinding contour should be made slightly concave by grinding on a convex ceramic tool with 95 grit.

Both sides of the planoparallel plates are given a semipolished surface. The surface quality should be one or two fringes or better. One surface will generally be convex and the other, concave because of the Twyman effect. The semipolished surfaces are checked in a Fizeau interferometer. (See Chapter 14.) With this interferometer two fringe patterns can be observed: first, the surface quality of the top surface; second, the Fizeau fringes of the parallelism between the two surfaces. Ignore the fringes of the surface quality and observe the bottom-fringe patterns. Several sets of Fizeau fringe patterns are possible. One is a number of equidistant parallel lines. The second consists of irregular oval fringes in an astigmatic pattern. If the second pattern is observed, the parallelism is good. The first set of parallel lines represents the residual wedge. To measure the wedge in either set of Fizeau fringes, the following formula is used:

$$\text{seconds of wedge} = \text{number of fringes} \times 1.5/\text{diameter}$$

for example; 3 sec. = 30 × 1.5 divided by 15. This would be good enough for most rhomboids prisms. If a more precise tolerance were required, the planoparallel sides should be made more parallel. Locate the direction of the wedge by placing one finger on a Fizeau fringe. Watch the expansion of the fringe. It will actually point toward the thinner area of the element. The thicker part of the wedge is directly opposite. Most opticians will finish polishing one side before attempting to remove the wedge angle.

The second side is polished and the wedge removed by placing a lead weight wrapped with masking tape on the polished surface. The amount of lead weight depends on the diameter of the planoparallel side and the sector wedge for its placement. This produces an astigmatic pattern in the surface quality after several wets which is removed by using a longer overarm

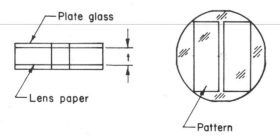

Fig. 10.2.   Pitch-cementing glass cover plates for protecting the polished surfaces.

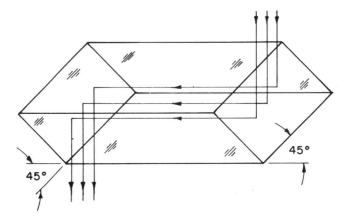

Fig. 10.3.   Tracing the light path through the rhomboid prism.

stroke with a slower rpm than the rotation of the work. Check the Fizeau interferometer again until the wedge tolerance is achieved.

The surface quality on both surfaces should be one-half fringe or better and the wedge angle, 3 sec or less. Plate glass covers with lens paper cemented between the polished surfaces are shown in Fig. 10.2. (See Appendix 3.) This will help to prevent scratching on the polished surfaces and chipping during the sawing and generating operations. The blanks are covered with masking tape when the two or more rectangular sections of the prisms are laid out. Figure 10.3 illustrates a typical Rhomboid prism. Caution: allow 3.2 mm for saw cutting the two 45° reflecting angles and even more if the prisms are large. The excess stock is easily removed by the generator. The glass rectangular slabs must be about 13 mm longer than the finished prism. See Fig. 10.4. for further details.

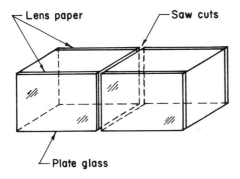

Fig. 10.4.   Angular saw cuts of 45° across the rectangular glass slabs for the rhomboid prisms.

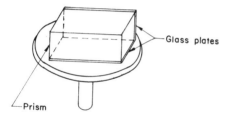

Fig. 10.5. Generating the 90° side angles of the rhomboid prism.

2. WORKING THE 90° SIDE ANGLES

In Fig. 10.3 there are two 90° side angles to the two 45° reflecting glass angles, which must be fine-ground with some degree of accuracy for pyramidal control and parallelism to fit in the holding fixture. A machinist's square can be used to obtain a desirable 90° side angle. The presawed rhomboid prism is finely ground with 250 grit on a flat-surface grinding tool. After squaring to a light-tight square pitch cement the prism on a flat aluminum holder and remove a minimum of glass stock. The reverse sides are generated down to 0.02 in. (0.5 mm) of the top tolerance for their width. The holding fixture on which the prism is mounted is shown in Fig. 10.5.

The final milling or generating of a rhomboid prism involves the two 45° reflecting surfaces. The prism is mounted on its hypotenuse face with the protective glass cover still in place. A special tool holder with a 45° aluminum angle block mounted on a flat holder is used. (See Fig. 10.6.) It is most important that the prism slabs be cemented tightly to the 90° side plate on the 45° angle block and the hypotenuse surface. After milling the

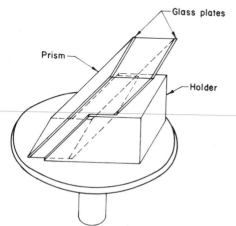

Fig. 10.6. Generating the 45° compound glass reflecting angles of the rhomboid prism in a special holder.

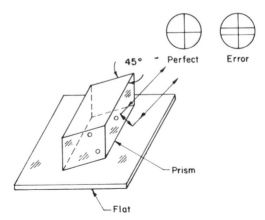

Fig. 10.7. The external autocollimation of the 90° side angles used for fine-grinding the prism.

first 45° face the blanks are reversed and the second face is milled to 1 mm of the top tolerance.

The next operation makes the prism's two 90° side angles parallel. Set up an external autocollimation device to be used during grinding. (See Fig. 10.7.) The collimator is set at an approximate 45° angle. Ground surfaces are placed on three prewaxed nylon balls on an aluminized flat. The external autocollimated image comes from the polished plate glass which is flat enough for this purpose. Because there are two surfaces, fine-grind each surface an equal amount. Any horizontal separation of the image denotes an error in the 90° side angle. The last grind is done with 145 grit and the images must be super-imposed.

In the final grinding the two 45° reflecting surfaces of the prism are rough-ground with 250 grit. The tools required for working these glass angles is a precision 45° and machinist's square. Bevel all sharp edges 0.02 in. (0.5 mm). After this operation the 45° glass angles are ground light-tight to the protractor and square.

3. WORKING THE TWO 45° REFLECTING ANGLES

Until this stage milling and grinding have been done with the cover glasses in place to protect the two polished surfaces and the sharp apex of the 45° angle. Before removing the cover glasses clean around them with running water and dig around the edges with a small wooden stick to remove any grit or wash with xylene. The cover glasses are heated to 80°C in a controlled oven and must be hot enough to part easily when lifting in an upward motion. When cooled, all sharp edges are beveled slightly with 145 grit. The cementing pitch is washed from the surfaces with xylene and again in detergent water.

Check the rhomboid prism again by external autocollimation, as discussed in Section 7.3. If there is error, it can be corrected by grinding with 145 grit for superimposed images. (See Fig. 10.7 for setup.)

In the next step the glass runners are pitch-cemented to the prisms. Here use can be made of the circular sectors cut from the glass disk. Care must be taken to cement the glass runners to the rhomboids to offset the unequal mass distribution. (See Fig. 10.8.) This procedure is described in Appendix 3. Three or four assemblies should be made up.

For pyramidal control of the 45° compound angle (45° and its 90° side angle) set up an external autocollimation check for this compound glass angle to test alternately the 45° reflecting glass angle and the 90° side angle. [See Fig. 10.9 and Section 7.3.] This applies to working the first and second faces of the rhomboid prisms. Briefly, the master 45° angle block is placed on a small high-quality planoparallel plate. The whole assembly is set on the three prewaxed nylon balls and the image reflected from the master 45° angle is located in the collimator's eyepiece. The collimator is securely tightened and the bifilar is set to the image. Next check the 45° angle, which is the polished hypotenuse surface. Then choose one of the preground 90° sides for the reference angle and mark it with a diamond scribe. This side will also be used for the fabrication of the second 45° angle. Present this 90° side angle to the collimator, and to make it reflective wetwring a small aluminized flat to its surface. Note that the 90° air angle is formed by the small flat on the prism and the optical flat, on which the assembly is resting. To uncover any errors in this compound glass angle locate the one pressure point during fine grinding that will simultaneously correction glass angle errors. (Review Section 7.5.)

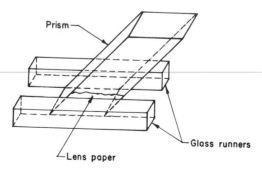

Fig. 10.8. Pitch-cementing the glass runners for working the unequal mass overhang of the rhomboid prism.

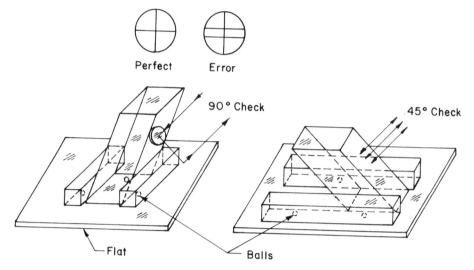

Fig. 10.9. The external collimation setup for checking the pyramidal angle and the 45° compound angle.

At this point in fabrication the total length of the prism must be controlled. The length is checked by measuring the two faces of the 45° angles with the micrometer. It is essential that a piece of lens paper be used to protect the polished surface. Allow for the thickness of the paper. Plan to remove one-half of the glass stock during the 250-grit grind; the remainder will be removed when the second 45° face is worked. The external autocollimation test is used during the 145- and 95-grit grinding. The images must be superimposed after the 95 grind.

Sometimes during the three fine-grindings the bevels are ground away and can be renewed with a microscope slide and 145 grit. First hone between the glass runners and the prism. A razor blade and 145 grit can also be used. Hone the glass runners and the apex edge of the 45° face and wash in detergent water before polishing.

The fine, preground surface is polished on a conventional pitch polisher. (See Chapter 3.) Check continuously for any drift in the compound glass angle autocollimation arrangement. (See Fig. 10.9.) The surface quality must be one-eighth fringe or better and be free of sleeks and digs.

During fine grinding the second 45° face is worked with the same glass runners and autocollimation setup shown in Fig. 10.9.

4. POLISHING AND INTERNAL AUTOCOLLIMATION

After several wets on the pitch polisher check the semipolished rhomboid prism with a new internal autocollimation test. (See Fig. 10.10.) The setup

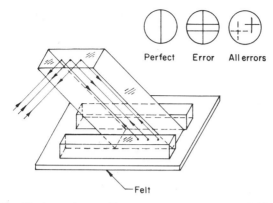

Fig. 10.10. The internal autocollimation mode for checking a finished prism.

actually follows the way the prism is used in the instrument. Note that the collimator is swung below the level of its platform and must be mounted at a height convenient for observation. If the observed images in the collimator's eyepiece are superimposed, the glass angles are good. It is most likely that there will be multiple sets of images. The correction of the glass angles during polishing is fully covered in Section 7.4. Briefly, there are eight possible sets of axial images when glass-angle errors are present. To locate the one pressure point during polishing for correction of the prism's glass angles first determine the direction of the pyramidal error (breath test) and whether the glass-angle error is positive or negative. Glass-angle errors are determined by two tests; the eyepiece is defocused inwardly toward the prism and the knife-edge test determines which horizontal image will disappear first. A midpoint must be located between the two pressure areas for correcting that error.

After correction of the rhomboid prism's glass angles to superimposed images the surface must be completely polished. An acceptable surface quality is one-eighth fringe or better and free of digs and sleeks.

EXERCISES

1. Make a small pair of identical rhomboid prisms.
2. Ray-trace the transmitted light path in a rhomboid prism. The centers of the entrance and exit faces are 3 in. or 76.2 mm apart in a 2-in. or 50.8-mm width prism. What is the total glass path?
3. Describe the Twyman effect.
4. Why are rhomboid prisms expensive?

5. Describe the autocollimation testing of the 45° optical face of the rhomboid.

BIBLIOGRAPHY

Hanna, G. D. "Making Rhomboid Prisms," in *Amateur Telescope Making—Book Three*, A. G. Ingalls, Ed., *Sci. Am.*, 264–276 (1953).

Smith, W. J. Modern Optical Engineering, McGraw-Hill, New York 1966, pp. 94–95.

See also bibliography in Chapter 8.

11

Precision Beam Splitters

Precision beam splitters, which consists of two identical right-angle prisms cut from one elongated tank prism with their opposite 45° angles cemented together, present a simplified procedure for their production. This is apparent because any right-angle prism with several minutes of error in its 90° faces (formed by two complementary angles), has one obtuse angle (plus) and one acute (minus). The pyramidal error is held to a minimum by a small aluminized planoparallel plate wet-wrung to one end of the tank prism during fabrication. With this method most of the pyramidal error can be nullified by a slight rotation of the pair of prisms cemented together as a beam splitter. However, the pyramidal error should never be more than 30 sec of arc.

1. GLASS PREPARATION

General procedures for the construction of tank-type, right-angle prisms are as follows: the glass disks are ground and semipolished as planoparallels plates (20 min of arc) to test for homogeneity strain, bubbles, and so on. (See Chapter 5.) The accepted glass blanks are covered with waterproof masking tape on both sides. A plan in which sufficient glass is allowed for sawing the tank prisms, generating, and grinding, (glass is cheaper than labor) is drawn for diamond sawing the rectangular slabs. It must be kept in mind that working a long right-angle prism depends on the base length of the finished beam splitters and the number of pairs of prisms cut from one tank prism. It is apparent that for prisms with a base length of 8 to 10 mm they cannot be made too long, but for the larger prisms of 25 mm or more they can be much longer. It is good practice to make the tank prisms long enough to allow for two or three pairs of smaller and four or more pairs of larger prisms (10 mm or larger). Always have some extra glass for fine grinding above the nominal tolerance. It must be remembered that diamond sawing causes polished surfaces on any optical component to twist (Twyman effect), and fine grinding and semipolishing of saw cuts are required to relieve the strains. Two things govern the total length of the tank prisms:

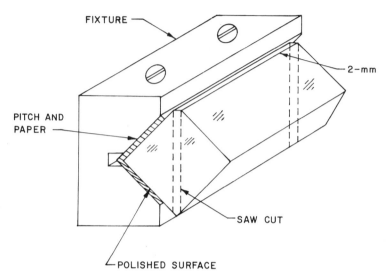

FIXTURE

2-mm

PITCH AND
PAPER

SAW CUT

POLISHED SURFACE

Fig. 11.1.   A vee-holding fixture for sawing the retangular glass slabs into two right-angle tank
prisms.

first, the diamond-saw splitting of long glass slabs into two 45° angular sections; second, the diameters of the channeled toolholders for generating and
fine-grinding the tank prisms.

The diagonal cutting or splitting of the glass slabs into two 45° right-
angle prisms is hazardous. It can be safely done by pitching both ends of the
rectangular slabs before placing them in a holder. (See Fig. 11.1.) Observe
the 2-mm bevel on one end of the slabs for guiding the saw blade and setting
the holder on the sliding table of the saw. An excellent, low-temperature
pitch mixture consists of three parts of water-white rosin, three parts of
beeswax, and two parts of Montan wax (by weight). After setting the
assembly for splitting place a rubber mat in which sectional holes have been
cut close to the holder to prevent breakage. After sawing the prisms and
removing them from the holder immediately bevel the dangerous sharp
edges on a flat grinding tool.

2. GENERATING, GRINDING, AND POLISHING

The rough tank prisms are generated in a 45° channel holder. Make sure
that any polished surface is bounded to the lens paper and the vertical side
of the 45° vee slot of the holder. Ignore any air gaps between the glass slabs
and the second side of the tank prism.

After generating the first 45° side generate the second 45° angle. Make sure that any semipolished surface is protected by lens paper. The first side is then tightly anchored to the toolholder. Next block and generate the hypotenuse surface.

A number of optical procedures have been devised for working the surfaces in turn which depend on the number of beam splitters required. If 10 or fewer are required, use glass runners cemented to the elongated tank prisms (see Appendix 3). The prisms should be long enough to furnish two or three *pairs* of right-angle prisms. (See Fig. 11.2.)

These techniques are similar to those described in Chapter 8. The hypotenuse surface is fine-ground and polished, shown in Fig. 11.3, with its glass runners on the·prism and its 45° compound glass-reflecting angle in an internal autocollimation procedure. It is preferable to cement the glass runners to the long side of the tank prism for easier handling. To control the 45° side and its 90° side angle (pyramidal control) a small aluminized planoparallel plate is wet-wrung to the 90° side of the tank prism for its 90° check by the autocollimator. The 45° face is checked by a preset 45° master angle placed on a planoparallel plate. (See Section 6.4 and Fig. 6.8.)

To mass produce these prisms the tank prisms are remounted in precision-made and well-annealed 45 and 90° cast-iron holders with extreme care and cleanliness in mounting, grinding, and polishing. All bevels must be carefully *renewed*. Lens paper must cover each polished surface of the prism. The temperature of the block must be raised to nearly 100°C (212°F) and the prisms firmly seated with a wooden block. The metal holders must have high quality and excellent parallel angular channels (checked by the master 45 and 90° angles).

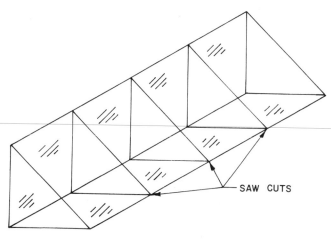

SAW CUTS

Fig. 11.2. A finished tank prism marked off into pairs of right-angle prisms for saw cutting.

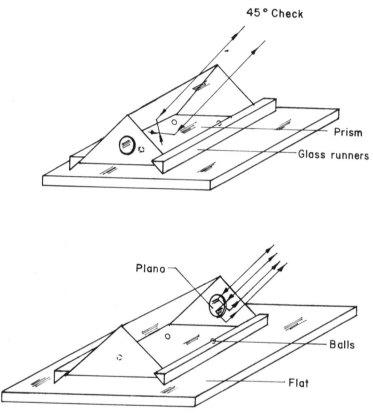

Fig. 11.3. The internal autocollimation test to check whether the 45° compound glass angle is plus or minus.

The dimensions of the tank prism (base length and vertex heights) are controlled by measuring with a micrometer height gage and checking at 120° intervals near the periphery of the holders. It is important that sufficient stock remain on the tank prisms for beveling and paralleling the prisms. Any bevels lost in this procedure can be renewed by using a shop razor blade and 145 grit. Grind between the prisms until a visible bevel is observed.

Note that at least 0.006 in. (0.2 mm) of stock should be allowed for the last two fine-grinding (145 and 95 aloxide grit) phases of the surfaces and sides of the prisms. Diamond-generated glass surfaces cause small deep-lying strains and craters in the surfaces. Beam splitters and other glass components having such faults cause background noise in electronic circuitry. Elements used in laser illumination should be more thoroughly ground in optical fabrication.

When blocking the tank prisms in the holders, the block must have a circular configuration and the smallest tank prism should be long enough to make a pair (not a single prism). The tank prisms are ground parallel and their height from the top surfaces of the tool should be checked with a depth micrometer. Polishing tank prisms is identical to polishing flat surfaces. The overall surface should be two wave or better (surface quality on a single prism will be proportionally less) and free of sleeks and scratches. Sleeks and fine scratches are caused by the dried polishing compounds that break loose from the bevels and edges and fall on the pitch polisher. The blocking, grinding, and polishing of the second face and hypotenuse surface are similar to the procedures already outlined.

3. SAWING AND ASSEMBLING THE BEAM SPLITTERS

After the prisms are finished mark the maximum number of pairs of right-angle prisms that can be cut from the tank prism with a felt-tip pen. Each pair must be placed together, one pair to a box section. Pairs cut from other tank prisms will be slightly different. If the sawing has been done carefully (each cut parallel), pitch them to a small flat holder and generate sufficient stock from the prisms to clean up the surfaces. Fine-grinding of the sides of the prisms removes about 0.2 mm of stock. For larger prisms the 90° sides must be semipolished. The blocked prisms are reversed and fabricated to their finished tolerances. Careful planning must be done to make the prisms parallel and 90° to the finished surfaces. If more than one tank prism is presawed, some consecutive numbering system should be penciled on each pair. Larger prisms may require the use of small strips of masking tape.

The finished prisms are retested and the plus and minus corners are marked with a red marking pen. Pairs of prisms are often mixed up because of carelessness during sawing more than one pair or a mixup in the coating room. This requires an internal autocollating setup. (See Fig. 8.4.) Here the glass runners were not on the prism because they were fine-ground and polished in the metal holders. Check the 45° glass angle and any vertical separation is an error in the 45° angle. Why not use the 90° check? Because it does not determine whether the 45° glass angles complement each other as needed. It merely establishes that a 90° glass angle really is a 90° angle.

Reverse the prism and observe the error in the second 45° angle; its separation is the same. This is to be expected because one glass angle is plus, the other, minus. How is it established that a glass angle is plus (obtuse) or minus (acute)? There are two rules to be remembered:

Rule 1. Defocus the eyepiece inward about 1 mm from the best focus; if the images tend to show more separation, the glass angle is plus.

Rule 2. Defocus the eyepiece about 1 mm toward the prism; if the images tend to come together, the glass angle is minus.

Rule 1 is subject to *verification* because a perfect glass angle will *expand* on *both sides* as the eyepiece is defocused. Therefore a minus glass angle will tend to close up the images when the eyepiece is defocused *away* from the prism.

If there is some confusion use the bifilar or the cross hairs to measure for identical separation. All 45° glass angles are identified as plus or minus and should be marked before the prisms are assembled. Strips of masking tape can be used on the sides. All large prisms should be checked in a Tyman or Hendrix interferometer to show that the optical path difference (OPD) is one-half fringe or better. Remember that the interferometer gives a double value. Most smaller prisms will meet this tolerance. Larger prisms may have to be localized. This is discussed in Appendix 8.

The tested prisms are cemented together into beam splitters after coating. (See Fig. 11.4.) Prisms are assembled in a parallel light arrangement and rotated or moved slightly laterally for superimposed images which must be found first with a collimating flat. Search for plus and minus glass angles and cement them in pairs.

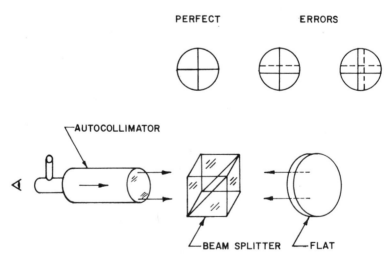

Fig. 11.4. Checking the assembly of a pair of right-angle prisms during the cementing into a beam splitter.

Vertical lines sometimes cannot be superimposed because of pyramidal error, in which case a slight separation of the horizontal lines is required to minimize the error for tolerance requirements. Rubber bands or masking tape are used on the assembled prisms while the cement hardens. The beam splitters must be checked several times before the cement becomes too hard and the prisms cannot be rotated.

4. CONCLUSIONS

This method of making prisms may not be new, but in a careful check of the literature as far back as 1905 no references were found.

If ultraprecise beam splitters are required, one tank prism at a time should be made by the methods used for right-angle prism.

EXERCISES

1. What is meant by the term "complementary glass angles" in beam-splitter assemblies?
2. Why is a small aluminized planoparallel plate wet-wrung at the tank prism ends during fine-grinding?
3. Name at least three optical instruments in which beam splitters are used.
4. Why is it necessary to prevent prisms cut from one tank prism from being mixed with other prisms?
5. Describe how the 45° reflecting glass angle is tested with an autocollimator. Is the pyramidal angle observered during this testing?
6. Describe how to determine when a glass reflecting angle is acute and when it is obtuse.

BIBLIOGRAPHY

See Chapter 8.

12

Corner-Cube Prisms

A corner-cube prism is a retrodirective glass element that returns a directed light beam by reflection from three 90° glass angles. These angles must be precise within several seconds of arc. The entrance face can have a 5-sec or larger error with no appreciable effect on its performance.

Early use of these prisms was made during World War II to outline secret airfields. The pilot simply held a flashlight near his forehead, and because the prisms return the light rays on-axis the airfields were located and landings made without lights. A more recent use for corner-cube prisms was established during the Apollo space programs. Arrays of 50 or more were left on the moon's surface. Astronomers then directed high-power laser beams toward them and the returning beam was picked up by their telescopes. The double passage of the laser light from earth to moon permitted the precise calculation of the distance. It takes light approximate 2.8 sec to complete this double passage.

Several methods of forming these prisms are outlined. Method A makes use of massive glass blocks in cubic or rectangular slabs. Eight corner-cube prisms can be sawed from one glass cube, one from each corner. Four prisms can be sawed from a rectangular solid. Briefly, the glass blocks are fabricated as follows: one corner of the glass block is made with three precise 90° angles done by external autocollimation during fine-grinding and internal autocollimation during polishing. This becomes the reference corner of the cube. The remaining sides are made planoparallel with a laser-illuminated Fizeau interferometer.

The preliminary 90° reference corner of the cube is fixed by external autocollimation which has a 4× amplification factor for the glass angles being fine-ground. Internal autocollimation is used during polishing and because it yields a 6× amplification factor the glass angles can be more precisely checked. Internal autocollimation is somewhat handicapped because all surfaces on the glass blocks must be semipolished. This can actually be a help because the Twyman-effect is prevented from twisting a polished surface when an adjacent preground surface is polished. These transmitting surfaces of the glass block need not to be fully polished (about

25% suffices) for internal autocollimation. The surface quality should be two fringes at the entrance pupil of the prisms and the edges can be turned if the autocollimator's objective interferes.

Method B is called the duplication process, by which the glass angles of corner-cube prisms are controlled interferometrically. To do this premade master optical squares or right-angle prisms are optically contacted with one polished surface. Because the glass angles on these master elements have been interferometrically controlled by Deve's method, the glass angle errors are less than 1 sec of arc. The total number of prisms that can be made is governed by the number of master elements available. Here, where one prism is made at a time, premolded prism blanks are often used after

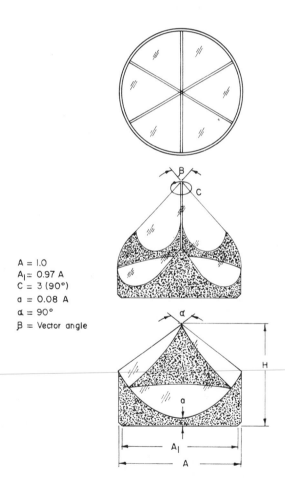

$A = 1.0$
$A_1 = 0.97\, A$
$C = 3\,(90°)$
$a = 0.08\, A$
$\alpha = 90°$
$\beta = $ Vector angle

Fig. 12.1.   Data for making the corner cube prism.

one side has been fabricated in plaster blocks. It is optically possible to contact master blocks on large glass blocks and control the glass angles by interferometric testing. However, this is not always practical because large prism surfaces are difficult to hand polish when glass angles are being corrected to the smaller area of the master angle prism.

The necessary data for making a corner-cube prism are given in Fig. 12.1. These data are based on the diameter of the entrance aperture, assumed to be 1.

1. PRELIMINARY FABRICATION

The sides of the glass block must be at least 2.8 times that of the required aperture; for example, if the aperture is 3.81 cm, this means that a glass cube has 11-cm sides. A little extra glass may be required for saw cuts. A preplanned layout should be made in for saw cutting eight prisms from the glass cube. Likewise, the layout for four prisms in a rectangular glass slab must be planned.

Glass (BK-7) for prisms comes in many forms: cubes, rectangular slabs, and flat disks. I have seen cubic glass blocks with 35-cm sides, all of which were being polished in a homogeneity test. The block was later cut up into penta prisms. These massive glass blocks are fabricated on lap-master machines, sometimes as large as 1 m in diameter.

In general procedure that can be followed by the optician small glass blocks are hand-ground on a rotating cast-iron flat tool. One corner with three 90° angles is chosen and the glass angles are coarse-ground with 250 grit until they are light-tight to a precision square. In the next step opposite sides are made parallel on a Blanchard-type generator and ground with 250 grit until the marks are removed. The result is the glass block illustrated in Fig. 12.2.

At this point check the glass block for homogeneity by using Schrelein-quality planoparallel plates with matching refractive index oil. The plates are placed on opposite sides during this test to make the glass block light transmitting. Glass blocks must be checked in three directions. (See Chapter 5.)

It is good practice to select one surface as the reference surface of the chosen corner on the glass element. The three 90° glass angles are referenced for this surface which is fine-ground by all the fine grits. The last 95 grit grinding should be done on a ceramic flat tile tool that is slightly convex to give a concave contour to the ground surface. The next two surfaces are then fine-ground and the 90° glass angles are checked by external autocollimation. (See Fig. 12.3.) The finished surface is placed on three balls which are tightly waxed to an aluminized optical flat. For further

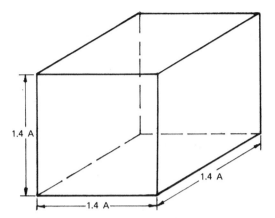

Fig. 12.2. One cube of glass used to make four corner cubes.

details see Section 6.A. External autocollimation is used to check each adjacent 90° glass angle at the reference surface of the chosen corner. Note the small aluminized planoparallel plate wet-wrung to the fine-ground surface. This makes the ground side reflective for checking the 90° angles by external autocollimation. (See Fig. 12.3.) Any horizontal separation of the images observed in the collimator's eyepiece is an error in the 90° glass angle and it can be obtuse (plus) or acute (minus). If there is an error, locate the direction of the wedge between the two surfaces that form the 90° angle. This is easily done by touching the glass element while viewing the area where the horizontal images come together in the collimator's eyepiece.

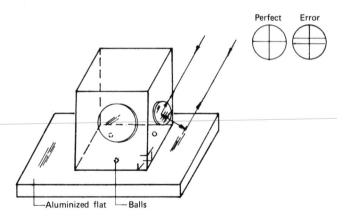

Fig. 12.3. The external autocollimation of the six 90° side angles of the cube while fine-grinding.

Mark this area with a felt pen or pencil and rotate the second side of the 90° glass angle. There are now two separated marks which were made in the 90° compound angle test. The point on the prisms at which pressure is applied in fine-grinding causes the horizontally separated images to come together (as in controlling a compound angle); the final pressure point is midway between the two predetermined points. This is due to three angular conditions: plus or minus glass angles and glass angles without errors. In any case, fine-grind the cube until all three 90° glass angles are super-imposed images when checked by external autocollimation. (See Fig. 12.3.) This means that the reference side must be checked against each 90° glass angle at the chosen corner. Beginners should number each surface. Surface 1 will check the 90° angle between 2 and 3, surface 2 will check the 90° angle between 1 and 3, and surface 3 will check the 90° angle between 1 and 2.

The next step is to render the three remaining sides parallel by grinding. The external autocollimation shown in Fig. 12.3 can be used. Some opticians like to change this setup slightly by placing the planoparallel plate on the top surfaces at the reference corner. The opposite surface is placed directly on the three prewaxed balls, and the autocollimator picks up the reflective image from the aluminized planoparallel. The glass block must be rotated until the vertical imagery is superimposed and the two horizontal images still show glass-angle errors. To determine whether there is a wedge angle use the collimator's eyepiece. If, when defocusing from the best focus toward the glass block, the separated images close up, the glass angle is plus; conversely, if the separated images move farther apart, the glass angle is minus. Because this setup produces a 2× amplification and the external autocollimation, a 4× factor, the latter is preferable. (See Fig. 12.3.)

Semipolish the six surfaces with approximately 25% polish to a surface quality of two or three fringes. There are two reasons for semipolishing the glass element: first, it prevents the Twyman-effect (twisting of polished surfaces when an opposite ground surface is polished), and, second, the new internal autocollimation setup allows the 90° glass angle to be checked with a 6× amplification factor. After this preliminary semipolishing one surface at the chosen 90° corner is completely polished to a surface quality of one-eighth fringe free of blemishes.

2. POLISHING THE CHOSEN 90° CORNER

A new internal autocollimation setup is shown in Fig. 12.4. In this illustration note that the waxed balls have been removed and a piece of felt placed on the base of the collimator. The parallel rays are directed at 45 or 60° toward the 90° glass angle formed by the reference side and the second sur-

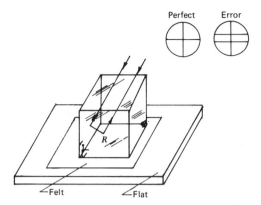

Fig. 12.4. The internal autocollimation of the 90° side angles used to correct the cube during polishing. One corner must become a reference corner and three 90° glass angles adjacent to the corner made perfect. (See Appendix 19.)

face. The finished reference surface is resting on the felt pad. The three types of image pattern are superimposed (a precise 90° angle) and horizontal separated images, which are plus (obtuse) or minus (acute). These patterns are discussed more fully in Section 7.4. If the two horizontal images tend to separate more when the autocollimator's eyepiece is defocused from the best obtainable focus, the glass angle is plus; conversely, if the horizontal images approach each other, the glass angle is minus. This is identical to Rules I and II (7.2). Here internal reflection and transmitted images with parallel light are dealt with.

Many opticians who are making prisms for the first time do not understand where to apply pressure while correcting glass angles. Figure 12.5 shows a cube with an error in its glass angles. One cube has a plus (obtuse) error and the other a minus (acute) error. The dashed lines in the figure show the position of the wedge and the pressure point for correcting the glass angle during polishing. Note that the pressure point is always at the minus glass angle.

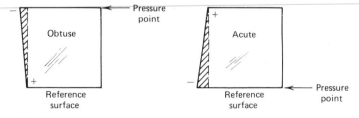

Fig. 12.5. The correction of glass-angle errors by using pressure at the correct point.

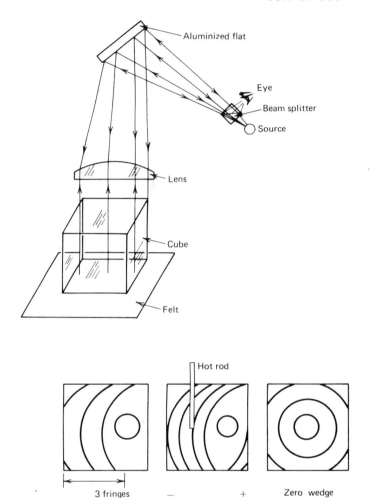

Fig. 12.6. The Fizeau interferometer test of the cube's six sides for parallelism. (See Appendix 19.)

Correct the 90° glass angle by working the second surface only. Never retouch the finished first side (place a piece of masking tape over it to prevent reworking). The 90° glass angle will be precise if the horizontal image is superimposed. When the polished surface is one-eighth fringe or better and free of sleeks and digs, it is considered finished. Modern electronic autocollimators may be used but care must be exercised when dealing with supplementary air angles.

The third side and the second 90° glass angle to the reference side (first surface) are polished as well. Use the internal autocollimation setup for checking the second 90° angle of the cube. (See Fig. 12.6.) Caution: place a piece of masking tape on both sides of the first 90° angle already worked. When the third side is almost finished and the 90° glass angle is approaching, place the piece on the side opposite the reference surface and recollimate the 90° glass angle. Another type of error can be observed in the glass angle when the reference side is down. Because this is a compound glass angle, correct the vector 90° angle in the second plane. Review Section 7.4.

3. MAKING THE OTHER SIDES OF THE CUBE PARALLEL

We should now have a precise reference cube corner on the cube with three 90° glass angles. Each of the three remaining sides is polished parallel to the reference cube corner. This is accomplished by checking the polished cube in a Fizeau interferometer. For large glass cubes the Fizeau interferometer must have a low-power, 0.3-mW laser source and the projected fringes viewed on a frosted screen. Mercury (isotope 198) sources can be used to check only to a 2.5-cm thickness. The placement of the cube is illustrated in Fig. 12.6. Note that the zero wedge exists when the ovals or circle patterns are dead center. Any patterns with equiparallel lines show a lack of parallelism. It does not pay to polish a large cube with three or more fringes on 2.5 cm of surface area. It would be best to fine-grind with 95 grit in the proper direction. How does one know the proper direction? Use a hot brass rod heated in running water and touch it on a fringe. (See Fig. 12.6.) Note the fringe break-up; it will point toward the thinner end. Of course the pressure point to remove the wedge is directly opposite the thinner edge. By regrinding and semipolishing the cube can be quickly figured to a precise planoparallel. A finished surface of a cube must be one-eighth fringe and the circular pattern of fringes, dead center.

The cube is ready for saw cutting when all the surfaces are planoparallel to the chosen reference corner. The procedures for each surface is identical to that described above. It is good practice to go back and recheck the 90° glass angles because the Twyman-effect is an ever-present problem that can cause the glass angles to drift as one surface is finished.

4. METHOD A: RECTANGULAR BLOCK FABRICATION

The rectangular glass-block method makes it possible to facricate four corner-cube prisms at one time with smaller glass blocks. The general procedures are identical to those already described. For easier handling of the rectangular glass block it is best to pitch-cement glass runners near each end. Four glass runners should be attached after the first 250-grit grind.

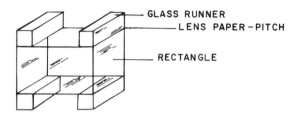

Fig. 12.7. A rectangular glass block with cemented glass runners for easier fabrication.

(See Fig. 12.7.) Before cementing the glass runners review Appendix 3. Caution: make sure that the bevels on the glass runners are strong and that there are small bevels on the glass block. Of course, any bevels lost during grinding can be renewed by using the edge of a microscope slide and 145 grit and beveling between the two glass surfaces.

5. CUTTING PRISMS FROM THE GLASS BLOCKS

Saw cutting the finished glass blocks must be done carefully because much time and effort have already been expended. The use of clean, waterproof masking tape is probably the simplest method of protecting the polished surfaces during the handling and sawing operations. A preplanned pattern is shown in Fig. 12.8. As one of many each glass block is a special case and should be laid out on the masking paper with a marking pen. Do not attempt to save glass because glass is cheaper than labor.

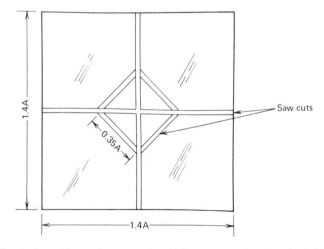

Fig. 12.8. A planned layout for saw-cutting the four corner cubes from the finished cube.

The first cut is made across one of the diagonals that divide the glass into four sections. For the larger glass blocks the diamond cutting wheel should be checked to prevent its holding washers from interfering during the sawing operation. The next saw cuts separate each corner-cube section. Because the corners of these sections are dangerously sharp, cut them off. This will also cut down the edging time. Save any surplus glass; these glass sections can be used for prisms and the polished surfaces can become reference surfaces.

6. FINISHING THE ENTRANCE FACE

The entrance face of the prism is not a 45° but a vector angle of about 54° 15″. Its precise value need not be known because it will be controlled later by external autocollimation. A special holding fixture is required for edging and generating this circular entrance face. (See Fig. 12.9.) It is made of round brass stock in two sections with 90° channel cuts to accommodate the three 90° glass angles of the corner in the recesses of the holder. Note the checking land which can be used to control the height of the prism during generation.

The prisms are pitch-waxed to the fixture by the method described for the glass runners. Briefly, the fixtures and prisms are heated together to a temperature of 80°C. Lens paper is placed between each surface of the prism and fixture as they are cemented together with pitch. Care should be taken that neither water nor breath condensation come in contact with the hot, uneven prisms.

Edge the prisms on an edging machine within 2 mm of the final tolerance. The next operation involves the generation of surplus glass to form the entrance pupil (entrance face). Leave a small surplus of glass, perhaps 0.05 mm above nominal tolerance, for grinding and polishing. Then edge the prisms to their final nominal tolerance. The reason why the prisms are not edged to their final tolerance initially is that some chipping may occur during the generation of the top surface. Next bevel the prism's entrance face with a small-radius tool and 145 grit.

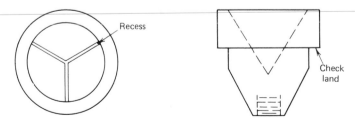

Fig. 12.9. A specially designed fixture for generating the entrance face of the corner cube.

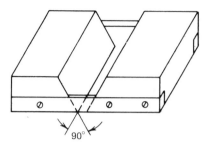

Fig. 12.10. A specially design fixture for beveling the 90° edges of the prism.

The elements are removed from the fixture by reheating to 80°C in a controlled oven. Allow the prisms to cool on paper toweling. Place one element in a sectional box that has four or more sections.

The cooled prisms are not cleaned at this time. Instead even the bevels if bevels are required. The lens paper and pitch cement will help to protect the polished surfaces during handling. Even bevels enhance the appearance of the prisms and should be done carefully. The paper and pitch are removed with xylene. Wash the prisms again in a warm, mild detergent solution and check all bevels for even widths. Regrind if necessary. Fine-grind the top of the apex corner with 145 grit. A beveling fixture is shown in Fig. 12.10.

7. WORKING THE ENTRANCE FACE

The autocollimator is setup as shown in Fig. 12.11. (See also Section 6.6.) Figure 6.9a shows the prism's entrance face resting on waxed balls. The collimator is preset to one of the reflecting surfaces of the prism. Set the bifilar or cross-hairs to the reflected image, and rotate each face of the element to locate the high and low areas. This is called drifting above or below

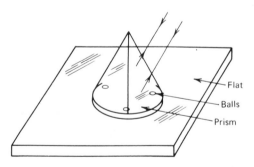

Fig. 12.11. The external autocollimation of the vector glass angle formed by the three 90° glass angles. We work for the same reading by the autocollimator's bifilar reading or the horizontal cross hair as observed in the eyepiece.

the null point. The entrance face is fine-ground until the three reflecting sur-faces read at the same null point, at this time the surfaces can be marred by sleeks and scratches incurred in handling. Masking tape can be used and a small window, cut into the tape before placing it on the prisms. When the prism has been ground to the nominal tolerance, it is polished to an optical quality surface of one-eighth fringe or better without digs or sleeks. Check during polishing that no drifting of the three sides has occurred by compar-ing the prism with the autocollimation in Fig. 12.11.

Several methods have been designed for testing the three glass angles of corner-cube prisms. Hopkins describes an optical layout for a Koester prism. A similar interferometric test is described by Drs. Sayce and Burch in *Interferometry, Symposium #11*, 1959, Her Majesty's Stationery Office, London. In essence both optical arrangements are the same. I have altered the Hopkins procedure to meet requirements that an optician can use. Twyman's method has been popular for a number of years and is described in *Prism and Lens Making*, 2nd ed., 1952, Hilger and Watts, London, pp. 447–448. Both optical arrangements are illustrated in Fig. 12.12.

In Fig. 12.13 another setup illustrates Twyman's method. The prism is placed in a triaxial adjustable holder that can be rotated and adjusted to the horizontal and vertical tilting plane. The prism is mounted with its round entrance aperture in the holder facing the parallel light from the divider plate. If the interferogram has equidistant and straight fringes, the 90° glass angles are well corrected. If the fringes are not straight but show several patterns of crisscrossed lines, there are errors in the 90° glass angles.

The general procedure for correcting these three 90° glass angles is com-plex. Use a partly coated planoparallel plate instead of a flat (Twyman method), for it will give the fringes a better contrast. Zero the fringes in one-half of the sector and rotate the prism in the second plane of the 90° glass angle. The fringes of the glass angular error are then viewed. To determine whether the 90° glass angle is positive or negative pull back

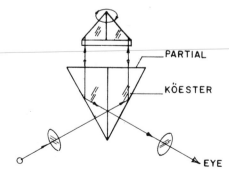

Fig. 12.12. The optical setup for testing interferometrically with a Koester prism for determining whether the 90° angles are within tolerance.

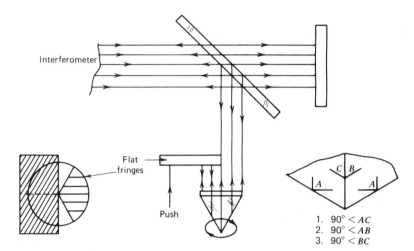

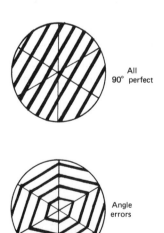

1. $90° < AC$
2. $90° < AB$
3. $90° < BC$

Fig. 12.13. Using a Twyman-Green or Hendrix interferometer for checking the three 90° glass angles of the prism.

slightly above the planoparallel to increase the light path. If the fringes move toward the center, the glass angle is obtuse or greater than 90°. If the fringes move outward toward the center, the glass angle is acute or smaller than 90°. (See Fig. 12.13.) Remember that the first surface becomes the reference surface and the second surface must be reworked.

Mark the reference surface (A) and rework (B) on a polisher to correct

All
90° perfect

Angle
errors

Fig. 12.14. Two observed interferograms showing a perfect corner cube and a corner cube with 90° glass-angular errors.

the glass-angle error. Use pressure while polishing at a predetermined position. Pressure is always exerted on the negative angle, never on the positive. (See Fig. 12.5.) The next step is to correct side (C) of the 90° compound glass angle. Because there are two planes on this 90° glass angle, determine the vector pressure point for polishing any glass-angle errors in both planes of the surface (C). Never retouch surface (B) because it is at 90° to the reference surface (A). If this procedure has been faithfully carried out, the 90° glass angles will be within 0.1 sec. (See Fig. 12.14.)

8. DUPLICATION METHOD

Duplication makes use of small, optically contacted, master right angles or squares to a reference surface on the element. (See Chapter 9.) Briefly, the method makes use of optically contacted, small master elements on a prepared reference surface of the corner-cube prism. This surface actually becomes one surface of the three 90° angles of the prisms and is generally plaster blocked when a larger number of items is worked at one time. The second surface is ground and polished and the 90° glass angle is duplicated and interferometrically controlled. One 90° side of the corner-cube prism is then finished.

Finishing the third 90° corner of the prism involves optical contact of the master element in a new position on the referencing surface. This is not too difficult and can be established interferometrically by rotating the master element to the newer position while observing through the planointerferometer. The third surface is then ground and polished in a mammer similar to that used for the second surface. Here the 90° compound angle is controlled by checking both planes in the interferometer.

It is obvious that by the use of master elements that are in optical contact with the larger glass cubes the 90° reference corner can be controlled during manufacture. Then each of the other optical surfaces is made parallel by a Fizeau interferometer. The drawback in this method is that the interferometer be laser illuminated. A low-power laser is sufficient for distances of several centimeters.

9. CONCLUSIONS

Little is available in the literature on step-by-step procedures for making corner-cube prisms. Several testing methods described by Murty, Puntambekar, and others can be used but only with caution. If the illumination is provided by laser, the optical surfaces must be worked to a higher degree of accuracy ($\frac{1}{16}$ of fringe). See the bibliography for these tests.

EXERCISES

1. Explain the finishing of the entrance and exit faces on a corner-cube prism.
2. Describe why one face of the three 90° angles must become the reference corner during fabrication.
3. What three angular conditions can exist in any glass reflecting angle?
4. Why are nonbeveled corners on corners-cube prisms so expensive?
5. Where are corner-cube prisms used?
6. Can a master right-angle prism be used during the fabrication of corner-cube prisms? Describe the procedure.

BIBLIOGRAPHY

De Vany, A. S. "Reduplication of a Prism Angle Using Master Prisms and a Plano Interferometer," *Appl. Opt.*, **10**, 1371 (1970).

De Vany, A. S. "Testing a Prism's Compound Reflecting Angles," *Appl. Opt.*, **14**, 2791–2792 (1975).

Hansen, G. and Kinder, T. "Corner Cube Prism Testing," *Optik*, **15**, 560 (1958).

Hopkins, H. H. "Discussion on Paper 3-5," National Physical Laboratory, *Symposium No. 11, Interferometry*, Her Majesty's Stationery Office, London, 1959, p. 177.

Johnson, B. K. *Optics and Optical Instruments*, Dover, New York, 1960, pp. 200–203.

Murty, M. V. R. K. "Newton, Fizeau, and Haidinger Interferometere," in *Optical Shop Testing*, Wiley-Interscience, New York, 1978, pp. 30–34.

Saunders, J. B. "Roof Prisms," J. Res. *Nat. Bur. Stand. USA*, **58**, 21 (1957).

Sen, D. and Puntambekar, P. N. "Corner Cube Prism Testing," *Opt. Acta*, **12**, 496 (1956).

Sen, D. and Puntambekar, P. N. "Shearing Interferometers for Testing Corner Cubes and Right-Angle Prisms," *Appl. Opt.*, **5**, 1009 (1966).

Twyman, F. *Prism and Lens Making*, 2nd ed., Hilger and Watts, London, 1952, 446–448.

13

Deviation Roof Prisms

A large number of deviation prisms with 90° roofs have been designed for use in sighting instruments. These roof prisms will revert and invert the image and at the same time deviate the line of sight through various angles, ranging from 45 to 120°. The most common deviation roof prisms are 45° (Schmidt), 60° (Frankford Arsenal #2), 80°, 90° (Amici), 115° (Frankford #1), and 120°. A penta prism is a special case. It is apparent that the angle of sight from the prism can be almost any angle of deviation, but it depends on the instrument and its design. It should be kept in mind that any deviation prism with a 90° roof will be an Amici (90°). The fabrication of this prism is identical to that of other roof prisms, but the entrance and exit included angles will be different. (See Fig. 13.1.)

1. HOMOGENITY OF GLASS

It is important that the glass be tested for homogenity, bubbles, striae, and so on. Methods of testing the homogeneity are discussed at length in Chapter 5. It is important to emphasize that glass blanks must be checked in three directions.

Preselected optical glass from manufacturers must be tested even if they claim that it is Grade 1-A and meets the specifications outlined in Jan. 174-A. I remember that a series of right-angle prisms had to be rejected after a fellow worker who failed "to know his glass!" had spent three weeks working on it. These hopeless prisms had small ripples of striae which could not be localized to meet the one-quarter wave OPD (optical path difference) under test by interferometer.

The glass blanks are made as planoparallel plates on a Blanchard or similar generator. The vacuum holder permits the reversal of the blank, and excellent parallelism can be achieved if care is taken. Chapter 5 describes particular fabrication methods for shaping glass blanks for three-way testing because most prisms have two parallel sides. The planoparallel plates should have two semipolished surfaces with a surface quality of two waves and 15 sec or better. For optical contact with master angle prisms the sur-

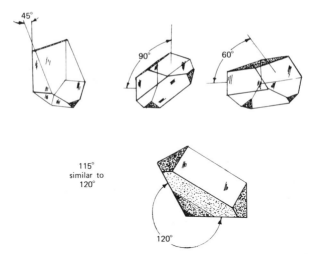

Fig. 13.1. Typical 90° deviation prisms: Schmidt (45°), Frankford No. 2 (60°), Amici (90°), Frankford No. 1 (115°) and 120°.

face quality must be one wave or better (single prisms will have proportionally better surfaces when saw cut from larger glass blanks). In any case, one of the better polished sides will become a reference surface for checking the 90° side angle to the reflecting glass angles of the prisms. The Fizeau interferometer is used to check the parallelism of planoparallel plates. (See Chapter 14.)

2. PROCEDURES: SAWING

The homogeneity test provides blanks or disks free of striae and other imperfections. In the first step two plate-glass square covers are cut to protect the two polished surfaces. Place a single sheet of lens paper between each polished surface with low-melting pitch. (See Appendix 3.) Briefly, the glass blanks are slowly heated in a controlled oven to a temperature of 85°C while resting on an asbestod pad. The plate-glass sheets are heated with the heavier glass blanks. The pitch is lightly smeared on one side and the lens paper is placed over it and topped by one cover glass. The glass assembly is turned over and the second cover glass is then cemented to the glass blank. Caution: wear cotton gloves for handling because the thick glass blanks are too hot to handle with the bare hands. Remove all water or damp toweling, for they can cause stress build-up and the glass disks may break.

After the glass assembly has cooled to room temperature it is covered with wide, waterproof masking tape. A template of the oversize prism is cut

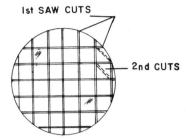

Fig. 13.2. A layout pattern for saw-cutting the prisms from semipolished planoparallels.

from cardboard or mylar and an ink drawing is made on the masking tape. Find the layout that will produce the greatest number of prisms with the least amount of glass waste. Allow for saw cuts. The first cuts, generally at right angles, divide the assembly into four sections. It is impossible to give a hard and fast rule because the size of the blank governs the size and shape of the number of prisms that can be sawed. Be sure to allow for saw cuts between prisms. If the form of the layout is such that all the sawing can be done in elongated slabs with a number of prisms in a series of square blocks (two prisms to a block), the sawing operation will be facilitated. Next each prism is separated by sawing through the hypotenuse side. Figure 13.2 illustrates one of the many ways of saw cutting prisms blanks from a planoparallel plate.

The cover glasses remain in place on the prism because they help to keep the breakage down in some of the ensuing operations. They also protect the polished surfaces of the prisms and offer a larger surface to work on during any rough hand-grinding operation.

3. ROUGH-GRINDING TO FORM

The prisms are rough-ground with 320 carborundum on all three sides to remove small glass humps remaining from the sawing operation. Large Amici prisms are generally hand-ground to final form at this stage. Small deviation prisms with 90° roofs must be generated to their form. The Amici prisms chosen as illustration have base diameters of 50.4 mm on their entrance and exit faces. Because the planoparallel plates control the thickness of the prisms, the 90° entrance and exit faces are shaped. The 90° roof is generated to form in a later operation. The tools required are a 90° toolmaker's square and a 45° precision brass template. (See Fig. 13.3 for final forms.) One face of the 90° angle is flat-ground; check the 90° side angle against the chosen reference side with the cover glass still in place. This should be light-tight to the square. The second face of the 90° glass angle is ground and must be light-tight to the first ground surface of the prism and

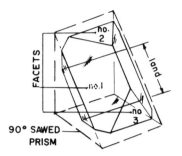

FACETS

no. 2

no.1

land

no 3

90° SAWED PRISM

Fig. 13.3. The various critical facets and reflecting angular surfaces on an Amici and similar deviation prisms with a 90° roof.

its 90° side angle; both glass angles should be checked by the square. Three 90° glass angles are light-tight in relation to one another. It is good practice to keep the same reference side (which is covered with the glass cover plate). Mark it with a diamond scribe so that it cannot be lost.

The hypotenuse face is trued by grinding on the surface plate with 320 carborundum grit. At this point the 45° angles are checked with the precision 45° angle brass template. Pressure is applied while final grinding on the necessary areas for correction. The 90° side angle must also be checked with the 90° square by using the marked reference side. Both 45° glass angles and their 90° side angles must be light-tight to the protractor and the 90° square. Bevel all sharp corners slightly. The result is a right-angle prism.

4. GENERATING THE FACETS

The first facet generated is at the apex of the 90° faces and becomes the reference face for checking all the dimensional heights of the prisms. Figure 13.3 illustrates the sequential facet generations. It should be noted that it is unnecessary to hold the tolerances of the prisms unreasonably close. The cost of a prism held to plus or minus 0.25 mm can be twice the cost of those of larger tolerances. Close tolerances serve no purpose when the prisms are assembled by the instrument maker because he uses cork shims and screw thread objective cells to adjust focal length changes due to long or short light path lengths caused by the glass in the prisms.

The ground prisms are mounted on a flat surface holder with a low-melting pitch. The amount of glass taken off is controlled by measuring the *width* of the land (a) of the prisms. (See Fig. 13.3.) The height of the 45° faces and the final 90° roof angle of the prisms will be controlled by the width (a) of this land. Any other method is time consuming. It should be kept about 0.025 mm short of the *low* tolerance. After generation the surfaces are fine-ground with 225 aloxide grit; check to see that the grinding of this facet is parallel to the base. Grind to the nominal tolerance for the total land width only.

The next facets are marked Nos. 2 and 3 in Fig. 13.3. These facets become equilateral triangles when the 90° roofs are finished. One of the 90° faces is pitch-cemented to a flat tool and the apex of the hypotenuse is generated until the height is 0.25 mm above the top tolerance. These facets are fine-ground with 145 aloxide grit while in the holder and checked for parallelism with a depth micrometer. Be sure that the pitch is removed from the areas on which measurements are being made. The prisms are reversed and similarly generated and fine-ground. Remove only enough stock in the fine-grinding to clean the facets and render parallel them to the flat surface of the tool.

Renew and rework any bevels lost during fabrication. The beveling of a prism is important because it adds strength against breakage and improves its appearance. Uneven bevels and corners are the result of the uneven widths and nonparallelism of the 145 grit grinding.

5. FINE-GRINDING THE ENTRANCE AND EXIT FACES

The prism at this point is in right-angle form with three facets. There are several methods of fabricating the 90° faces. In one the hypotenuse surface is fine-ground to an excellent 90° side angle by external autocollimaation. (See Fig. 13.4, first position.) Because this gives a four-time amplification factor, the 90° side angle becomes a reference angle for controlling the

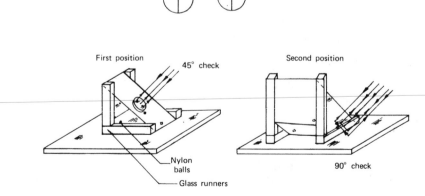

Fig. 13.4. Two external autocollimation modes used alternately to check the 90° entrance and exit faces by rotating from position 1 to position 2.

pyramidal error. Note also that the premarked reference surface is involved. The preground surface is placed on three balls waxed to an aluminized flat.

In this preliminary fine-grinding it is not necessary to fine-ground all the pits out because this surface becomes the 90° roof angle. Fine-grind the surface flat and stop when the 90° side angle is perfect, shown by the autocollimated superimposed images.

The new method is a little more direct and is preferred by most opticians because no master 45° angle block or pregrinding of the hypotenuse surface is required. The external autocollimation of the 90° entrance and exit faces during fine-grinding is illustrated in Fig. 13.4, which shows one 90° face of the prism resting on three waxed balls on the aluminized flat. The collimator is set approximately to a 45° angle, and the 90° side angle to the first 90° side of the prism is placed on the three balls. Note that the surface of the prism being ground faces the collimator and no compound glass angle is involved.

The second 90° glass angle uses the same external autocollimation setup. Here a compound glass angle is involved. The first preground 90° face is made reflective by using an aluminized planoparallel plate (wet-wrung) on its surface. During the fine-grinding of the second 90° face there are two 90° glass angles whose surfaces must be presented to the collimator. These are the first preground 90° angle with the aluminized flat and the 90° side angle to this compound angle. Always make sure that the first surface is not fine-ground twice because of lack of identification. Mark it in some way.

Any separation observed in the horizontal images of the cross arms in the external autocollimation (Fig. 13.4) indicates positive or negative errors in the glass angle of the prism. It is easy to determine in what direction the prism must be ground to correct any two glass-angle errors simultaneously. Place a finger on an area of the prism during each external autocollimation and watch for the horizontal images to approach each other. Mark this area on the prism with a felt pen. Examine the premarked areas. The midpoint between the two premarked areas is the single pressure point that will correct both glass-angle errors simultaneously. It is obvious that one glass angle can be corrected and the other positive or negative when either one of the external autocollimating setups in Fig. 13.4 is used. The entrance and exit faces must be carefully and continuously checked for even widths.

The size of the prisms governs the procedure to be used. The prism's entrance and exit faces range from 0.5 cm to 15 cm in diameter. The smaller diviation prisms of 0.5 cm are generally pregenerated to their final shape and are then fine-ground and checked by external autocollimation methods. Three surfaces of a deviation prism with a 90° roof are worked in plaster-of-paris blocks and the last surface of the 90° roof is worked by cementing glass runners to each reflecting surface of the prism. The second method makes partial use of the glass runners in the later phases of fabricating the 90° roof and depends entirely on the base diameter of the prisms. If the prisms are 10 mm, they can be mass produced by plaster blocking and using glass runners on the 90° roof. Prisms with a 50-mm base are worked free hand or with glass runners.

6. USING GLASS-RUNNERS DURING FABRICATION

The use of glass runners cemented to large deviation prisms with 90° roofs is the approach taken by many opticians. The procedures are discussed at length in Appendix 3. Briefly, the glass runners and prisms are heated to 80°C and pitch is flowed around each heated unit. Lens paper is then placed between prisms and runners and the heated units are allowed to cool to room temperature. The use of glass runners does not materially effect the autocollimation of the 90° side angle and causes only a slight loss of intensity of the image. Some care must be exercised when placing the prism assembly of the three prewaxed balls on the aluminized flat to maintain a stable area for the external autocollimation of both glass angles of the prism's compound angle. (See Fig. 13.5) Figure 13.5 also shows the external autocollimation setup in which the prism assembly rests on the three balls for each compound angle check.

Both faces of the Amici prism are alternately fine-ground and checked by using the setup in Fig. 13.4. Fine grinding must extend fully over the prism's surface and glass runners. An underground area is called a "witness mark." It may be better to use external autocollimation of the 90° faces of the Amici prism because a 4× factor is given, whereas the external autocollimation of the 45° angle is only 2×. Always fine-grind on a convex glass tool because this produces one- to two-wave concave surfaces which are more stable for a autocollimation check as well as for polishing. Remove all pits with 95 grit.

The fine-ground surfaces are semipolished before one surface of the prism is finished. The 90° side angle is the only one that needs to be checked at this time. This prism face is completed when its surface is one-fourth of fringe and free of scratches and sleeks. The second face of the Amici prism

Glass runners

Fig. 13.5.   The method of cementing glass runners to an Amici prism (90° roof not yet generated) for easier fabrication of the 90° faces.

is polished next. Here the surface being fine ground is placed on three balls as shown in Fig. 13.4. Check the compound glass angles in both external autocollimation setups: first the 90° face (finished surface) and then the 90° side angle. Work for well-corrected, 90° angles in this optical check. Because it is an external autocollimation arrangement, determine the one pressure point that will, if possible, correct any two glass angle errors simultaneously. The method involves touching the prism while watching in the eyepiece of the collimator and finding the area on the prism in which the horizontally separated images tend to close up. This must be done for each external autocollimation setup. The midpoint of the premarked areas on the prism is the one point at which pressure should be applied during polishing. The surface quality of this second side should likewise be one-quarter fringe or better and free of surface faults. The glass angles of the Amici (90°) prism are often within 2 min or better because of the external autocollimation check on the 90° angles.

7. GENERATING THE 90° ROOF OF THE PRISMS

The 90° roof angles of the semifinished prisms must be generated with the most careful workmanship. Before cementing the prisms to the 90° channel holder a middle mark which determines the ridge of the 90° roof should be scribed on several of them with a height gauge. One is cemented into the precisely made jig or fixture to serve as a guide during generation of the roof angles.

The prisms are warmed in an oven to 80 to 85°C. Heat the fixture on a heating plate covered with paper toweling. If the paper begins to scorch, the temperature is too hot. Smear the warmed fixture with pitch-cement until it flows freely. Place one of the premarked prisms in the middle of the fixture with the semipolished reference surface *down* on the 90° milled cut. Apply pressure on each prism but make sure that the facet land at the apex of prism, as well as the reference side, is anchored securely. Replace the prisms in the oven and heat to 85°C. Recheck to see if all the prisms are evenly cemented in the fixture. A straight wooden stick helps to press the prisms into place. Large prisms are generated singly in a specially designed fixture, and heated prisms must be kept away from draughts. Glass of uneven cross-sections will fracture when subjected to excessive thermal shocks. (See Fig. 13.6.)

When a number of heated assemblies have cooled down to room temperature, mount the fixture tightly to the generator's spindle and mill within 0.5 mm of the premarked prism in the middle of jig. Record the micrometer setting for future milling operations. Check the parallelism of the blocked prisms with a micrometer at 120° intervals. If it is within 0.1 mm, the

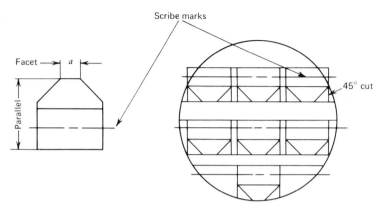

Fig. 13.6. The partially fabricated Amici prisms placed in a tooling fixture for generating one-half of the 90° roof angle.

prisms are of near-equal size. Place these blocks of cemented prisms aside and mill the remaining blocked assemblies. If the parallelism is off by 0.1 mm, the tool fixture has become wrapped or is not precise.

Check the milled assemblies to see if any milled surface widths are larger than average. Mark these prisms with a pencil for separate handling later.

Prism assemblies with parallelism of 0.1 mm are fine-ground with 225 aloxide grit and made parallel to width 0.05 mm on a rotating flat tool. The cleaned prisms are placed in a degreaser or reheated in an 80°C oven. The second side of the 90° roof is similarly reblocked and generated.

Excessively wide prisms must be handled separately because of the uneven thickness of the plates. Occasionally the operator overshoots when milling and passes the prescribed mark (these cases are more difficult to correct) and special handling is required. The prisms are reblocked and the second side is milled 0.1 mm short of the ridge of the roof. Check the previous reading of the generator's micrometer setting. The prisms with the wider lands (caused by overgenerating) may be rescued by fine-grinding the 90° angle to the lowest tolerance.

8. FINE-GRINDING AND POLISHING THE 90° FACES

Some of the smaller deviation prisms (12.5 mm and less) are fine-ground and polished with the first surface of the 90° roof angle remaining in the holding fixture or plaster-blocked and then fine-ground and polished. If the smaller prisms are worked in the metal fixture, they must be cemented, with strips of lens paper between each surface. The plaster blocking is described in Appendix 11. The last surface is fine-ground and polished with cemented glass runners and lens paper is placed between each surface. Before working the last surface the first surface must be aluminized or silvered for internal autocollimation. This point is discussed again in later sections.

For the larger prisms cement the glass runners with lens paper between each surface of the 90° roof. (See Fig. 13.7.) Because the ridge of the 90° roof must be 90° to the entrance and exit faces, make another external autocollimation arrangement. (See Fig. 13.7.) The prism rests on three nylon balls, prewawed to an aluminized flat. These balls should be in a stable configuration on the narrow surface of the prism. The collimator is preset approximately to a 31° angle and the reflective image from one of the 45° faces is picked up. This is a vector angle of the prism's compound angle which is formed when the prism rests on the waxed balls. The collimator is tightened and the cross hairs, set to the reflective image. Check the width of the roof.

In the procedure used during external autocollimation the reflective image is set with the bifilar on the horizontal line of the cross hair of the collimator. In the next step the prism (same side) is picked up and the second face of the prism is rotated. Determine whether the new reflective image is below or above the previous bifilar setting. If the 90° roof ridge is perpendicular to both faces, there will be no drift of the image, either above

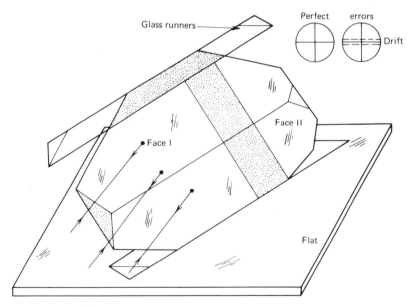

Fig. 13.7. Externally autocollimating each face of the 90° roof for controlling the ridge-roof of the Amici prism. Each 45° face I and face II is alternately placed on the nylon balls.

or below the first setting of the bifilar. Autocollimators often lack bifilars. In this case fine-grind toward the midpoint of the horizontal drift. In any case a null point must be achieved on fine-grinding.

The surface is polished on a flat polisher and the surface quality must be one-eighth of fringe or better and free of surface faults.

One polished surface of the 90° roof can be coated with aluminium or silver but generally it is not. The glass runners and lens paper are cemented to the 90° roof with the coated side of the 90° angle in an upright position.

The second side of the 90° glass angle is also fine-ground. At this point in the fabrication the 90° roof-angle is checked by external autocollimation of the prism's compound glass angle. In this test the prism is picked up and repositioned on the three nylon balls. This 90° glass angle is presented to the collimator setting shown in Fig. 13.7. It is necessary during autocollimation to have a leveling fixture for the prism and aluminizing flat. The reflected image of both settings (Fig. 13.7) must be at the same position as the autocollimator's image. The images need not be superimposed. The desired point on the prism for pressure while fine-grinding is the one that causes the horizontally separated images to come together. This is easy to determine; touch the prism during both collimating checks and watch for the horizontal images to approach each other. If two angles are involved (as in controlling a compound angle), the final pressure point is midway between two predetermined pressure points.

9.   POLISHING AND THE INTERNAL AUTOCOLLIMATION

The second face of the 90° roof differs slightly during polishing because it is required that glass-angle errors be corrected. The prism with glass runners is semipolished and checked by internal autocollimation. (See Fig. 13.8.) There will be three images in the eyepiece of the collimator if the prism has glass-angle errors. Three images occur because the first is a reflective image from the front surface of the prism and the second and third images are internal reflections transmitted after crossing back into the eyepiece of the collimator. The multiplication factor is 12×. This is an extremely sensitive test; consequently the latent heat absorbed by the prism during polishing must be considered and sufficient time must be allowed before checking. If the prism is perfect, the three images will be superimposed in the collimator.

When the prism is properly set with the autocollimator and the images are centered in the field of the eyepiece, the next step is a study of the images.

Assume that one side of the roof angle has been corrected (by external autocollimation) and that the surface is one-tenth fringe. It is possible, however, that secondary or ghost images may be viewed with the eyepiece.

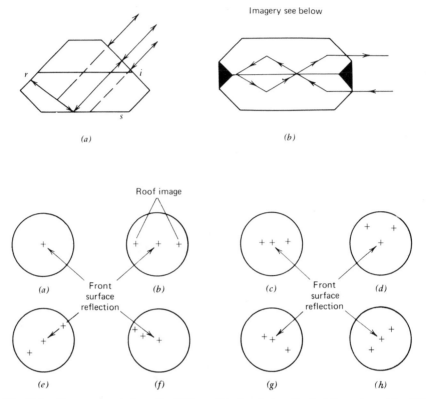

Fig. 13.8. The upper part shows the $12X$ amplification factor of a deviation prism with a 90° roof when internally autocollimated. The lower part contains several sets of image patterns that may possibly be observed in the collimator which shows a 90° angular condition.

Disregard these blurred images and work with the three brightest. Because it is impossible to anticipate the pattern of images that will be observed, take some hypothetical conditions for analysis.

In Fig. 13.8a–h observe the relationships among the multiple images that can be observed for glass-angle errors.

Many beginners in the optical trade have difficulty understanding the meaning of acute (minus) and obtuse (plus) glass-angle errors. (See Figure 13.9a–c.) Figure 13.9a shows a glass-angle (minus) error in the 90° angle; (b) is a 90° glass angle with an obtuse (plus) error; (c) illustrates a slant angle of the roof's ridge with an acute angle error; and (d) is an obtuse slant angle. It must be kept in mind that these glass-angle errors are small—not more than ±1 min of arc. The dashed lines show the wedge of glass that must be removed by polishing. The optician cannot add glass—only a computer can do that! Note that pressure must *always* be exerted on the acute or minus side of the prism when error occurs. The arrow pointing down is the pressure point for correcting glass-angle errors.

Several methods will determine whether glass angles are acute (minus) or obtuse (plus). Here the breath test is used to locate the reflective image from the first 45° face and the image that generates the internal autocollimated images from the prism. This is done by breathing heavily on the 45° face farthest from the autocollimator. Watch for the reappearance of the double images in the field of the eyepiece as the moisture evaporates. The image that does not *disappear* is reflected by the front surface. The reappearance of double images results from internal autocollimation of the 90°

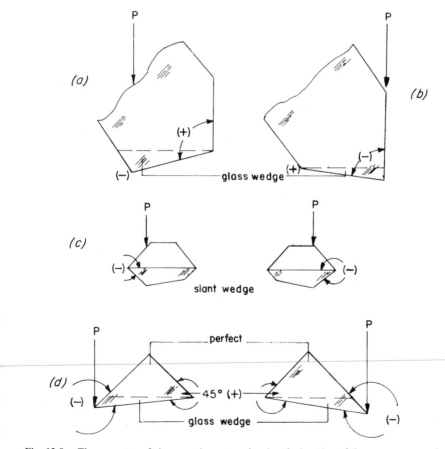

Fig. 13.9. The geometry of glass-angular errors showing the location of the one pressure point that is applied during hand polishing for correction of the angular errors in the 90° roof. Note that one always applies pressure on the minus glass-angle.

roof; (x) the separation of the two side images near the referencing image (reflective image from the prism's front surface) is the error in the 90° roof and (y) is the slant angle of the 90° roof. It is important that the referencing image be centrally located in the field of the eyepiece. To determine whether the glass angles are obtuse (plus) or acute (minus) make use of two rules.

Rule 1. Defocus the eyepiece inward from the best focus, and if the separated images tend to close up with the reference image the glass angle is acute (minus).

Rule 2. Defocus as in Rule 1, and if the glass angle tends to separate further it is obtuse (plus). Another method called the knife-edge test can be used to verify the glass angles. Shade from the bottom at the eyepiece with a machinist's ruler. If the top of the horizontal image disappears, the glass angle is acute or less than 90°. If the bottom of the horizontal image disappears, the glass-angle is obtuse or greater than 90°. In Fig. 13.8.a the triple image is superimposed into one for a perfect prism.

In Figure 13.8b the center image is reflected by the first 45° surface and the two side images come from roof. This is verified by the breath test in which two of the three images reappear. A number of possibilities (actually four) relate to this doubling of the images from the roof's ridge. If, however, the center one and one on either side fade out, they will be images from the roof. Assume that the two outside images have faded out. Now because all images are in the same plane horizontally, this means that the slant angle of the ridge in relation to the reference face (first 45°) is correct. The separation of the two outside images represents a glass-angle error in the 90° roof, which can be determined by using Rules 1 and 2. It is pointed out here that the two outside images may not be equidistant from the central reference image. (See Fig. 13.8c.) One may be close to the reference image and the other at a greater distance. This would mean, in effect, that the roof is slanted in relation to the chosen reference side of the prism and that one of the angles formed by the first 45° surface and its related roof's ridge was plus 135°, the other minus 135°. (See Fig. 13.8c.)

Another possibility is a combination of errors in which the slant angle of the ridge is not 67.5° in relation to the surface that forms the reflective reference in the collimator and the slant angle of the roof is in error. (See Figure 13.8d.)

Figure 13.8e indicates an error in the slant angle of the roof's ridge, a tilt that relates to the chosen reference side, and the 90° roof angle is positive or negative. Figures 13.8f, g, and h are similar.

As in Figure 13.8e, there are possible left or right conditions in the horizontal plane and two conditions that could exist if the two roof images were completely displaced to one side or the other of the reference image, above or below its horizontal plane. (See Fig. 13.8h.)

In summary, other combinations can occur, but in general the examples outlined cover most of the possibilities if the work has been done by external autocollimation. It is obvious that if correction of the 90° glass-angle is achieved (no doubling on the horizontal lines) little can be accomplished by polishing. It would require regrinding the second 90° face of the Amici prism to bring the other set of images to the reflective referencing image from the first 90° surface. This offset of the referencing images from the second set is not necessary for most deviation prisms with a 90° roof if the 90° is well corrected.

It must be assumed that any optician attempting to make a prism of this type has had enough experience to be familiar with the autocollimator and considerable experience with high surface quality and critical glass angles. If, however, the worker studies the glass-angle image formations (especially Fig. 13.8.) and checks the change in the image position in relation to the reference image, he will quickly learn what to do. It is an excellent idea when moving the eyepiece to defocus in one direction only. Thus confusion is avoided.

It is good practice to remember, when checking a glass angle that forms two images that if the point of intersection of the two images is exactly in the focal plane then, by moving the eyepiece *in* the images will separate, thus indicating a possible obtuse (plus) angle. To test this procedure defocus the eyepiece *outward* by an equal amount; the images will again separate by the amount noted when the eyepiece was defocused inward. Keep in mind that by defocusing from the best focus the images separate *outward* on the movement of the eyepiece in *both* directions.

The simplest way of determining whether the angle is good within the limits of the resolving power of the collimator is that with the eyepiece at the best focus and only one image seen the glass angle is well within 2 sec or better with the help of the $12\times$ factor. After polishing to make angular corrections be sure to reestablish the one-quarter fringe surface quality before testing to allow for the latent heat absorbed by the prism to dissipate. This is important.

10.   OTHER TEST METHODS

Another method used in some optical shops involves a target of ruled horizontal and vertical lines some distance away—approximately 6.5 m—with a sighting telescope of 20× at right angles to the target. The prisms which are placed in front of the objective of the telescope direct the image of the target to the observer.

For prisms with well-corrected roofs the superimposed horizontal lines will appear as a single line. If the flat surfaces of the prisms are free of astigmatism, the vertical lines will not double. Striae in the glass of the prisms will effect both vertical and horizontal images. The glass-angle error in the 90° roof causes doubling of the horizontal image similar to that seen in internal autocollimation.

Prisms with errors of 2 sec or more will cause the horizontal lines to appear with a slight swelling or doubling. This is the result of one line from each face of the roof transmitting the image in a different plane. If all surfaces are one-eighth fringe or better and free of astigmatism, then all horizontal and vertical lines of a well-corrected 90° roof will be equally sharp. Understand that this does not necessarily mean that the reflective referencing image is superimposed as observed in the autocollimation test, even for an excellent prism. The target position does not relate to this referencing image. In the case of 90° roof prisms made by internal autocollimating setups this is unimportant. Prisms smaller than 1.8 cm base width cannot be tested by the target method at this distance and internal autocollimation is used.

As discussed for internal autocollimation, the eyepiece test determines whether the glass angles are acute (minus) or obtuse (plus). The eyepiece is defocused inward from the best obtainable focus and the crossed lines will spread equally before going out of focus for a well-corrected prism. If both horizontal and vertical lines form a small square, the prism is well corrected. If the image is rectangular in the vertical plane, the glass angle of 90° roof is obtuse; if it is in the horizontal plane, the roof is acute.

11. INTERFEROMETER CHECK OF THE 90° ROOF

In this section testing the 90° roof in an interferometer is described briefly. (See Fig. 13.10.) The interferometer is setup in the conventional manner with two end mirrors spaced for observation of the fringe patterns. The prism is placed inside the light path of the stationary mirror. Because this is an Amici (or 90° right-angle prism with a roof), it is placed in its normal assembly position; one of its 45° faces receives the parallel light from the retrodirective flat nearest the prism (m_3). The second 45° face is positioned near the stationary mirror which has been rotated from its original position. The support for the prism must be a proper height when placed on the base of the interferometer. The object is to bring the ridge of the roof prism

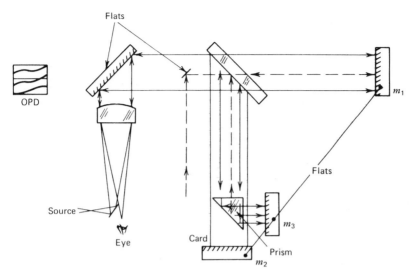

Fig. 13.10. The interferometric testing of the 90° roof angle and the movement of the fringes away from or toward the roof ridge. OPD fringes are often observed in this interferometric setup from the larger prisms.

centrally into the parallel light beam. This support can be a heavy block of metal (plane and parallel) with masking tape on the top and stable at the base but need not be attached permanently.

Occasionally two moving lights are observed in the eyepiece, and one of the light sources must be superimposed on the stationary light source by rotating the prism or using the fine adjustments of the holder with its end mirror. Rotate the eyepiece out of line and look for fringe patterns, which are often observed because the wrong light source was chosen. Some interferometers (Hendrix and others) have an angular-wedge divider plate; therefore no back-reflected image is observed. The first fringes observed are usually multiple, narrow, and parallel. Move the sliding mirror until there are three or four fringes on the roof of the prism. They are oriented in the vertical plane and an equal number occurs on each half roof of the prism. If the 90° roof angle were perfect, the fringe pattern would be like Fig. 13.11a, rectangular instead of circular because of the foreshortened effect of the prism's hypotenuse face.

If the 90° roof angle were acute (minus) or obtuse (plus), the fringe pattern would be like (b) or (c) in Fig. 13.11. The slant of this fringe pattern is due to two conditions: first, the 90° roof angle is not a perfect 90° angle and, second, the fringes are actually circular arcs caused by the vertical

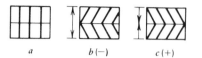

images (from the two plane mirrors) arriving in the focal plane, one behind the other. When the two coincide, the lines of intersection is marked by straight fringes, which may slope to left or right.

To determine whether the roof line has plus or minus error on the 90° roof, pull back slightly on the flat mirror nearest the prism being tested, thus introducing a longer light path. If the fringes move toward the ridge of the roof, the glass angle is obtuse or greater than 90°. (See Fig. 13.11b.) If the fringes move away from the ridge of the roof, the glass angle is acute or less than 90°. (See Fig. 13.11c.)

Other irregularities may occur in the fringe patterns. This applies particularly to the condition of the fringes, which may be slightly curved or the ends bent up or down (s pattern). caused by polished surfaces with turned down or turned up edges or by refractive index variations in the glass of the prisms. This can be established by rechecking the polished surface with an optical flat. If no fault is found then it results from the (OPD) nonhomogeneity in the glass of the prism. If the fringe change is small, localize one of the 45° faces. (See Appendix 8.)

Remove the cemented glass runners from the prisms by heating to 80°C in a controlled oven. When the prisms are cooled, they should be hand washed in xylene, followed by a warm detergent solution. If the coating on the roof is silver, it is removed by concentrated nitric acid. Aluminum can be removed by concentrated lye or hydrochloric acid.

EXERCISES

1. Where is the pressure point for correcting a plus glass angle and a minus glass angle?
2. Why is a third mirror helpful in testing a deviation prism in an interferometer.
3. Name five deviation prisms with a 90° roof.
4. Describe the method to control the 90° roof ridge of a deviation prism.
5. What is the amplification factor for a deviation of an Amici prism and a Schmidt prism?
6. Why are the 90° faces of an Amici prism finished before the 90° roof?
7. Describe the setup for testing the 90° roof a target.

8. How are the small facets and glass angles of prisms smaller than 0.5 base length formed?

BIBLIOGRAPHY

Brown, M. "Making 90° Deviation Prisms," *Optical Shop Procedures and Theory*, **8**, Northronics—Northrop Corporation, 1955.

De Vany, A. S. "Deviation Prisms: Controlling the 90° Roof-Angle Ridge," *Appl. Opt.*, **15**, 1104 (1976).

De Vany, A. S. "Deviation Prism Interferometric Testing," *Appl. Opt.*, **17**, 7 (1978).

Ferson, F. B. "Prisms, Flats, Mirrors," in *Amatuer Telescope Making—Advanced*, A. G. Ingalls, Ed., *Sci. Am.*, 89–93 (1959).

Moffitt, G. W. "Autocollimation of Prism," *J. Opt. Soc. Am. Rev. Sci. Instrum.*, **7**, 41–51 (1923).

14

Precision Planoparallels and
Laminated Windows

Planoparallels and optical windows have many areas of use; for example, in bubble chambers, guided missiles, and window tunnels. Quality assurance ranges from a low quality in a bubble chamber to the highest in space optics. Multiple-laminated windows in tanks and elsewhere can be made cheaply to provide a safety factor.

The earliest guided missle with an optical guiding system was probably the SNARK, made by the Northrop Corporation. Basically, three telescope lens systems of small aperture in an equilateral triangle guided the missile by selected stars sighted through a 20-in. (50.8-cm) window. This chapter discusses the construction of small as well as large planoparallel plates, some as large as 20 in. (51 cm).

The general specifications for any planoparallels are 5 sec or better parallelism and an optical pitch-polished surface to meet the 40–10 dig and polish code specification outlined in Mil-0-13830 (ord.). The call out is surface quality of one-half fringe over 10 in. (25.4 cm) in diameter, less than 0.25 astigmatism, and spherical power less than one. The general speciations are outlined in Jan-G 174 A.

The homogeneity of the glass disk, which is generally felt polished as it comes from the glass manufacturer, can be verified by Murty's test method (See Chapter 5.) Here Scherlien-quality planoparallels with a matching oil because of the poorly polished surfaces must be used. With this method stones, bubbles, and striae can be detected, but because of the large apertures in the glass blanks only one section can be treated at a time.

Fabrication involves the grinding and polishing methods common to smaller planoparallel production. The parallelism is checked by a large reflector-type Fizeau interferometer. The laminated windows are sawed from finished polished glass disks and cemented together in a unique way which cancels glass-angle errors.

1. FUNDAMENTAL OPERATIONS

After being tested for homogeneity the blanks are parallel gererated on a vacuum chuck that permits the back to be made parallel to the front with a high degree of accuracy. A 45° bevel with a 0.5-in. (12.5-mm) width can be done at this time.

Because the glass disks are generally circular edged by the manufacturer, this operation is not required. It can also be done by using a grinding wheel on the generator. (See Chapter 1.)

Two methods are in common use to strengthen the edged circular disks: in one the bevel and peripherial edge are ground and felt-polished on a specially built machine before the parallel surfaces are ground. In the other all surfaces are acid-etched with hydrofluoric acid in special plastic or wavetanks. This takes only a short time because only the fumes are involved but it is dangerous. Good ventilation with special handling equipment is required. The etching fumes eat into the diamond grit swirls (about 0.01 mm deep). It has been shown that glass blanks so treated are actually stronger. I recall seeing some of these 51-cm blanks broken up after the SNARK program was discontinued. The glass disks were removed from their potted holders with a 20-lb hammer and came out without being broken. These were the disks with polished edges and bevels; some of the older fabricated windows with unpolished edges were smashed in the process.

The acid etching of the pregenerated surfaces can be problematic unless enough time is taken to grind the bottom of the diamond generator marks. I recall that a number of Schmidt-Cassegrain cameras were returned from the field to have scratches removed from the corrector plates and their coatings redone. After the removal of the coatings with a concentrated sulfuric and borax mixture the diamond swirls could be seen on the flat side but not on the correcting side because that side receives four times more polishing than the flat side. It appears that some of the polishing action is like spreading butter on toast to fill the cracks.

To find grind the tested glass blanks for a homogeneity test place them on a light aluminum holder of the same diameter with a piece of rubberized felt underneath the disks. To prevent the disks from sliding several layers of masking tape are wrapped around their edges. The grinding tool is made of bathroom tile pitch-cemented to an aluminum holder. (See Figure 14.1.) The cemented tiles are separated by the width of one tile. One of the reasons a tile tool is used is that it can be controlled to give a concave or convex ground contour on the glass element; a cast-iron tool cannot be easily changed. Also the separated tile will give a finer grind and, unlike a cast-

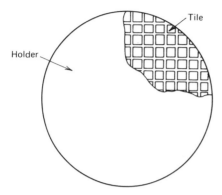

Fig. 14.1. Grinding facet flat tool made up of bathroom tiles cemented to a flat aluminum holder.

iron tool, will not be subject to seizure when the grinding wet becomes dry. Because these planoparallel elements are thin (1/20 ratio of diameter), fine-grinding is always done with the glass disk on top and the tile tool on the bottom spindle. Generally the bottom spindle rotates at 20 rpm and the eccentric arm, at 10 rpm, with a third stroke.

The surface is controlled by checking with a 12-in. (30-cm) spherometer. The desired reading should be a slightly concave 0.001 in. (0.0025 mm).

After fine-grinding one surface with 250 aloxide grit check the parallelism with a micrometer at eight sectors of glass disk. The second side is also fine-ground with 250 grit. The high side is marked with a felt pen before grinding. To grind out a wedge added lead weight is placed on the aluminum holder. Equation 1 serves as an estimate of the amount of weight needed to grind out the wedge in the glass disk.

Circular flat, no center hold:

$$D = r(W + L)/4L \tag{1}$$

where D is the distance from the center of weight on the holder, r is the radius of the glass disk, W is the weight of the glass disk and backing plate, that is, the total weight, and L is the lead weight to be added eccentrically.

Example in pounds and inches: a change is to be made in the wedge of a 20-in. (51-cm) diameter glass disk whose weight with toolholder is approximately 30 lb. Where is the 15-lb lead weight placed? Calculation: $D = 10 (30 + 10)/4 \times 15$ or $400/60 = 6.6$ in. (16.8 cm). Assume that it is desirable to change the wedge angle without removing more glass than necessary. In practice the wedge is to be zero opposite the lead weight and maximum where the lead weight has been placed. This is not often achieved and a ground witness appears, in which case the lead weight must be placed nearer the edge than the calculation shows. The use of several strips of masking tape will keep the weight in place. It is important that the parallelism be within the tolerance before proceeding with the next grit.

Each side is alternately fine-ground with the next two grits (145 and 95 aloxide). The surface must be flat within the limits of the 30-cm spherometer. The amount of wedge removed by the 145 and 95 grits is small and before the 145 grit grind it should be within tolerance. It is essential that the micrometer differences in the readings at each sector be well within 0.01 mm over the aperture before the 145 fine-grinding.

The added weight can be used further to remove any wedge angle because this weight will help to speed up the grinding with 145 grit. It should be removed after several wets, because it can cause astigmatic surfaces to be ground in the glass disk.

Because concave surfaces are desirable before polishing, the 95-grit grind should be controlled by slowing down the bottom spindle's rotation and speeding up the overarm. The stroke of the overarm must be shortened, perhaps to 15 cm. Here the 95 fine-grind must be thoroughly carried out on each side of the planoparallel element. Check especially for any unground witness marks where the wedge was last found.

2. POLISHING AND FIGURING

A pitch polisher is described in Chapter 3. Here a full-size-diameter aluminum, lightweight holder is required. A large polisher is not difficult to make and only a little heavier to handle. Pour the hot pitch mixture on a warmed aluminum holder with a dam formed by masking tape 1 in. (2.5 cm) above its surface. The pitch should be about 0.5 in. (1.5 cm) thick. After the pitch has cooled (test with a slightly wet finger) it is ready to be flattened if a finger imprint is visible. This is accomplished with one of the pressing plates, which should be thoroughly wetted with polishing compound. Remove the masking tape before gently placing the glass disk on it. Next move the glass plate over the polisher and decide whether the entrapped bubbles are getting smaller and the pitch is even around the glass disk. Cool the glass disk quickly and wet the aluminum holder with a sponge dipped in cold water, especially around the peripherial areas. All entrapped bubbles need not be removed and even some areas on which the pitch did not become flattened can be ignored. They can be filled later with melted pitch.

Following the method described in Chapter 3, cut the facets into the pitch with 1-in. (25-mm) facets. Briefly, the pitch lap is wetted with a detergent-water solution, and with a 1-in. (25-mm) brass strip make the first cut channel across the lap. Move the brass strip over to allow for the full width of the channel and repeat the process. Next emboss the pitch lap with smaller facets called ripples. (See Fig. 14.2.) It is necessary to make these small facets by the "cold press method," which involves laying a wet nylon or screen netting over the precut channeled lap wetted down with polishing compound. The ground disk is placed on top and a number of lead weights are added to the aluminum holder. Generally this will take several hours and can be done overnight without the added lead weights. Examine the imprint of the netting by sliding the glass tool off the polisher a short distance. Caution: do not allow it to impress itself any more than one-half the thickness of one of the threads or pitch will be pulled out of the lap. It often helps, however to remove the glass disk and its holder by squirting water into the pitch channels before removing the assembly. The capillary action causes the water to creep underneath the glass.

One surface is now polished perhaps with 20 rpm on the bottom spindle, 15 rpm on the overarm, and a 15 cm stroke over the polisher. After a 20-min run check the progress of polishing. If the center is brighter than the edges, the polishing can be continued for several hours, but if the edges are brighter change the length of the stroke to 3 in. (7 cm) and speed up the top spindle to 20 rpm. After 1-hr check with a 10-in. (25-cm) optical flat. The surface should be no more than three fringes, either concave or convex. If it is concave, polish another hour or two, but if it is convex turn it over and polish the second side. The reason for this is that during fine grinding one side will approach the convex and the other will be concave. In any case, semipolish only one side for a short time and then turn it over and finish *that* side to tolerance. If the mistake is made of finishing one side and not turning it over, the Twyman effect will cause the finished side to become astigmatic by wrapping.

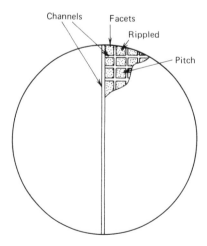

Fig. 14.2. A flat pitch lap makeup with channel cut facets and with embossed smaller facets made by screen door netting.

This surface to be finished is polished by changing the length of the over-arm stroke: in one third diameter of the element, then one-third and one-fourth at 20-min intervals.

In a like manner the second surface is completed by polishing. Perhaps here some astigmatism will be uncovered by the plate test and can be the method described below.

3. FIGURING

Often the weight of the aluminum holder plus the weight of the glass plate will produce unequal polishing, thereby causing a cylinder to develop in the element. Three-legged, spiderlike drivers, which have rubberized supports resting on the glass are the answer. These spider drives will permit the glass disks to float on the polisher and bring the higher edges down to the level of the lower areas.

As the figuring progresses, it is often desirable to contour the polisher with a pressing plate of the opposite contour to the configuration on the work. These pressing plates will generally be two to three fringes concave on one side and a like number of convex fringes on the second side. If the glass surface is convex, it is apparent that the concave side of the pressing plate is needed and a convex surface is pressed on the pitch polisher. Conversely, a convex pressing plate will contour the pitch polisher with a concave surface, a process that can be speeded up with a heavy cast-iron grinding plate on top of the pressing plate and a rubberized felt pad to protect the glass surface. This will generally take about 15 min if the small embossed facets are still present.

To attempt to give the various rotation rates and stroke lengths of the bottom and top spindles is impossible because of the number of variables. This can be learned only by experience.

The testing of the parallelism of the polished planoparallel involves a large f/3.5 spheroidal mirror. The planoparallel is placed inside the focal point of the concave mirror to form an autocollimation setup. Mercury sources were originally used with a limit on the thickness of the glass element that could be tested. Illumination with an 0.3-mW laser can be used and the projected fringes, observed on a frosted screen. The Davidson 12-in. (30-cm) planointerferometer will check the surface quality and the Fizeau fringes of parallelism. Figure 14.3 shows a glass disk worked by the techniques in this chapter. Observe the four fringes with diameters of more than 51 cm. Because rough calculations reveals 1.5 sec per fringe and the planoparallel has a 50.1-cm diameter, this is 0.12 sec of arc:

$$\text{seconds of arc} = \frac{1.5 \times \text{number of fringes}}{\text{diameter of piece}} \qquad (2)$$

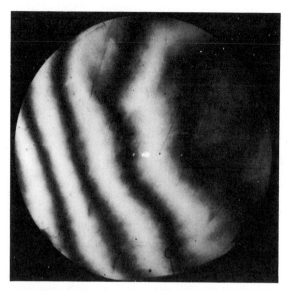

Fig. 14.3. An excellent 22 in. (55-cm) window made by the techniques described in the text. Note four fringes overall or 0.26 sec.

As stated in Section 1, the specifications of the window include a surface quality of one-half fringe over a 1-in. (25.4-cm) diameter, less than one-quarter fringe astigmatism, the spherical power to be less than one, and 2 sec parallelism. Question: Figure 14.3 is a photograph of a 20-(50.1-cm)-diameter planoparallel. What is the wedge arc? Answer: 0.12 sec of arc. Do not make the mistake of assuming that the off-center circular pattern of fringes is surface quality; the are Fizeau fringes formed by interference between the light paths of the two polished surfaces. The overall surface quality of this glass disk is probably in the neighborhood of 4 × ½ or two fringes, plus, over 51 cm.

4. SMALLER PLANOPARALLELS

Smaller planoparallels are made like the larger ones, but it is a little more difficult to obtain the degree of parallelism required because we cannot with a micrometer measure the high degree of accuracy as with larger glass disks. However, with a special three-legged fixture and a Cleveland electronic measuring probe we can obtain a very high degree of accuracy, perhaps 0.001 mm of wedge measurements. Fine-grinding must be done by hand to remove the wedge from the glass disk. As in the larger elements, we must remove most of the wedge with 250 grit. The fine grinding with the last two grits can be done by machine working.

The polishing of the smaller elements is similar: first one surface is semi-polished (25%); then the second side is completed, but before beginning to polish the second side check the element in the Fizeau interferometer. Count the number of fringes and use formula (2). If it is within tolerance, continue the polishing. If it is not, determine the high point of the wedge so that correct pressure can be applied by hand polishing to lower the wedge. This is done by placing a finger on one of the fringes for several seconds. (See Fig. 14.4.) The fringe will expand slowly in a fingerlike area and point directly at the thinnest part of the glass wedge. When the wedge becomes small (parallelism), the fringes will often be circular or fuzzy ovals at dead center which indicates a zero wedge. (See Fig. 14.3.)

If the wedge is rather large, the Fizeau fringes will have parallel lines equally spaced; this method of identification is the same as that described above. If the wedge is above 10 sec, it would be best to find the highest point and regrind with 95 grit. Pressure is applied at the highest area during this grinding. A 10-min polish is required before rechecking.

The ever-ready Twyman effect is present on polished surfaces to give an astigmatic fringe pattern (called saddleback). Hand polishing must be

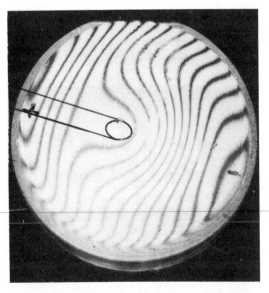

Fig. 14.4. Checking the direction of the wedge angle by placing a finger over several fringes. The fringes will actually point toward the thinner area of the wedge angle.

applied on the highest areas (the location of the fewest fringes of the flatter two areas). (See Chapter 4, Fig. 4.2.) On the smaller elements it does not pay to use lead weights to correct for saddle-back fringes.

5. LAMINATED WINDOWS

The making of precision laminated windows is similar in many respects to making optical windows of high quality, the only difference being that the parallelism need not be high-precision.

Examine Fig. 14.5 and note the parallel fringes of an optical window which indicates that a wedge exists in the planoparallel. To fabricate a laminated optical quality window saw cut the two sections from one glass window with this wedge angle. The element is cut-along the direction of the Fizeau parallel fringes. (See Fig. 14.6.) If the two proper sides are cemented together, the assembled cemented unit becomes a planoparallel. (See Fig. 14.7.) Observe the wedge angle that shows a vertical parallel where the saw cut was made to divide the window element into two sections. Also shown are two alternate methods of cementing sections (a) and (b) to nullify the wedge angles.

In addition, it is good practice to select the two optically superior flat surfaces of the element carefully to minimize the power of the optical surfaces which contribute to the assembled laminated window. (See Fig. 14.7.)

What is meant by power given as a number of Newton rings or fringes? It is generally observed that one side of a large optical window may have a surface quality of several fringes convex; the other side will have almost that

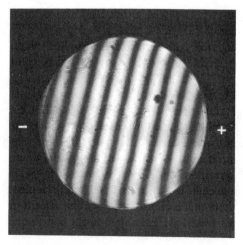

Fig. 14.5.   A 55-cm planoparallel ready to be saw-cut into sections for a laminated window.

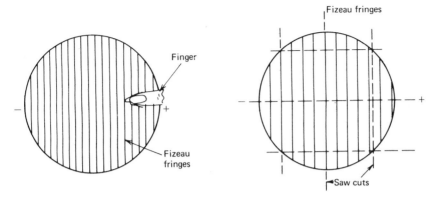

Fig. 14.6. The saw cuts made along the parallel Fizeau fringes into sections before cementing them together.

quality (not necessarily) concave. This does not mean that the apparent surface quality is such that no choice need be made. It will always involve a fraction of a fringe or ring.

It is important that wedge angles in seconds of arc (measured by Fizeau fringes) remain unmixed with surface quality which is the sole contributor

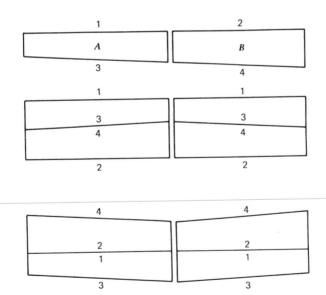

Fig. 14.7. Several alternate methods to cement the glass sections together, thus eliminating the power worked into the respective glass sections.

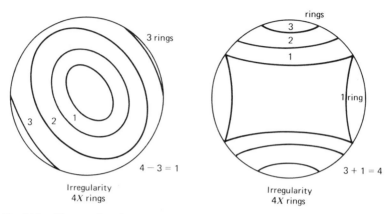

Fig. 14.8. The meaning of power (X term) and its calculation of the polished surfaces.

to power. Figure 14.8 reveals what is meant by power and irregularity (two differences). An optical shop print will often enter these notations under tolerances. Unless otherwise stated, the rings are counted for the entire aperture of each polished surface. For large elements a 10-in. (25-cm) test flat was used and the specifications were outlined. Caution: Fig. 14.3 represents but *one* side; the other side may be quite different; the power then is the algebraic sum of both surfaces.

6. SAWING AND ASSEMBLY

It is important after sawing any polished optical window that the sawed edges be fine-ground and felt-polished or deeply acid-etched with hydrofluric acid (HF). This means that all chipped edges and bubbles are removed. I remember that a very large window carried in the ALOTS camera plane to cover the Apollo launchings was broken after several landings and had to be replaced. When the broken window was examined, it was discovered that all the cracks running across the window had a common origin in a large chip (25-mm) diameter on the peripheral edge. The circular cracks showed fine annealing (not ragged), which would indicate poor annealing. The size of the crown glass window was approximately 152 × 76 cm × 9 cm. As described above, the 51-cm glass windows 2.54 cm thick had to be hammered out with a 20-lb sledge with six or more blows, yet they did not break in two. They actually popped out whole and unbroken from the sealing compound and their holder. These SNARK windows had polished edges and bevels.

Returning to window sawing saw cut the rectangular or squares shapes from the circular optical window and split them down the middle along an inked line of the Fizeau fringes. It would be best to mark the wedge direction on the edge of the window with a piece of waterproof masking tape. The polished surfaces must be protected during sawing with wide masking tape; a planned drawing should be made. The waterproof tape can be left on during the fine-grinding and polishing phases. A simple method uses flexible rubber disks that carry various grades of carborundum paper for grinding the edges. These edges are then polished with a crocus cloth.

For acid etching the edges on the sawed contour the polished surfaces are protected by brushing with warm paraffin. Check to see that all polished areas are well covered with the hot paraffin, but be careful to remove it from the edges with xylene.

The acid etching is a hazardous operation because the fumes are corrosive and poisonous. Beeswax or vinyl plastic containers are uneffected by the acid and highly vented draw fans must be used to cover the enclosures. Generally a 20-min period is sufficient if the acid is 50% commercial grade. Care must be taken to prevent breaks from occuring in the paraffin because the glass surface will surely be marred by etching. As stated before, this acid etching digs deeply into the strains setup by the diamond generating and sawing operations.

Several methods of assembling the finished element into a laminated window after the etching and grinding and polishing the edges are shown in Figure 14.7.

EXERCISES

1. How is the wedge angle determined in a Fizeau interferometer?
2. Name several methods of strengthening windows.
3. Explain what is meant by power and irregularity.
4. In what direction is the glass section to be used as a window section for a laminated-window?
5. How is power eliminated in cementing the window sections for laminated windows?
6. Calculate the wedge angle for a 24-in. diameter window with two fringes per inch.
7. How is the floating action polishing carried out while polishing windows?
8. Describe the makeup of the flat tile tool.
9. Describe how the measured wedge angle is removed during fine-grinding.
10. How are the edges polished on a window?

BIBLIOGRAPHY

De Vany, A. S., and Griffth, T. "Working Large Plano Parallels," *Optical Shop Procedures and Theory*, **13**, Nortronics—Northrop Corp. (1955).

Forman, P. F. "A Note on Possible Errors Due to Thickness Variations in Testing Nominally Parallel Plates," *Appl. Opt.*, **3**, 646–647 (1964).

Halderman, P. H. Private communication, Nortronics, Northrop Corp., ALOTS Window.

Twyman, F. *Prism and Lens Making*, Hilger and Watts, London, 1952, pp. 386–388.

Welford, W. T. "Bubble Chamber Optics, and Bubble Chamber Windows," *Appl. Opt.*, **2**, 981–1012; 1037–1042 (1963).

15

Spherical Test Plates

Interferometric control for precision working optical surfaces is carried out by observing fringe patterns from precisely measured master test plates. Most test plates are spherical but in a few cases aspheric test plates are used for small secondaries in multiple-mirror telescopes. Test plates are matched pairs of concave and convex surfaces of low expansion glass generally pyrex or a like material. With a matched pair of spherical surfaces in close contact with one another Newton rings or fringes are formed. All test plates for checking len's surfaces must have slight air spaces between the surfaces for the fringes to be seen. The number of fringes or rings observed depends on the amount of air wedge formed by dust particles and the difference between the two radii in wavelengths. If there is no air wedge and the surfaces have identical curvature, no rings will be observed; a uniform color will be visible, depending on how close they are.

The fabrication of high-quality spherical test plates with a surface quality of one-tenth fringe and 0.005 mm tolerance is expensive and time consuming. Where there is quality assurance control the test plates are made of tighter tolerances than those used in optical shops. This means that the radius of curvature must be precisely measured on an optical bench and the surface quality must be one-tenth fringe or better. These test plates are sometimes used for low-quality surface work such as eyepiece fabrication in which the tolerance might be two or more rings. Here the tolerance is still maintained because these test plates can be used for other lenses. By counting the number of fringes of departure per centimeter diameter, new test plates need not be made if the curvatures are similar. Firms often exchange test plates for use in making tests of their own or for a short run of lens elements.

Test plates can have short radii (microscope objectives) and long radii of curvatures. An optical proof plane (flat) actually has an infinitely long radius of curvature. Test plates diameters range from 6 to 254 mm. I once made a pair of test plates with 10-in. (25.4-cm) diameters, 212-in. (238-cm) radii, and a surface quality of two rings. This large diameter is rarely used because of the danger of scratching or causing digs in the polished surface of a finished lens.

Test plates with 100-in. (2.54-cm) radii require no compensator if they have an R/D number of 6 or more, where R is the radius and D, the test-plate diameter. A test plate with a low R/D number or ratio requires a compensator radius on the opposite side. The convex test plate requires a concave compensator radius (not the same radius) on its opposite side; conversely, the concave radius needs a convex compensator curve.

1. COMPENSATOR CURVES

Deve's book, *Optical Workshop Principles*, (Hilger & Watts, London, 1954) describes some of the difficulties of using test plates for the control of surface quality and radii of lens work, chief among which is the fringe test pattern for test plates with low R/D numbers when the usual broad mono-chromatic mercury or helium sources are used in standard viewing boxes. These extended sources are useful for flat surfaces and the longer radii, but break down for test plates with R/D numbers of 6 or lower.

Most beginners and even experienced opticians become confused by the foreshortening effect caused by the fit of a concave test plate on the convex test plate of a lens surface. The foreshortening effect is caused because the convex half of the test plate's match is small and not fully illuminated. Therefore for accurate fringe measurement the line of sight must be perpendicular to the surfaces. This is not always possible for low R/D ratios. The best that can be done (without auxiliary equipment) is to polish the two matching surfaces until they can be pressed into near optimal contact and only one color shows over the entire spread of the surfaces. Caution: small differences in radii between the two test plates cannot be measured by counting fringes from either side. Fringes are caused by dust particles and a wider separation of the surfaces is the result. Another problem is often encountered in the optical shop. A discrepancy can exist between the fringes observed with the convex surface uppermost and those with the concave surface on the bottom. This is especially evident when the surfaces are brought to a close tolerance; perhaps one-half fringe when viewed on one side and then two or more fringes viewed on the other. This discrepancy may be due to several conditions; for example, two surfaces are rarely in absolute contact, for particles of dust can hold them apart by several fringes; in addition, scratches and digs on a test plate in near contact on a lens or test-plate surface. Much thought has been given to the use of auxiliary equipment and compensator radii. (See Section 4.)

In conjunction with Carl Short of Northrop Corporation I examined Deve's method of having compensator radius of opposite curvature for each type of test plate. Deve did not give complete formulas for these compensator radii. The following formula was developed for the convex compensator radius on the concave test:

$$r = R - d/[(n^2 - 2n(1 - h^2/R^2 + 1)^{1/2})^{-1/2} + 1] \qquad (1)$$

where r equals radius of the compensator curve, d, the maximum thickness of the test plate, h the semidiameter, n, the refractive index (1.5), and R, the prescription radius.

The concave compensator radius on the convex test plate is different from formula (1). It is approximately equal to

$$r = R/2(n - 1) \qquad (2)$$

where r is equal to the concave radius on the convex test plate, R is the prescription radius, and n, the refractive index (1.5).

It is apparent from formula (2) that the radius of the compensator curve is about equal to one-quarter of the radius of the test plate. From both formulas it can be seen that the compensator radii are always of sharper radius than those of the prescription radii of the test plates. The viewing distance for the test plate fringes is at least 10 or more diameters.

Formula (1) is rather difficult to handle without an electronic hand calculator. The same perplexing problem occurs with low R/D numbers near 1.0. The formula proved useful for R/D numbers greater than 1.5.

2. GRAPHICAL RAY TRACE OF COMPENSATOR CURVES

A newer method of approaching this problem is graphical ray tracing of the compensator curves was proposed by Daniel McGuire in 1964. This can be done easily by optician or draftsman. Only a compass and a parallel rule are required. The draftsman has a drafting machine and can produce a full-scale drawing and accurate compensator curves can be determined. The method of drawing is straightforward. See Fig. 15.1 for a graphical ray trace. Both concave and convex test plates with compensator curves are shown. The R/D number of the test plates is 4.2 and the viewpoint is 20 diameters away (near parallel light distance). The following steps are taken to produce a drawing:

(a) Draw the axis line.
(b) Draw viewpoint lines 20 diameters away from the test plates.
(c) Draw two radii of 127 mm R_1 from T, where the thickness is ⅛ of plate.
(d) On the concave test plate extend the radius line some distance.

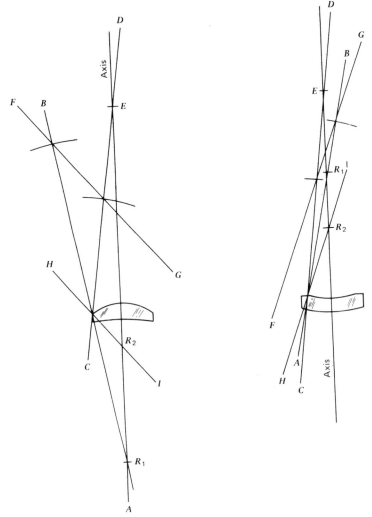

Fig. 15.1. A graphic ray tracing of the compensator curves of the concave and convex test plates.

(e)   On the convex test plate extend the radius to the axis line.

(f)   Swing arc₁ at any convenient radius at thickness T (right) on *view-point line*.

(g)   Swing arc₂ which is approximately 1.52 times larger than arc₁ on *radius line*.

(h)   Draw lines between arc₁ and arc₂.

(i)   With a parallel ruler bring line arc₁ and arc₂ over to T and draw
       from T across the axis line. This is the radius of the compensator
       curve on the test plate.

Now check the radius on the 50-cm (127-mm) concave test plate. It is
approximately 1.5 in. (37.0 mm) (scale measurement). The convex test plate
is 1.85 in. (47 mm).

In this section some of the complexities involved in the use of test plates
when they have low R/D numbers have been discussed. Although it is not
usually the optician's job to design test plates with compensator radii, it is
helpful to know why the foreshortening effect occurs. The next section deals
with auxiliary equipment for better viewing of the fringe fit.

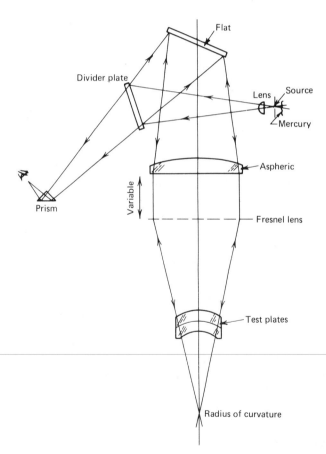

Fig. 15.2.   A planointerferometer setup for checking test plates. Fresnel plastic lenses can be
interposed for test plates with low R/D numbers.

3. AUXILIARY EQUIPMENT

In many optical shops auxiliary viewing equipment is available for viewing the test plates in parallel light. Most of the equipment is based on the interferometer principle. Several types of instrument are described. The interferoscope principle makes it possible to reflect a small source of monochromatic light normal to the test plates or lens surfaces under test. The fringes are easily observed for any R/D number. The makeup of the Fizeau interferoscope is described in Fig. 15.2. It has an aperture of 6 to 12 in. (15 to 30 cm). The lens is often an $f/4.0$ planoconvex aspheric or a two-element objective connected to the source. In the planoconvex aspheric the master flat is the plano surface and in the objective lens a second wedge element (20 min) with a one-tenth fringe surface serves as the master flat. The source of illumination is a mercury reed with a 0.004-mm pinhole. Often for brighter light sources a short focal lens (18-mm planoconvex) is used. Its planosurface must be fine-ground to act as a diffuser. A beam splitter of planoparallel glass with a 50/50 coating reflects the light from the source to the test plates. One of the best planointerferometers of this design is Davidson's (Optronic) Corporation, West Covina, CA.

Some shops will use cellophane tape (0.05 mm) on their test plates to standardize the separation. This has several advantages; first, it keeps the test plate from being scratched by dust particles; second, the resilient nature of the tape permits an even uniform airspace; and, third, it eliminates the partial vacuum that makes the removal of the test plates difficult.

4. THE ZIMMERMAN INTERFEROSCOPE

A simpler interferoscope based on the use of cheap plastic Fresnel lenses is described by J. Zimmerman of Itex Corporation in *Applied Optics* September 1971, pages 2216–2217. As shown in Fig. 15.3, this instrument consists of a diffused mercury source with a Fresnel lens inserted into slots at various heights above the test plate on the lens surface. A beam divider with a 40/60 partial coating reflects the fringe patterns and gives a Maxwellian view over the full aperture. The test plates must be graphically designed with compensator curves for low R/D ratios. Fresnel plastic lenses with various focal lengths can be purchased from Edmund Scientific Co., Barrington, NJ. At times a Fresnel lens must be used (Fig. 15.3) to help to cover the aperture of the test plates.

5. WORKING PROCEDURES

Test plates are generally made from low-expansion glasses such as Pyrex. The graphic ray trace is made (see Fig. 15.4) and the prescription concave

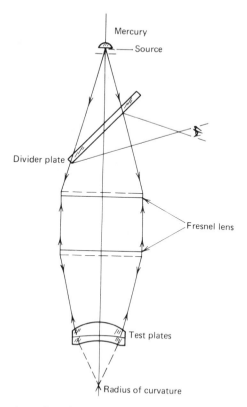

Fig. 15.3. Zimmerman's version of the interferoscope made with two plastic Fresnel lenses. The lenses positioned in variable slots give a Maxwellian view of the test-plate fit.

and convex radii, each with its compensator curve, noted. Check for proper thickness which is roughly ⅙ ratio to the diameter of the test plates. Test plates are always slightly larger than the diameter of the lens.

The grinding tools for the compensator curves are generally green plate glass and have the same diameter as the test plates. All the necessary information that can be obtained from the ray trace should be listed. It will be needed later when aluminum test plate holders and pitch polishers are ordered.

Generate all four radii on the test plates and grinding tools being sure to place the compensator curve on the correct plate. Do not try to generate the radius any closer than a 0.1-in. (0.25-mm) reading of the spherometer. During the fine-grinding work for the higher precision reading on the spherometer. Bevel all edges to a sharper radius with a cast-iron tool and 145 grit.

Because aluminum holders for the test plates and later for the lens elements will be needed, they should be ordered in advance. The test plates have the same diameter and equal radii of curvature and a set of four holders is required. Another set of four must be supplied for the pitch polishers; their machined radii differ from the radii of the test plates and their compensator curves. Holders for concave radii must be 0.2 in. (5 mm) longer because of the thickness of the pitch. For convex radii the holders must be 0.2 in. (5 mm) shorter. Generally toolholders for test plates are larger than the lens elements; they can be turned down on a lathe for lens fabrication.

Depending on the type of spherometer available, always use the largest pin setting (radius of ball stems) that the lens surface will permit. Among the electronic spherometers on the market one is the Strasbaugh Model 18-N which has various pin-setting plates that range from 0.75 in. (19.05 mm) to 8 in. (203.2 mm). The sensing device is a Sheffield Model 51 C Accutron (electronic) which can read to 0.00001 in. (0.0002 mm) differences. The Accutron is used to set the spherometer to a value found in a table of settings for any curvature and the diameter of its mounting. (See Fig. 15.5.) It

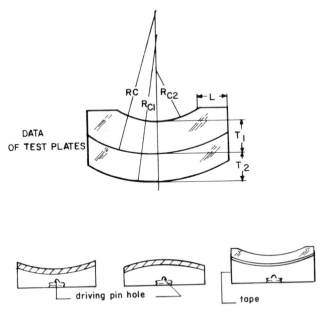

Fig. 15.4. The data for test plates obtained from the graphic ray trace. Here RC is the prescription radii, R_{C1} the radius of the compensator curve of the concave test plate, and R_{C2} the radius of the compensator curve of the convex test plate; T_1 and T_2 are the respective thicknesses of the test plate. The lower part of the figure shows the basic design of the pitch lap radii holders; an equal thickness of the pitch should be a requirement.

Fig. 15.5. The Strasbaugh Model 18AS spherometer with a number of pin-setting plates.

is simple to interpolate for any radius not found in the table and to find the radius of a lens surface from the sagitta reading. The precise spherometer readings are taken only after fine-grinding with 95 grit or during the polishing phases.

When grinding the pregenerated glass surfaces with 250 grit, the radius of curvature of the glass elements can be rapidly changed by several methods. If the ground surface is too steep on the test plate, the setup on the machine must be reversed. The test plate is then placed on the bottom and its grinding tool is worked on top. This is generally done by hand until the radius is within 0.004 in. (0.1 mm) of the correct sagittal reading. If the ground radius is too shallow or low reading, the grinding tool is placed on the bottom spindle and the test plate is worked on top.

Fine-grinding is continued through 145, 95, and 50 grits. During the 95- or 50-grit grinding use the finer calibrated scale of the Accutron device. Some optical companies do not use the 50 aloxide grits because it grasps the two glass elements too tightly as the slurry dries. This can happen in other fine-grit grinding but there is no difficulty in separating the elements with a shop razor blade by placing it under the bevel and striking its edge with a rawhide hammer. Talc or pumice is often added to the finer grits to prevent grasping.

6. POLISHING AND FIGURING

The compensator curves are semipolished about 40%. The polishing machine is set to run at approximately 150 rpm with a one-third stroke at

50 rpm. The steepness of the radius of the compensator curve governs the overhand stroke and its rate of rotation during polishing. A harder pitch is used and the polisher is often subdivided into four unequal quadrant sectors by cutting across two diameters.

To polish the precise radius surfaces on the test plates, make up facet-type polishers. (See Chapter 3.) Briefly, these polishers have 0.5-in. (12-mm) square facets cut into the pitch and are embossed with smaller facets formed by nylon or view-screen netting to give better contact. Polish for a period of about 20 min. Determine whether both surfaces are even. If not, work the test plates by hand for a short time by maintaining uniform pressure and good contact during polishing.

Check the test plates together in a test setup and note the fringe departure. It is not important at this time if they are separated by two to three fringes. It is better to have a number of rings than one fringe and a turned-down edge. Enter the fringe departure and its concave or convex fit in your notebook. Remeasure the concave test plate with the Accutron and check for correctness of the reading on the spherometer, using the *finer* scale reading. Some optical shops do not stock this type of spherometer, in which case the concave test plate is measured on the optical bench. Some shops coat this concave surface with aluminum or silver.

Some opticians pick up the convex test plate for polishing, thus destroying the reference surface. Of course, this is experienced by those who have measured the test plate on the optical bench. Others who have the use of the electronic measuring device can polish both elements intermittingly. Before measuring with the finer scale on the spherometer allow the latent heat to normalize to room temperature or an error will occur.

Those who are interested in the measurement of a concave test plate should review Chapter 25. Briefly, a series of 10 readings is made at two positions of the tester and their average is taken; for example, a set of test plates 101.6-mm in diameter with a radius of curvature of 255.27 mm and a tolerance of plus or minus 0.08 mm. The R/D number is 255.27/101.6 or 2.51, which is used to bring the test plates to tolerance. (See Table 15.1.)

Because the pair of test plates is within 1.5 fringes plus, the concave test plate is brought to a near zero fit with the convext test plate. It can be seen that the semipolished convex test plate was within the nominal tolerance of 0.003 in. (0.08 mm). This particular example shown in Table 15.2 demonstrates how important it is not to rework the convex test plate while the concave test plate is being measured. It serves as a referencing standard. Now, because the concave test plate is 1.5 fringes plus, it must be reduced by polishing to make it more concave or to shorten the radius. This is done by polishing the concave test plate on top of the polisher with a one-third diameter stroke on the overarm.

TABLE 15.1

Test plates diameter 4 in. (101.6) radius 10.042 in. (255.07 mm)	Average Readings
R/D Number 2.51	255.27 nominal tolerance
From Table A (2.5 R/D) #10	255.17
740 fringes per cm	0.10 minus error
A. 7.4 moving decimal two places	0.08
B. For 0.02 mm over 101.6 mm diameter is .02 × 10.0 cm × 7.4 mm = 1.5 fringes	0.02 minus error
C. Previous value in Notes 3 fringes convex or plus	
D. Total change requires is 3 minus 1.5 or 1.5 fringe change	
E. Make concave test plate 1.5 fringes more concave	
F. This reduces the match fit now to 1.5 fringes overall (convex).	

The example listed in Table 15.1 was close to the nominal tolerance, and it is not uncommon for test plates to be off by 10 or more fringes (plus or minus) before measurement.

If the concave test plate was on the plus side by 0.005 in. (0.12 mm), we would have to flatten the concave test plate or give it a longer radius. This is done by working the concave test plate on the bottom spindle with the polisher on top.

Some optical firms are now using the Accutron device with the spherometer to produce measurements as precise as 0.0002-mm. Other firms require an optical bench measurement to verify the tolerances.

In this section another simpler method is illustrated. Here the ratio is based on a measurement taken on the optical bench. It is not uncommon for test plates to be reworked three or more times before they are acceptable

TABLE 15.2

Number of fringes from last movement	=	X change
Difference of radius of previous measurement		difference in last measurement

Example: $\dfrac{4.5 \text{ fringes}}{0.5 \text{ mm}} = \dfrac{X}{0.1 \text{ mm}}$ or $0.45/0.5x$ $x = 9$ fringes for nominal tolerance. Of course, the change can be plus or minus and for that reason a notebook should be kept for reference.

When the concave test plate is within nominal tolerance, it is matched with the convex plate. It must be kept in mind that the concave test plate will now have to be reworked in conjunction with the convex test plate—*but* it is a waste of time to try to remove all the irregularities in both plates until the concave plate is within tolerance. The surface irregularities become more pronounced as the surface quality improves on test plates. As the pair of plates comes closer together (color matching) the latent heat absorbed during polishing and handling must be controlled.

A Tropel vertical interferometer is a useful instrument for measuring and testing the surface quality of a concave mirror. It has the capability of recording the radius within 0.01 mm in a 10-cm radius range. Knowledge of the surface quality of the concave test plate within one-tenth of a fringe helps in matching it to the convex plate. In the older method the center of the concave test plate is offset about one-fourth of its diameter to give a composite of the zonal slopes in both surfaces with relation to a good surface on either one. By shifting pressure with the finger the fringes can be centered at will on each element. The zonal departures (hills and holes) can be detected by bobbing the head or by exerting pressure. (See Figure 15.6.) Tropel Model 4000 vertical interferometer is shown in Fig. 15.7.

It is difficult to explain how to correct each glass fault and much depends on the experience the optician has gained. It is noteworthy that if the edge is turned down the whole area of the test plate must be removed down to the beginning of the slope. This is the reason that the polishing pitch is always harder on test plates. Most turned edges occur because the plates were handled or worked on before the absorbed heat had left them. It is good

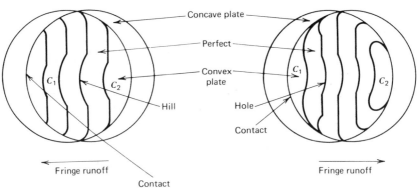

Fig. 15.6. The off-setting of a pair of test plates to check for zonal departure; always use the best area of the test plates because it becomes the referencing zone.

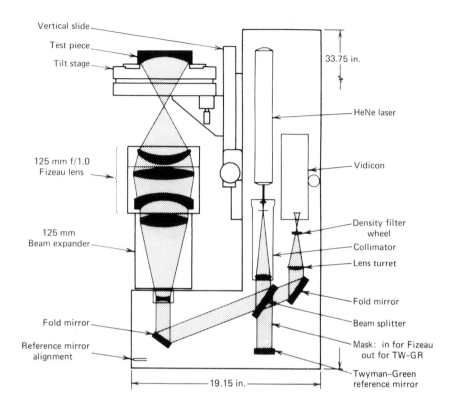

Tropel model 4000 vertical interferometer

Fig. 15.7. A Tropel model 4000 test-plate vertical interferometer.

practice to place the test plates on the polisher for several minutes before reworking them. Adding a small lead weight helps to contour the polisher. Make sure also that the small embossed facets have not disappeared into the body of the pitch. These facets must always be present. Check that the channel groves have not filled up and renew the channels with a razor blade. Work the test plates by hand for a time and then let the polishing do the smoothing. Holes and hills are generally eliminated by a third diameter stroke across the polisher. Stubborn holes or hills are often the result of channel cuts and an uneven rotation of areas of pitch. A center set at dead center will cause a central hole.

Some opticians will make use of the thumb or middle finger to correct high rings or turned-up edges and then work out the remaining irregularities with the machine. A watery slurry of polishing compound is better than a heavy one. When retouching with the fingers extra care must be taken not to over do.

Remember that after any polishing a concave surface comes toward the convex surface. A convex surface never comes up (some crystals are an exception) but always has more convexity.

When the surface quality is accepted for the test plates, the concave surface must be remeasured. Test plates will depart from specification during figuring and this is one of the many reasons why they are expensive. The measured value is often marked on top with a diamond scribe for identification.

EXERCISES

1. Make a pair of test plates 2 in. (50.4 mm) in diameter and with a prescription radius of 4.025 in. (102.235 mm) and a 1% nominal and two-ring tolerances.
2. Describe Zimmerman's test for checking test plates.
3. What is the correct distance for viewing test plates?
4. Graphical-ray trace the compensator curves for these test plates.

BIBLIOGRAPHY

Biddle, B. J. "Testing Steep Radius Test Plates," *Opt. Acta*, **16**, No. 2 (1960).

Cooke, F. "Aid to Viewing Test Plates Interference Fringes," *Appl. Opt.*, **10**, 2216–2217 (1971).

De Vany, A. S. "Compensator Curves of Lens Test Plates," *J. Opt. Soc. Am.* **TD75** (1959).

De Vany, A. S. "Testing the Sphericity of Short Radii Convex Surfaces," *Appl. Opt.*, **13**, 229 (1974).

Deve, C. *Optical Workshop Principles*, Hilger and Watts, London, 1954.

Johnson, B. K. "Test Plates," in Optics and Optical Instruments, Dover, New York, 1960, pp. 183–187.

McQuire, D. "How to Make Test Plates of Precise Radii" unpublished (1960).

McQuire, D. "Designing and Making Air-Spaced Test Plates," unpublished (1963).

Short, D. "The Use of Fringes in the Determination of Spherical Test Plates," Nortronics Division Memo, Northrop Corporation (1955).

See also Chapter 4.

16

Lens Equipment

Lens making has an ancient origin. Lenses of natural quartz which date back to 1200 BC, have been found in Troy and Egypt. Objective lenses as large as 40 in. (101.6 cm) were made for astronomical research by Alvin G. Clark in 1897. This telescope is at Yerkes Observatory, Williams Bay, Wisconsin, which is connected with the University of Chicago. As soon as the 40-in. objective was figured Clark expressed his readiness to begin work on a 60-in. (152.4-cm) objective, but this never came to pass. A 50-in. (127-cm) objective was made in France and was mounted horizontally at the World's Fair in Paris in 1896. The light of the moon and planets were reflected into the objective by an optical flat on a celostat. However, it was never shown in a conventional manner and its present use is known only to the French government. The three testing methods applied by the Clarks are of interest: first, they used a knife edge to set the lens in an equatorial mounting that followed a star; second, an artificial star was devised in an underground tunnel some 75 m in length; third, they developed an autocollimation setup with a large, silvered, optical flat that doubled the irregularities of the figuring.

The methods of lens fabrication for the beginning optician are limited perhaps to 10. Use is made of glass grinding tools and quality control of radii and polished surfaces by test plates. Recent progress in mass production with pressed-molded lens blanks is not discussed here. This method requires five-diamond grit radius tools and lenses mounted on precision holders. These lenses are generated and polished rapidly in less than 6 to 8 min.

The methods covered here have been in use for more than 100 years. Diamond generation had its start around 1914. The production of prototype lenses for a newly designed instrument follows these techniques.

1. SAWING AND EDGING

Lens blanks larger than 8 in. (20 cm) should be checked for homogeneity by the methods outlined in Chapter 4. If smaller lens are to be cut from larger blanks by diamond sawing, the number of smaller circular lens blanks that can be sawed from the larger must be laid out. Always allow sufficient glass for sawing and edging. Cut off the surplus around the prescribed circular patterns to form an octagon-shaped figure and reduce the edging time.

Glass blanks of approximate diameter are cheaper than glass slabs of larger diameter. In any case, these blanks are edged to a diameter 0.8 in. (2 mm) larger than the call-out on the shop print.

The generation or edging of these glass blanks depends on obtainable shop equipment. The modern Strasbaugh generator edges and generates on a vacuum-chuck mounting. (See Chapter 2.) In another method the first radius surface of the precut octagonal blank is generated; the second surface is treated next. The edges are then made circular. This method keeps the two generated radii surfaces from forming appreciable wedges between the glass surfaces. If the glass blank is edged and the first and second radii surfaces are generated, one large wedge will occur between them. The reason is that the generated glass blank cannot easily be recentered for the generation of its second radius surface.

A third method makes use of precisely made thin brass holders with outer edges that run concentrically to the threaded screw hole. Octagonal or circular glass blanks are preheated on a hot plate with a sheet of white paper toweling on the metal base. Do not allow the temperature to become hot enough to brown the paper. Keep the hands dry and do not allow water or damp fingers to touch the heated glasses. Flint glasses are more susceptible to thermal shock than crown glasses.

Insert the wooden handle with its metal screw into the recess holder. Heat the brass holder over a Bunsen burner and coat with the inside and outside peripheral areas with stick pitch. (See Fig. 16.1.) The heated glass blank is picked up with a fresh piece of paper toweling and set down on another. Next, the preheated brass holder is placed near the central area of the octagonal glass disk which is determined with a small ruler and checked by the scribed pencil mark. The heated brass holder can be moved about easily at this time. Make sure that the pitch encircles the peripheral area of the holder. If not, add more heated pitch. Reheating the brass holder will permit the pitch to flow around both edges and over the glass by capillary action. Three holders with roughly centered glass elements can be made up so that as one assembly is cooling another assembly can be fabricated.

2. PRELIMINARY EDGING

The generated octagonal blanks are more difficult to edge than the circular, the reason being that the protuberant edges of the octagonal blanks are not uniform. Hand feed the edger slowly into the glass blank until the first click

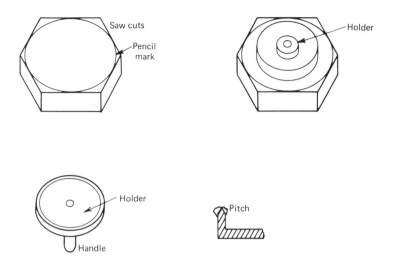

Fig. 16.1. Precut glass blanks pitch-cemented to a holder for edging and generating. Both glass blank and holder are preheated before assembly.

of the glass striking the rotating diamond wheel is heard. If it strikes too hard, it will be knocked out of the brass holder.

Before setting the automatic feed make sure that the glass blank will remain on the full travel of the width of the rotating wheel. If it rides over either of the two edges, it will chip the edges as the glass disk translates over the diamond wheel while being fed in. The setting is made by adjusting the length of travel by the eccentric arm. (See Fig. 16.2.) A rubber mat may be placed between the glass blank and metal surface to prevent damage to the disk if it breaks away from the holder. Edging of the periphery is checked by marking it with a pencil while the disk is rotating. If the pencil mark remains, the edging is not circular. The glass-grinding tools are generated and edged in a similar manner.

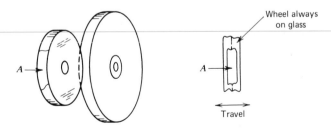

Fig. 16.2. The setting of the eccentric arm of the edger by keeping the grinding wheel on the work continuously.

3. GENERATING THE RADII ON THE ELEMENTS

The generation of glass surfaces and prescribed radii have been discussed but not the methods, although a brief description is given in Chapter 2. Table 16.1 lists the angle of tilt of the generator's diamond cutting wheel for concave or convex surfaces. Here the listed calculated values are found by using formular (1):

$$\sin T = D/2R \qquad (1)$$

where T is the angular setting of the wheel, D, the diameter of the measure cutting edge (outside diameter for concave radius and inside diameter for convex), and R is the prescribed radius.

Because the operation of the Strasbaugh generator is described step-by-step in Chapter 2, it is not discussed here.

Two areas must be closely watched: first, the spherometer radius reading must be within 0.25 mm of the book value of the pin's diameter setting; second, the generation of the second radius must be similarly controlled with respect to the spherometer setting and the thickness of the glass. This is accomplished with a dial indicator and a small stud fixture set to zero before measuring the thickness between the two radii. Generally about 0.008 in. (0.2 mm) above the highest side tolerance will suffice if both glass radii have been carefully generated according to the techniques outlined. (See Fig. 16.3.)

The use of a vacuum holder on the Strasbaugh generator involves the careful centering of the glass blank after the generating and edging the first surface. Here examine the edged blank with a dial indicator on the vacuum holder. The final setting may involve several movements of the glass disk on the vacuum holder before its run-out is zero. After this centering the second surface is treated in a similar manner.

TABLE 16.1

	Too steep	Too low
Concave surface on elements		
Grinding tool	Top spindle	Bottom spindle
Lens elements	Bottom spindle	Top spindle
Convex surface on elements		
Grinding tool	Bottom spindle	Top spindle
Lens elements	Top spindle	Bottom spindle

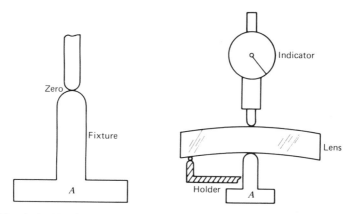

Fig. 16.3.   A checking fixture for determining the thickness of the generated glass elements.

The glass-grinding tools for each surface must be generated in turn. Some beginners may wonder why the grinding tool cannot also be generated with this setting. The reason is that the cutting edge of the diamond wheel has a different tilt angle for opposite surfaces.

4. LOOSE GRIT GRINDING

After the diamond generation of the lens elements a fine spiraling of scratch marks can be seen over the surfaces of the curve. These surfaces are actually ellipsoidals because of the geometry of the setting. Small central hills and holes are also often present.

If the lens elements are larger than 2.0 in. (5 cm), they are fabricated one at a time. Most of the generator marks are removed with a 250-grit grind and for laser illuminated optical systems about 0.008 in. (0.1 mm) of glass must be removed. The reason is that the diamond generator cracks extend far below the surface of lens and will cause electronic background noise in electronic circuitry. The smaller lens elements are generally pitch-blocked in multiples of three, seven, and so on, depending on their diameters. Figure 16.4 shows several methods of blocking lenses. In the first a blocking tool is made up with an opposite matching curve and wax paper to hold the lens to a fine-ground and polished. It should be noted that the matching curves of the metal holders are screwed into the radius tool (convex). The opposite surface is also fine-ground and polished. The second method is the conventional one of precasting pitch mallets to mount the lens to a convex or concave radius holder.

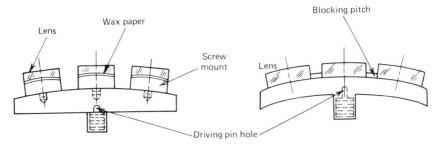

Fig. 16.4. Two different methods of blocking optical components.

Working the larger elements requires checking the parallelism of the two radii on the element. Place it on three ballbearings soldered to studs separated by 120° and anchor three plastic pins close to the studs to act as stoppers. A dial indicator checks the run-out as pressure is placed on one side. This is all hand done while the wedge angle from the first surface of the lens is ground. Apply pressure on the highest side of the wedge while fine grinding. One way to hold the lens during fabrication is to mount it on an aluminum holder with felt between the surfaces. Two layers of masking tape encircle the lens. This assembly is used during machine grinding and polishing. A check of the run-out and assembly of the lens to the holder are shown in Fig. 16.5.

Loose-grit grinding of blocked or single lenses requires a spherical surface to be maintained on the elements at the prescribed radius. In the sequential grinding with the finer grades of grit the radii are harder to change. The spherometer reading must be definitely *on* at the end of the 145 grit before proceeding to the 95 aloxide grit. The spherical surface grind by the finer grits is well covered by W. Rupp of Itex Corporation in an article published in *Applied Optics*, December 1972, pp. 2797–2817. It is possible to loose-

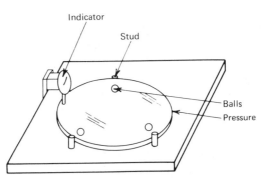

Fig. 16.5. A three-ball checking fixture for paralleling optical elements during grinding.

grit grind an optical surface to one-eighth wave. The concave test mirror had a 25-cm diameter with an $f/2.2$ ratio. The mirror was worked on top of the grinding tool with a random stroke (asymmetrical) of 0 to 100 mm, the grinding tool rotating at 30 rpm, and the eccentric arm rotating at 5 rpm. Random-motion grinding and polishing machines are made in the United States by Strasbaugh. It was found that there is no need for sophisticated feeding of the slurry of grit to ensure a constant feeding ratio for best results. A fine-ground concave mirror can be given a Ronchi test by applying a light coating of paste wax and buffing with a nylon cloth ball.

The grinding of any lens elements to a prescription radius of curvature requires an interchange from top grind to bottom grind. (See Table 16.1.) Of course, too steep means that the spherometer reading is too high or a larger value than required; conversely, too low means that the spherometer reading has a lower value than required.

Caution must be exercised when grinding a steep radius on top of a concave grinding tool. This grinding must be done by hand and not by the conventional use of machine grinding and polishing. Another factor not discussed is the length and rate of the stroke over the glass elements. A longer stroke will change the concave and convex surfaces to a steeper radius with the grinding tools in proper relationship. (See Table 16.1.) A short, fast stroke will lower the radius with the grinding tools in the relationship shown. Steep radii on the elements are slow to respond to any change in rate or position.

5. POLISHING LAPS

As pointed out previously it is necessary to consider the pitch thickness of the aluminum radius holder, especially for the steeper radii lens surfaces. See Fig. 16.6 for the particular design of the pitch tool and method of holding a single element during fabrication. As shown, a driving pin is drilled in the screw hole and should be designed for fastening a spindle or for inserting

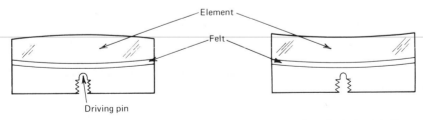

Fig. 16.6. A design lens holder for holding the glass elements with strain during grinding and polishing. Note the driving pen placement close to the work and the easy reversal of the work from top to bottom on the polishing machine.

the driving pin during machine grinding and polishing. For longer radius work this is not too important. Still it is good practice to have the driving pin as close to the work as possible. Too much tilt of the polisher and work is not desirable.

The formation of steeper convex and concave polishing tools is not a simple matter. Generally the polishing pitch is allowed to cool to the consistancy of molasses, and the convex curve tool is heated in hot water or with a Bunsen burner. Before dipping the convex holder into the pitch a small amount of soft wax or turpentine is smeared over the metal. The holder is then dipped a number of times into the pitch and allowed to cool slightly until the pitch is about 0.2 in. (5 mm) thick. The next step is to form the pitch radius on the blocked elements or the single large lens. The blocked lenses are presently a problem which can be overcome by the use of another radius glass or cast-iron tool of approximately the same radius (plus or minus 0.04 in. or 1 mm). For another solution use an aluminum foil sheet and place it on the blocked elements. The foil is thoroughly wetted with polishing compound and the warm, roughly shaped convex-polisher is formed over the foil. The foil is easily removed from the blocked elements and the warm pitch if both surfaces are wet. Grinding glass tools are often used for making test plates if they are large enough. During the formation of the pitch lap iron out any recesses caused by sinking the blocked elements into the pitch.

In the next operation deep channel grooves are cut into the pitch to form 1-cm squares. Again, iron out and contour the polisher by reheating the pitch in hot water and working it until it is uniform. Again cut away any overflow of pitch in the channels. The polisher should be embossed with smaller facets with nylon mesh cloth or wire netting. This finishes the concave surface polisher.

To make a convex polisher pour the pitch into the recess and impress the contour with another tool close to the desired radius or with a special glass tool generated to the radius required. The pitch lap is then cut with 0.4-in. (1-cm) square facets and the lap is contoured to the elements by warming the polisher by hand as the blocked elements are rotating on the bottom spindle. This requires careful workmanship because the blocked elements can become distorted if too much heat is applied. In the long run it often pays to generate another glass contouring plate to form the polisher. In any case, it would be best to regrind the blocked elements with two sets of 95 grit to reestablish them.

6. POLISHING

The polishing pitch laps are run on top of the glass elements to conform with the surface. First check the lapping by hand to see that there is no grasping by the polisher during its stroke. After the polish has been underway for a short time check the general appearance of the semipolished elements and the radius of the lens with a test plate. If the grinding has been careful according to the Accutron spherometer, each lens ought to be within two or three rings. (See Chapter 15.) If the lenses are concave to the test plate, the polisher must be run on top with a longer sweep and more rapid stroke. If they are convex, with longer or flatter curves, polish the blocked elements on the bottom with a shorter stroke. If short-radius lenses cannot be easily worked on the bottom because of drag or a loose lens the length of the stroke and rotation of the bottom spindle must be varied. If the departure of the radius on the lens under test-plate check is two or more rings on small blocked lenses, it would be best to regrind and verify with the spherometer and the Accutron's more sensitive scale reading. This is the principal reason for examining the lens in a semipolished state.

If the lens elements are well within the test-plate fit, polish the lens to the necessary degree. Blocked lenses are seldom worked to better than two-fringe (0.5 wave) test-plate fit over the entire block.

After several hours of polishing it might be well to retrim the polisher and determine whether the smaller facets have totally disappeared or the channels have overflowed with pitch. The shorter radius elements often remain too convex to the test-plate fit and cannot be adjusted by shortening the stroke or speeding up the bottom spindle, in which case make the polisher smaller, perhaps of 0.7 diameter, and work it on top with a short, rapid stroke.

As a general rule blocked elements should be finished in one day because lenses have a propensity to shift overnight. A good practice is to make up the polisher the day before and have the elements ready in the morning for one wet regrind with 95 grit.

Astigmatism is often the result of uneven polishing or lateral shifting of one or more blocked elements. The test plate reveals it as a saddleback or elliptical fringe. Several methods will correct astigmatism:

1.   For saddle-back fringes find the area in which the fringes are the closest fit to the test plate (flatter); then by hand polishing, with the polisher on the bottom rotating around 50 rpm, the blocked elements are worked across the flatter plane (in two directions) but allowing no rotation during the stroke. (See Fig. 4.2.)

2.   For oval fringes all weight should be removed to speed up the polishing action.

3.   The polisher can be dragging because of the close fit between the glass surface and the polisher; determine whether the smaller facets are still visible. If not, reemboss the polisher.

7. INSPECTION

Most beginners have difficulty in knowing when polishing is complete. Look for gray, sleeks, and scratches. Most optical shops use a dark box with a 100-Watt light bulb with a bare-wire filament (very few are available today), but a frosted bulb can be substituted. The magnifier should be a 5*X* binocular type. The observer should start from the center and sweep to the outer edge. Use a cotton swab with a small amount of cleaning fluid and treat each unit being checked. Watch for gray marks (unpolished surface) near the edge of the elements. Spots that do not move may be a dig (testplate mark), gray mark, or a small bubble. Ammonia-based window cleaners are also used. Pick up the reflected image and examine several areas. Many opticians use a finger to check spots. Dried polishing compound and pitch mixture often found along the extreme edge can be removed with xylene. This build-up should not be permitted, however, because it is a source of sleeks. Particles will fall on the optical surface and become entrapped in the pitch. The lenses should be cleaned with a cotton swab and xylene from time to time when searching for unpolished gray surfaces.

For unblocked single lenses a fluorescent bulb can be installed in the black box at eye level to check the lens surface.

8.   REBEVELING AND CENTERING

By working with only one side of the blocked elements they can now be removed by placing them in a freezer or refrigerator for a short time to release the lenses from the pitch or wax paper. The elements are cleaned in xylene. Check the bevels again and rebevel those that require it with 145 grit by using a small cast-iron radius (concave). Be sure that the radius is a sharper than that on the element.

Check the thickness with a micrometer and use a piece of lens paper between the polished surface of the element and the anvil.

Fine-grinding and polishing the second lens surface are identical to working the first side. After finishing the second side recheck the elements and wrap them in lens paper until they can be edged.

Centering the lenses is necessary to remove the prismatic wedge that lies between the two surfaces. At least three methods for centering lenses can be used; the method depends on the available auxiliary equipment and the type of edging machinery. Some shops favor the old method of centering in which the lenses are centered on small brass holders with cementing pitch and placed on the spindle of the edger. In the next step the lenses are moved with a small wooden stick until the two reflected images of a small light source perhaps 15 ft (5 m) away are superimposed. (See also Chapter 2.)

The second method makes use of auxiliary equipment by which the demountable quills can be placed alternately in the auxiliary apparatus and edger. This is Strasbaugh's method, which is convenient for edging large or small lens for zoom-type systems. (See Fig. 2.5.) The lens is mounted on a brass holder with a piece of lens paper pierced by a central hole, which extends slightly beyond the brass holder (to protect the polished surface). When the assembly has cooled sufficiently for handling, it is placed on the quill in an upright position. Two methods are used simultaneouly, a dial indicator checks the lens on top for runout, and a 20X microscope is used with a small collimated source which has an illumination recticle. If the image remains stationary and the run-out is zero, the lens can be edged. Of course, if the cemented lens on the brass holder wobbles, then it is shifted in the holder. The quill and its assembled lens and holder are than placed in the edging machine.

The third method makes use of the LOH (Wetzler) or Strasbaugh bellcentering (automatic centering) for small or medium-sized lenses (150 mm is the largest). (Review Chapter 2.) These automatic centering machines have two precision bell-like cups which, when released, grab the lens surfaces and by pressure automatically center them. For long radius surfaces the machines have extra spindles to center the lens. These machines can be set up to complete several operations at one time. Edging and beveling the lens with precise bevels for mounting takes place in one operation. (See Fig. 2.21.)

9. MASS PRODUCTION METHODS

When several thousand or more lenses are required for optical sights such as binoculars, microscopes, and artillery sights, the lenses are produced by special techniques. Molded or preslumped lens blanks are used, as are special recess holders for the curves generated by diamond radius tools and for polishing after the finely generated lens blanks have been completed. The polishing is done on two types of polishing material—polyurethane and transelco. A thin layer (0.5 mm thick) is expoxied to an accurately ground spherical holder.

The polishing cycle is approximately 8 to 15 min with high-speed methods. Automatic LOH (Wetzler) tray machines are used for centering.

Anyone interested in the aspects of mass production methods should buy *Optical Workshop Notebook*, Vol. I, edited by J. B. Houston, Jr., 1974–1975, Optical Society of America, 2000 L Street, N.W., Washington,

D.C. 20036. John L. Plummber of Plummber Precision Optics describes these modern methods of mass production. This notebook is written in the simple language of the beginning optician. It is also valuable for optical engineers.

EXERCISES

1. What are the two causes of astigmatic surfaces during polishing?
2. How is a polished convex surface test changed to a concave fit for the test plate?
3. How is a polished surface tested to ensure that it is completely polished?
4. Describe one method of blocking lenses elements.
5. How are large lens elements fine-ground in parallel?
6. How is thickness of a lens element checked?
7. How is a lens element centered on a Strasbaugh generator for generating the second radius on the lens element?
8. Describe centering lenses with flat radii.
9. Can cylinder lenses be centered by automatic chucking?
10. Why should the grinding wheel remain on the peripheral during edging?

BIBLIOGRAPHY

Ferson, F. B. and Lenart, P. "Lens Production," in *Amateur Telescope Making, Book Three*, A. G. Ingalls, Ed., *Sci. Am.*, 163–199 (1953).

Horne, D. F. "Production of lenses in Quantity," in *Optical Production Technology*, Crane, Russak, New York, 1972, pp. 163–199.

Strong, J. "Laboratory Optical Work." in *Procedures in Experimental Physics*, Prentice-Hall, Englewood Cliffs, NJ, 1943, 29–93.

Twyman, F. "Production of Lenses in Quantity," in *Prisms and Lens Making*, Hilger and Watts, London, 1952, pp. 191–250.

17

Optical Elements from Hard Crystals

Making optical elements from medium and hard materials calls for techniques that differ in several ways from that using glasses. Calcium fluoride from Eastman Kodak, Rochester, NY, is called Irtran 3. This company also supplies other optical crystals of chemical compounds, such as magnesium fluoride (Irtran 1), zinc sulfide (Irtran 2), zinc selenide (Irtran 4), calcium oxide (Irtran 5), and cadmium telluride (Irtran 6). These crystals are medium hard, yet vary in working techniques; for example, calcium oxide withstands thermal shock and calcium fluoride does not. Lithium and calcium fluoride crystals are representative of large ranges of other crystals worked into optical elements.

1. GENERATING

Lithium and calcium fluoride elements can be generated, if certain precautions are followed: reanneal the crystal blanks for 3 hr at 100–125°F in a controlled oven. Do not preheat the oven. Place the blanks on sheets of paper toweling.

Modern generators with vacuum holders simplify the generation process, but if metallic holders are required several sheets of lens paper are placed on the metal holder and both brass holder and crystal elements are heated in an oven at 130°F. A low melting pitch (see Section 2.8) is placed on the peripheral edge before heating. Do not set the brass holders near the blanks because they may be hotter than the crystals. Raise the temperature to about 140°F (but no higher) until the pitch begins to liquify. Pick up the brass holders and place them on crystal blanks. Use a pair of cotton gloves to lessen the thermal shock that might occur if the elements were touched. When the assemblies have been made, do not remove them from the oven until they have cooled to room temperature.

The diamond cutting wheel of the generator is generally 180 to 220 grit. Automatic feeding of lithium is a problem because the wheel burns the crystal. It is better to use the hand feed. Generate 0.01 in. (0.054 mm) at a time and wait 10 sec for cooling the generated crystal before starting another cut. Calcium fluoride should also be generated slowly without the automatic feed.

In the modern generator with vacuum holders it is easy to place the lens blank on the "O" ring of the holder and to check a dial gauge for run out before generating the second radius on the lens. For the brass holders place the assemblies in the oven and reheat them until the lens blanks can be removed by the gloved hands. Check to see if the brass holders have enough pitch on them for attaching to the curved surface of the lens. If it is insufficient, the pitch must be renewed on the holders. Allow them to cool in 140°F oven before pitching them to the curved surface of the lens. If the stock of blanks for the final edging diameter tolerance is limited, care must be taken.

Because both radii of the prescribed curves on the generator were checked by a dial indicator, the wedge should be small. The elements on the brass holders will surely have a wedge in them because of the difficulty of setting the blanks precisely while hot. Do not take the assemblies of these elements on the brass holders from the heated oven because they will be subject to thermal shock. The cooled blanks are edged by a 320-diamond grit or 120-grit carborundum wheel large enough to edge them.

2. GRINDING AND BEVELING

Before they are ground the lenses should be beveled with 95 aloxide and a strongly curved concave glass tool. This tool should be on top and the lens element, on the bottom spindle. The lens is held by a glass radius tool of equal curvature and diameter during this operation. In this way the emery slurry flows down more easily on the lens and a finer bevel results. Wash in water and detergent at room temperature and wipe dry with a towel.

Polish all bevels with a felt concave tool until a bright sheen is observed. Caution: this should be carefully done at this stage because sleeks and scratches in fine-grinding and polishing can be traced to chips of broken crystal that keep falling on the grinding or polishing tool.

Grind all surfaces with three grades of aloxide grit: 145, 95, and 50. This means that the thickness of the lenses should be near the top tolerance when generated because little stock is removed. With the 145 grit all wedge is removed by using a three-post tooling fixture with a dial indicator that reads to 0.01 mm. (See Fig. 17.1.)

Wedge is removed by pressing on the highest reading edge. Caution: always maintain constant pressure against two posts when measuring.

Generally 1 hr of 145 grinding is enough for each surface. There is little difference between grinding calcium and lithium fluoride. Lithium fluoride

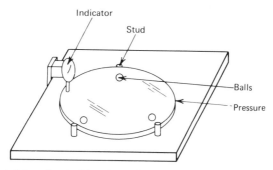

Fig. 17.1. A test fixture for checking the wedge angle of a lens element. Pressure is applied while checking toward the two stops under the dial indicator.

is softer and fine-grinding lines appear. Use glass grinding tools with facets sawed into the impress.

The 95 and 50 aloxide grits grinding phases are similar; each grit must have at least 30 min duration. Caution. The optician must be honest with himself because any fine lines seen will *not* polish out. No. 50 grit grinding ought to be done by hand and not be allowed to become too tight because it is extremely difficult to separate the crystal blank from the fastened tool. If this occurs, the only recourse is to place a sharp razor blade under the crystal blank and tap it with a small wooden mallet. Make sure that no fracture has occurred. If it has, regrind with 145 grit until it is removed. I have never chipped or fractured an optic by using this method if there were sufficient beveling. (See Fig. 17.2.)

During any grinding it is always essential that the ground optical surfaces be checked constantly by an excellent spherometer such as the Strasbaugh electronic. It is good practice to check this instrument by measuring the opposite radius test plate. Of course, by alternating the top and down posi-

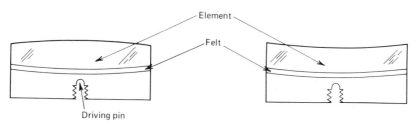

Fig. 17.2. Mounting the lens elements on an aluminum holder with rubberized felt between the element and holder. Masking tape encircles the assembly.

tion of the grinding tool the radius reading will be changed on the spherometer. If the reading is high, place the grinding tool on the bottom and conversely, if low, place the grinding tool on the bottom, with crystal on its opposite side.

3. POLISHING

The pitch radius polishers are made of three parts Burgundy pitch and water-white rosin by weight. Other pitches that melt at about 140°F can also be used. Zobal pitch and soft Swiss pitches (Gugolz No. 64) are reliable because they are quality controlled. Zobal needs no softeners and can be used without. The aluminum holders for the pitch tools should have an approximate difference in radius equal to pitch thickness or about 0.125 in. (3.2 mm). The pitch tools are facet-cut lapping instruments with facet widths equal to 0.1 of the diameter of the polished lens. This can be a rippled facet with nylon or plastic screen netting and the opposite glass grinding tool for pressing the warmed pitch. *Always* use preground glass tools.

The polishing agent is Linde powder types A and B in a 5% sodium bicarbonate water or vinegar solution; four parts solution to one part powder. Type B is used until the lens surface quality is within one-half wave of matching the test plates and there grinder A is used.

After 4 hr of polishing with a little weight (1 to 3 lb) the rotation is slowed down to 50 rpm and the surface whould be within one- or two-fringe convexity. The figuring should be done with a two-thirds polisher and the stroke, increased. Caution: hand work should not be used. Sleeks are often caused by hand figuring because of a poor fit (loss of contact in some area due to dried up polishing action). They do not show when a watery slurry is used continuously with pressure and under rotation.

A strange phenomenon occasionly appears on surfaces under the testplate test; the cleaved plane of the crystal's surfaces is often lifted up and spread out as oil on water. Around the digs and sleeks the fringe pattern is broken up and the fringes run off around the imperfections. If the crystal is not one continuous homogeneous mass (follow the growth structure), it should be rejected before too much work is done on it. If the homogeneity tests were made as discussed in Chapter 5, this should not happen.

Any polished surface should be covered with Opticoat or a similar collodin coating material to protect it when the second surface is worked. It can be removed simply by pulling it off.

The edging of lithium fluoride is done with a close-fitting brass radius holder of smaller diameter than the finished edge lens. It is mounted on a turned brass radius holder on a demountable spindle (Strasbaugh horizontal fixture). Three strips of waterproof double-coated tape are laid down along

$120°$ radial lines and the crystal lens is placed lightly on top and then zeroed with a dial indicator 1 mm inside the extreme edge. The gap between the tape and lens surface is filled with soft lab wax and each surface is coated with Opticoat or sprayed lacquer to protect the surfaces during edging with the grinding wheel.

The lens is beveled again with 95 aloxide grit. Care must be taken if semi-polishing is necessary.

4. FLATS OR PRISMS. HAND WORKING LiF AND CaF_2: PRESHAPING

Prisms and slabs can be cut by the conventional diamond-bonded circular or band saw. Slabs 5 mm thick can be diamond scribed and broken into several sections. This is done by scribing the ground surface with 120 aloxide grit (but not carburundum) on a cast-iron flat tool (carborundum grinding on cast-iron tools causes a splintering action). The scribe is made once across the section with a straight edge and a small steel rod 1 mm in diameter is placed under the scribed mark. Pressure is exerted over each section. Prisms precut by sawing are preground to their final form and 0.5 mm or 0.20 in. of excess stock is allowed above top tolerances.

The final 225, 145, 95, and 50 aloxide grits grinding is accomplished with two flat, facet-grooved glass tools. One is used for the 225 and 145 grits and the second is made two fringes convex by grinding with another tool over it. This convex tool fine-grinds with 95 and 50 grits; No. 50 grit can be used if precautions are taken. It is important to remember, however, that sleeks or scratches cannot be removed by polishing.

5. POLISHING AND FIGURING

A pitch polisher about three times the base diameter of the work is made up of soft Zobal or three parts Burgundy pitch and one part water-white rosin and perhaps Guzolz pitch No. 64. A wax lap consisting of two parts beeswax and one part water-white rosin with a melting point at $145°F$ is often used. This is a solid polisher formed or shaped with pressing plates and a small buffing glass. (See Fig. 17.3.) The lap is contoured by two pressing plates: a two-fringe concave pressing plate and a one-fringe pressing plate.

The polishing slurry is Linde A and a bicarbonate or vinegar-water solution, similar to the polishing slurry described in the preceding sections. The polishing slurry is placed on the pressed polisher. A small area is removed from the central region to allow for the flow of the pitch. The polisher is lightly scratched with $90°$ parallel lines to form squares of medium size. The

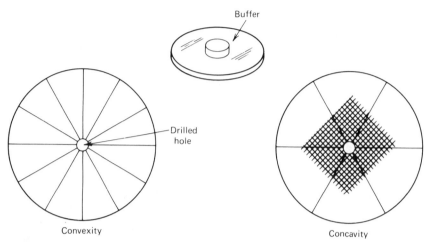

Fig. 17.3. A typical solid pitch lapscribed by a razor blade for correcting a flat surface with faults. The glass buffer is often used to change the contour of the polisher or to charge it with polishing compound.

wet polisher is buffed with a flat buffer to break down the Linde slurry and charge the lap.

The polishing action is "W" strokes over the full aperture with medium pressure. The polishing of thin optics such as planoparallel elements requires an assembly of the component mounted on a plastic holder of equal diameter and a cork or rubberized felt between the two solids. The assembly is held together with waterproof masking tape around the periphery. A fast polishing stroke is used with pressure. Nothing can be done with a loose fit and slow stroke.

6. OPTICAL FIGURING

Two methods are used for figuring flat crystal components. First, the pressing plates contour the polisher to its opposite configuration; that is, a concave pressing plate forms a convex surface on the pitch lap. Second, the convex plate forms a concave contour. In the other method the lap is scratched with a razor plate and pitch is removed from certain areas. (See Fig. 17.3.) Again it is possible to iron the polisher with the glass buffer to a certain contour to fit the required surface fault found by the test plate. With a combination of contour polishers, stroke, scratching away pitch, and stroke, optical figuring to one-quarter fringe surfaces can soon be learned.

7. IRTRAN CRYSTALS AND SIMILAR CRYSTALS

Kodak's publication, No. U-72, contains detailed information about two crystals; for example, soft lithium fluoride and a harder calcium fluoride which is sensitive to shock. In this publication Kodak illustrates many techniques of polishing their Irtran crystals and use is made of them in the following sections.

7.1. Irtran 1. Magnesium Fluoride

This technique resembles glass working but the polishing is done on a 140°F flow wax polisher (rosin and beeswax) with a 1-μ diamond compound. It is finished with an 0.5-μ diamond compound (Hyprez fluid and diamond dust). Irtran 1 can be core-drilled with glass dummies cemented on both sides with a wax mixture of one part beeswax and three parts rosin. It is generated with a 180-grit diamond wheel.

7.2. Irtran 2. Zinc Sulfide

This technique is also similar to glass working but Linde powders are used for polishing on a 140°F flow wax polisher. The mixture is two parts beeswax and 1.5 ronin (rosins vary). Start with Linde C (coarse) and finish with Linde A. The same core drilling is used. Like glass work, Irtran 2 can be preshaped by diamond generating or rough-ground with coarse emeries.

7.3. Irtran 3. Calcium Fluoride

The technique remains the same. Irtran 3 can be core drilled like Irtran 2.

7.4. Irtran 4. Zinc Selenide

This crystal is probably one of the most temperamental to polish because an occasional grain "pullout" will occur. There is no recourse except to continue polishing below the pullout. The initial polishing slurry is Linde C and finishing is done with Linde A on a wax-lap, as described above. It can be core-drilled (see No. 1) and diamond-generated.

7.5. Irtran 5. Magnesium Ozide

The techniques are those of glass working for generating, blocking, and grinding. Irtran 5 can be fine-ground with No. 50 aloxide if a little talc filler is added to the slurry lubricant. Its ground surface must be free of sleeks and pits for they will release pull-outs that will drop on the polisher and cause scratches. A typical time for polishing a 1-in. (25.4-mm) on a lens surface is 20 hr. The polisher is the same as that already described. Linde C is

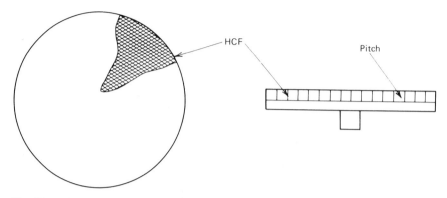

Fig. 17.4. A special combination polisher made up of honeycomb impressed into a soft pitch lap. The contour of the polisher is controlled by the use of concave or convex pressing plates. Additional weights can be added to speed up the contouring of the polisher.

the initial polishing slurry. Finish with Linde A and core-drill. (See Irtran 1.)

7.6. Irtran 6. Cadmium Telluride

This material must be handled with care in the polishing and cleaning sequences to produce an excellent polished surface. Two soft wax-laps are required—one for Linde C and the second for finishing with Linde A and B. (See Fig. 17.4.) Surface cleaning requires a new, clean solvent. Xylol, used first, is followed by rubbing alcohol. Pat dry with fluffy diaper cloth; *never* rub. Use a 220-diamond grit wheel for generating surfaces and 220 aloxide for rough-grinding. Core drilling is not recommended.

Hand polishing and figuring flat work follow the techniques described for lithium and calcium fluorides except for the differences in polishing and drying Irtran crystal materials. Because it would be impractical to hand polish calcium oxide crystals like planoparallels or prisms, it would be best instead to mount them on special parallel tools for windows. The prisms would also require runners of the same crystal material cemented with a shellac wax. Polish their surfaces on a rotating wax-lap with Linde powders. (See Figs. 17.3 and 17.4.) For further information refer to Chapter 8.

8. SILVER CHLORIDE, ARSENIC TRISULPHIDE, KRS-5

In this section some of the special techniques applicable to a number of crystals that can replace other crystals are discussed. Some of the representative crystals in this category are silver chloride, arsenic trisulfide, thallium bromide, iodite, and lithium fluoride.

The poisonous effects of thallium and arsenic are similar to lead poisoning because of the slow, cumulative effect built up in the worker. It would be like a sloppy painter getting lead poisoning by eating or smoking with lead paint on his hands. Because of the relative insolubility of KRS-5 (thallium salt), no poisoning effects are to be expected. It should be pointed out, however, that the slowly accumulation of minute quantities on the fingers can be transferred to the mouth by smoking or eating with unwashed hands.

As in all crystalline materials, operations such as sawing, lathe turning, and core drilling should be performed under liquid coolants or in vehicles in which the materials can be entrapped. Working with cuts or scratches on the hands should be avoided because such cuts will heel very slowly. Wear thin rubber gloves as protection from prolonged exposure to coolants that contain these materials.

Headaches are often the result of working with the arsenic trisulfide because of its fumes.

Silver chloride is nontoxic and can be worked with the bare hands except in surface figuring.

9. REANNEALING

When working toward one fringe or less, the reannealing of the crystals is important. Because cleaved and thermal strains are setup by the several operations, such as the sawing, grinding, and blocking, they will, if allowed to remain in the material, often gradually relieve themselves over a period of time and cause surface painting and twinning.

Strains are likely to be introduced in the process of sawing smaller pieces from a larger blank. The English method of sawing with smaller diamond grit wheels as slow as 200 rpm should be used rather than the conventional diamond sawing with larger wheels and higher speeds. Saw fractures or surface twinning will appear if sufficient grinding (0.20 in. or 0.5 mm) has not been done.

To remove most top surface strains from the preshaped elements caused by the cutting and rough-grinding processes the crystals are reannealed in an controlled oven heated to 150–175°F. This reannealing must be carefully done. Place the element in a cold oven and start at normal room temperature. Raise the heat 20°F every 15 min until 150°F, plus, is reached. The crystals are held at this temperature for 2 hr. It is then lowered in a similar manner.

10. MOUNTING AND BLOCKING

The crystals can be affixed to a chuck with a low melting wax (three parts rosin, three parts beeswax, and two parts montan wax by weight). It is often desirable to increase the thickness of the wax between the chuck and the work with several layers of paper. The metal holder should be heated slowly in an oven no hotter than 150°F. Take care not too touch the crystals with the hands. Use paper toweling or cotton gloves when setting the holder on the material.

Start with cold parts and place them in the oven at normal temperature. Raise the temperature 20°F every 15 min until 150°F is reached. When heated, smear the low-melting wax on the material and place two sheets of lens paper over it. (See Section 8.6.) The hot holder is also covered with soft wax before it is placed on the material. Work on a wooden board protected by toweling that has also been heated. Work in the oven. After the assembly is complete lower the temperature of the oven every 20 min until it reaches room temperature.

It is often necessary to move the crystal when centering it, in which case heat the brass holder rapidly without allowing the flame to touch the crystalline material. The element can be moved before it becomes too hot. This is true for all materials; for example, lithium, barium, and calcium fluorides and thallium, silver, and other chlorides.

To remove the elements from this sandwiched assembly repeat the procedure described earlier. Never try to remove the element from the holder with a "shocker" or by tapping the brass holder with a hammer. In prism work it may be desirable to cement (using the wax described) two prism blanks together to reduce flexing on the edges and vibration effects from creating odd shapes.

11. GENERATING

These materials can be generated with a 280-diamond grit wheel if certain precautions are taken.

The wheel loads up quickly with the soft material. Use a wire brush to remove engrained material in the rotating wheel and a soft emery stick to dress it.

Make small cuts no more than 0.01 in. or 0.25 mm, and let the crystal run free for 10 sec. This allows the coolant which washed out some of the crystal grains on the wheel to cool the crystal. Use a 220-diamond grit wheel.

12. CORE DRILLING AND MACHINE WORK

Core drilling with diamond drills can be accomplished if window-glass sheeting is cemented on top and bottom of the slab crystal according to the procedure described in § 3. Be sure when drilling through the crystal to use very light pressure.

Crystals of chlorides and thallium (KRS-5) have been drilled by the conventional method and lathe-turned but extreme care must be exercised.

Use plenty of soapy or generator coolant. The tool must have a 60° rack angle and good clearance.

The speed of the lathe is about 1000 rpm for turning and a mirrorlike finish, especially on KRS-5, can be obtained. Silver chloride can be lathe-turned but it melts is the coolant is not jetted under the cutting tool.

Soft crystal is drilled with a sharp drill at 300 rpm and rather slow feed. These machine operations must be done before fine-grinding and reannealed as described.

13. GRINDING

All grinding, rough or fine, should be done with 120 aloxide grits on glass tools because many crystalline materials react with cast iron. With silver chloride free silver becomes embedded in the material. In addition, a bad odor and taste is encountered when cast-iron tools are used. The lubricant in most cases is a dishwater solution of Joy or Dreft.

It cannot be stressed too strongly that a crystal is a boundary oriented with layers of parallel sheets of molecules, not unlike mica sheets, and that fractures from carborundum on cast iron will implant deep "weldlike humps." To achieve a ground surface free of cut strains grinding must be done slowly with gentle pressure. Do not take off too much material at one time. A "W" or elliptical stroke that covers the whole aperture is the best method. Check the parallelism continuously by one of the several methods available.

It is preferable to fine-grind from 145 aloxide grit to 50 grit on a flat tool with a four-fringe convex contour that makes the element slightly concave. Test this surface quality by placing a small optical flat on the fine-ground surface and holding the assembly at a grazing angle to the monochromatic light. Observe the fringe contour.

14. HINTS

When fine lapping, use only slight pressure or the 95 or 50 aloxide will grab and anchor to the grinding tool and will have to be removed with a shop razor blade. This is done by placing it under the bevel of the material and tapping with a small hammer.

If the fine-grinding is not correctly done, polishing by hand or machine will be impossible.

If runners cut from the same crystals that surround the work can be used, it will expedite later grinding (strong bevels are required) and polishing. (Use the low-melting wax mentioned in mounting crystalling elements).

The slow cutting of crystals by a diamond grit wheel is relatively new in the industry. An English firm, Metals Research, Ltd., grows crystals by the Melbourn process and markets the Macrootom II for slicing at slow speeds and the Miroslice annular saw. Their American subsidiary is Image Analysing Computers, Mosey, NY. This method of cutting crystals into sections tends less toward reintroducing strains or toward hardening areas in the crystals.

Always cover all fine-ground surfaces with masking tape or Opticoat when fine-grinding or polishing another surface because moisture from the fingers will etch the surfaces and fingernails will scratch them.

If these instructions are carried out on all crystals such as thallium (KRS-5), arsenic trisulfide, silver chloride, sodium iodine, or sodium chloride the optician will have few problems. The less polishing done on crystals, the better.

15. SILVER CHLORIDE

Silver chloride is best preshaped with polished metallic dies in various forms, such as a lens or prism if mass production of these elements is required. Where a limited number of experimental elements is required, however, it is possible to grind and polish this material into reasonably functional shape.

Crystals can be sawed by conventional methods. Flats, windows, and radius work are best done on glass tools with a slurry of aloxide grits and a soapy solution as the lubricant. The material grinds slowly and 225 grit is the coarsest that can be used because it becomes embedded in the plastic-like material.

The material is sensitive to light and the last wets of fine-grinding should be undertaken in subdued light or darkness or the crystals will turn purple.

The best polishing is done on a honeycomb beeswax lap which is laid over a 140° melting-point pitch. (See Fig. 17.4.) Some care must be taken to make sure that the pitch is not too hot to melt the wax honeycomb. When the polisher can be touched with a finger and the pitch does not cling to it, the temperature is correct. The honeycomb is pressed flat in the lukewarm pitch with a soapy pressing tool. It should be serrated into small finger-width facets. The diameter of the polisher is 154 mm.

Before the crystal is polished, the polisher must be well buffed with a flat brushing glass tool which has been wetting with a solution of ordinary photographic hypo solution in a Linde B mixed at a ratio of four of liquid to one of powder. The contour of the polisher should be made slightly concave with pressing plates.

The polishing stroke is a zigzag or "W" across the small polisher and must be done in a faintly lighted room or in complete darkness. It takes practice to polish in the dark but a red darkroom light is an alternative. The optician must discipline himself to add only the top liquid of the hypo solution in a few drops after each dozen or so polishing strokes. A slurry of 20% (by volume) butyl amine in ethyl alcohol can replace the hypo solution.

Two actions are taking place—mainly the dissolution of the crystal and the charged Linde powder and the dried crystal's polishing action.

Wash well in a detergent-water solution and dab dry with paper tissues. Never scrub dry.

Silverlike deposits sometimes form around the bevels and must be removed to prevent scratches.

Caution! Any surface ground or polished must be protected by 202 masking (3M) tape or it will become stained or pitted by fingers wetted with hypo.

16. ARSENIC PHOSPHIDE AND ARSENIC TRISULFIDE

These crystals have similar optical techniques for grinding and polishing. They also have a high coefficient of expansion and will stand little in the way of thermal shock.

All elements should be reannealed by being reheated, starting at room temperature in a controlled oven and raising the temperature every 20 min until 180°F is reached. Hold at this temperature for 12 hr. Then reduce it in a similar routine until room temperature is regained.

Plates, windows, and other elements to be blocked after the reannealing process can be treated for fine-grinding by heating them in water until 118°F, the temperature of the low-melting wax melts, is reached.

Do not place elements directly on the bottom of the pan. A raised platform of mesh wire should be used. The elements can be removed one at a time and placed directly on a 111°F, heated metal holder in the generating or grinding configuration. The elements are wiped dry and the low-melting wax is spread uniformly over each of them. Work on paper toweling and do not allow water to touch the heated elements.

These crystals grind about twice as fast as glass because they are quite soft and will scratch easily if mishandled. Ordinarily glass tools and emery no coarser than 225 aloxide should be used. Cast-iron plates can emit a bad odor and taste if used during grinding. Wash in a detergent water and use prewashed diaper towels. All bevels should be strongly ground on all edges and semipolished before grinding on a flat felt polisher with cerium oxide.

17. POLISHING

The makeup of the wax polisher is described under Section 18.7.

Eighty-hour-milled cerium oxide has been used successfully as a polishing compound in a slurry of 5% sodium bicarbonate-water solution. TK68 from Universal Shellac can be similarly mixed. Soft felt will give a good polish and can produce test-plate quality. Another pitch mixture supplied by the Universal Shellac and Supply Co., is the No. 450 polishing pitch, two parts (by weight) pitch to one part beeswax. This lap, if kept wet will produce interferometer quality in 15 to 30 min.

The elements are removed after polishing by reheating them in a vacuum oven in which the air cannot tarnish the surfaces. Follow the procedures described for reannealing. The only difference is that they can be removed once the melting temperature of the wax has been reached. Leave the elements in the oven until they have returned to room temperature. The optical elements are washed in xylene and then in a detergent-water solution. Dry by dabbing with diaper cloth; do not rub.

18. EDGING

These crystal elements can be edged if parallelism has been carefully maintained during lens-element grinding. Parallelism can be achieved by the use of the three-post fixture. (See Fig. 17.1.) The cool elements can be placed on the trued brass holder provided that two sheets of lens paper lie directly between the surface of the lens and the rim of the holder. The brass holder on the spindle is heated slowly while on the Strasbaugh centering holder. The wax on a holder should have reached the melting point and the heat evenly distributed with a Bunsen burner. A dial indicator with a strip of masking tape on the stem is used to prevent thermal shock and scratches on the polished surface.

Edge very slowly because the crystaline materials tend to clog the grit edging wheels.

Caution! As described for Irtran 4 (Zinc selenide) these crystals often develop a pull-out that will scratch or sleek the polished surfaces. These pull-outs can never be predicted and the only recourse is to keep on polishing below the point of appearance.

19. SILICON AND GERMANIUM

These elementary crystals grind and polish like crown glasses. No special techniques have been devised for working them. They can be cloth or pitch polished by Linde A and a sodium bicarbonate-water solution.

Silicon polishing requires a wax polisher similar to that used in Irtran crystal polishing but it must be loaded with Agroshell or a nutshell ground to superfine grains. Generally, a 1 to 1 mixture of ground shells to pitch is adequate, but because of the foaming action the shells must be added slowly to the melted pitch mixture. This is required because silicon polishes slowly without added weights. More weight is used than is normal for glass polishing. It is apparent that the wax-pitch polisher will retain its softness yet will not flow so freely. Hand polishers often rely on walnut shells. (See Chapter 3.)

For the final pitch-polished surface X-OX polishing compound is the choice for figuring silicon and germanium crystals.

20. METAL GRINDING AND POLISHING

Like crystals, a large range of soft and hard metals is used in many areas. For missile-guiding optical systems the metallic surfaces are in contact with the glass surfaces and must be lapped to withstand many G's of shock.

Finely ground metal or glass laps for ground lens surfaces are often used to give a semipolished effect on the narrow-rim metal holders. This is accomplished by dry lapping, the metal holder to the required radius. If a wedge angle must be removed from a metallic circular-rim holding fixture (a reflecting surface), diamond dust is needed on a cast-iron or copper-aluminum lapping tool. These tools are often serrated with small facets.

The technique is similar to hand polishing flat surfaces on a post polisher. Use the various grades of diamond dust in a grease-carrying vehicle smeared uniformly over the lap. The final polish is done with finishing diamond grits. As in fine-grinding clean up after each grit during surface lapping.

Lapping tools are kept flat by using a series of three and the sequential grinding of No. 1 on No. 2, No. 2 on No. 3, and No. 3 on No. 1. Flatness is checked with a precision metal knife-edge. The grinding slurry is 145 aloxide and kerosene.

Surfaces as large as 2 in. (51 mm) have been lapped by this method with a bright finish on aluminum and invar.

21. MACHINE LAPPING OF SOFT AND HARD METALS

Large-aperture flat and radius work requires a special technique. Stainless and aluminum flat work as large as 8 m has been figured optically to yield good images. Many metallic optical elements such as paraboloids and ellipsoids 1 m in diameter working at a $f/0.16$ ratio have been made by Ramsom Physics Laboratories, Los Angeles, CA, with an image quality equal to a 10-mm circle of confusion. Stainless-steel, off-axis paraboloids 3 m in diameter and a short $f/$ratio are also produced.

Many methods of preshaping and grinding metal elements requires large vertical lathes that turn the metals to form. These lathes often have punched-hole paper tapes which direct the tool cutting. The preliminary grinding is accomplished by several methods. Cooke's method calls for a narrow sanding belt controlled by the tape. Ramsom's method makes use of carborundum honing sticks of 600 grit cemented to circular sponge-rubber pads. These pads are never larger than 1/20 of the diameter of the work. A reciprocating hydraulic arm carrys the flexible grinding tool during the roughing or final-finish grind as the work piece is slowly rotated. The coarsest carborundum honers are usually 220 grit, the finest, 600 grit.

Polishing is done with a flexible polisher and a honeycomb foundation or hard pitch on sponge rubber. Optical figuring is similarly carried out. In the figuring of these $f/0.16$ paraboloids the rapidly changing central zone vertex creates most of the problems.

Because aluminum surfaces require nickel for reflectivity, all figured surfaces must be fine-ground again, repolished, and figured to the final tolerance. Stainless-steel elements require only the Linde C preliminary polish on the pitch lap, this is followed by a Linde A finishing polish. The Linde powders are also used for metallic polishing medium-hard materials such as beryllium.

22. COPPER AND COPPER ALLOYS

Copper alloys and copper substrates are finding application in high-power infrared lasers.

Disks as large as 12 in. (31-cm) have been figured to a surface quality of two waves or better. These disks are ground on flat tile tools made of bathroom hard tile. The semiground surface is often nitric-acid etched before being polished to remove strains and grit embedded in the copper surface. The softer natural grits like those of the American Optical Company are used—mainly 302 1/2 and 303 1/2.

The fresh-feed method of bowl polishing in which the slurry is recycled on a wax polisher produces a brilliant surface with a quality of two waves or better. Some beryllium-copper surface flats have been figured to one-quarter wave. The polishing agent in Linde A powders is continuously broken down from repeated use. A sodium-bicarbonate solution with 5% water is in the lubricant.

The bowl-feed method is preferred by many firms for polishing crystals because the polishing machines can be left unattended.

23. RUBY, SAPPHIRE, LITHIUM NIOBATE AND SULFIDE

These hard crystals are best ground and polished with diamond dust mixtures and oil or grease on rapidly rotating cast-iron flat tools and a lead–tin soft lap for polishing. Optical elements such as windows of sapphire with diameters as large as 9 in. (25.4 cm), laser rods, and lenses are being installed in missiles and high-power lasers.

24. PROCEDURES

Rough-diamond-generated crystal elements are cemented to a toolholder by cellulose caprate or a mixture of six parts of shellac and three parts of oil by weight. These elements are generated and edged as in any conventional glass operation.

1. The elements are removed from their holders by heating them with a large Bunsen burner. The cement used is easily removed from the holders with acetone.

2. Heat the element sufficiently to cement it to the flat or radius metallic holder.

3. Bevel the edge with a cast-iron radius tool.

4. Semipolish the bevel with No. 6 (orange Elgin) with a micarta 90° including angle tool.

One of the three cast-iron tools is used solely for rough grinding (Nos. 45 through 51 Elgin diamond grits). Begin with No. 45 compound by squeezing about 0.5 in. (12.5 mm) on the lap and adding thinner (hyprez) to spread the compound smoothly over the entire surface. Use an eccentric arm rotating at 40 rpm and a spindle (machine grind) rotating at 60 rpm and carrying a 5-lb weight. Add a spray of thinner as necessary. Do not allow the compound to become too thin or scratches will appear. After the compound has broken down and thinned out, stop the machine, take the work off the lap, and collect the diamond around the side of the lap on your finger. Respread it over lap and grind as before. When only very little diamond is left, add fresh compound; No. 45 compound is used to clean up generator marks and other defects.

The second cast plate for flat or radius elements is used with No. 30 Elgin compound to remove diamond-grit pits.

The third tool is used for removing the No. 30 pits with No. 15 compound. Wash all laps with detergent and warm water and make sure that such lap is numbered for future use.

25. POLISHING

Three lead-tin laps which are generally two parts lead and one part tin (by volume), are required. Some commerical solders contain this mixture. The manufacture of these laps to a close tolerance is typically not the opticians task. The flatness of the surfaces should be within four waves or better and the surfaces should be checked with an optical flat or test plate. Because the lap surfaces are reasonably bright, to view the fringes they should be coated with a 45/45 partial coating of aluminum. This coating is removed by using a strong lye solution (NaOH) for conventional testing later on.

One lap is used for each grade of compound. It has been found, that for intermediate polishing powdered diamond dust is best. The lap is serrated in thin lines with a razor blade. The pattern of serration is not important because many patterns give equivalent results. However, serrations should be about the width of a fingernail.

Check the three lead-tin laps and determine which lap departs by the greatest number of waves from the test plate. Mark No. 1; second best is No. 2, and No. 3 for the best lap.

Place two or three drops of No. 20 motor oil on the No. 1 lap. A small amount of powder is removed from the No. 9 bottle (6–12 micro compound from U.S. Industrial Diamond Co., New York) with a toothpick and mixed with the oil. When the mixture appears to be uniform it is spread over the entire lap with a finger in a circular motion. The machine can be used for polishing by adding a 15-lb (7-Kg) weight. After running the machine for a short period add a drop or two of No. 20 oil. When it becomes necessary to add more oil, stop the machine, collect the diamond around the lap, and spread it evenly. Sometimes a little soft grease is added. The grease can also be renewed. The reason that grease is used in place of oil is that the lap can be charged with diamond grit and needs only grease thereafter. Remove the work from the machine and examine the surfaces. (Grease can be removed by washing with warm water and detergent. The surface should be semi-polished but it will still have some scratches. If they are not deep, disregard them; they will be removed with the other grits.

The next lap makes use of No. 6 (4 to 8 micro grit). Repeat procedure No. 3. Scrub the lap with a warm water and detergent and a stiff brush. Use this lap again for No. 3 powder (1 to 5 micro grit).

Check the surface quality with the test plates after Procedures 1 and 2 and determine whether the tolerance requirement has been reached. Some limited control can be exercised over this tolerance by exchanging the positions and stroke on the element on top or bottom on the tool. (See Chapter 16.) On flat work the elements working on top tend to become concave and on the bottom, convex. The third lap is used with No. 1 compound and the work is held in the hand. The lap is actually placed on a 1750-rpm electric motor, which is controlled by a knee or foot switch. The work is held with some pressure against the lap until it is dry; then grease is added to the lap and the procedure is repeated until the surface is even. Clean in warm detergent water and check with test plates. If it is too concave, move directly across the center area to the lap; if it is too convex, work near the peripheral area. The surface will have a bluish tint and will be sleeked. The lap will eliminate the surface defects with No. 1/2. The grit will have been washed off with a brush and detergent water. The surface should be bright and shining with some small sleeks but no scratches.

26. SUMMARY

Polishing the bevels is important because there is no breakage of the crystalline materials in any of the grinding and polishing phases.

These crystals must be held on their mounting tool with a high-melting mixture (cellulose caprate or the shellac mixture) because high temperatures are generated in the final polishing phases.

27. EPON-ROSIN METHOD OF WORKING HARD CRYSTAL

For precise lens elements, one makes radius surfaces from epon resin compounds with loaded 95 aloxide grit. These surfaces are then generated to the precise radii required after curing. It is obvious that metallic tool radii are difficult to change by the conventional up-and-down method. Precast ceramic tools can be worked like glass tools.

The data in the following section will serve to explain the methods of casting radii laps for grinding and polishing.

28. CASTING OF MATERIALS

1. Epon resin: #828: Shell Chemical Co., New York.
2. Catalysts: DTA or D.
3. Aloxide No. 95, paper tubes.

   Mixing Formula:

   a. Use eight parts (by weight) of epon-resin to one part of curing agent DTA or D.
   b. Weigh the desired amount of epon-resin and mix in as much 95 grit as the epon-resin will absorb without becoming too thick to pour.
   c. Add curing agent DTA or D.
   d. Seal off one end of two tubes for six tools with tape.
   e. Pour the mixture into paper tubes and bake in an oven for 2 to 8 hr, depending on the curing agent: DTA produces faster curing. The oven temperature should rise slowly from room temperature to 150 to 200°F.
   f. Remove the mixture from the oven when satisfactorily cured, cool, and remove the paper from the casting. Saw in six pieces.

29. GENERATING AND EDGING

Generating and edging crystals and epon-resin radii tools for grinding and polishing are discussed in several areas. (See Chapter 16.)

   Crystals do not require high-temperature melting wax for generating or edging.

30. GRINDING AND POLISHING

Attach the crystal surface not being worked to a metallic radius holder with a radius opposite that of the crystal with a high-temperature blocking pitch (six parts shellac, three parts pine tar by weight). Also attach the proper grinding tool to the metallic holder. Caution! Do not make assemblies of holders and crystals too high.

   Separate grinding tools are used in all grinding or polishing operations.

   Short wets are necessary because of the heat generated by the high speed (1750 rpm) of the tools.

   All radius work of small diameter [1.5 in. (3.8 cm)] should be worked at high speed and held by hand. Maintain good contact at all times.

   Generally 3 to 4 hr are required for polishing and figuring. Use of variably controlled motor is helpful during figuring to match test-plate tolerances.

   To avoid scratches always make use of paper tissues to dry the optics.

Flat work of small diameter can be similarly worked. The epon-resin flat tool for polishing can be kept on the convex side by grinding it on a second epon-resin tool by the conventional upside-down methods.

Epon-resin radii tools are superior for casting turned tools or for soft-solder polishing tools and are more easily made to precise radius of curvature.

30.1. Method for Laser Rods. Hacker Instruments, Inc.

This company publishes a manual *Technical Manual and Applications Notes* for working crystals for laser rods and other elements. It can be obtained from Hacker Instruments, Inc., P.O. Box 646, Fairfield, NJ 07006.

The use of their machine and jigs permits these elements to be made with high precision.

BIBLIOGRAPHY

De Vany, A. S., et al, *Optical Shop Procedures and Theory* Nortronics, Northrop Corporation, 1955.

Eastman Kodak Co., *Kodak Irtran Infra-Red Materials*, Bulletin U-72, 1971.

Exotic Materials, W. Loucks, private communication, 1965, Costa Mesa, CA.

Harris, L. B. "Cleaving and Sawing Crystal elements," *J. Phys. E*, 432–433 (1969).

Holmes, S., Klugman, A., and Kraatz, P. "Copper Mirror Surfaces for High Power Ir Lasers," *Appl. Opt.*, **12** 1743 (1973).

Horne, D. F. "Dioptric Substances," in *Optical Production Technology*, Crane, Russak, New York, 1972, pp. 74–125.

Optovac, Inc., "Optical Crystals," *Bulletin 50*, North Brookfield, MA, 1971.

Paul, H. E. "Building a Birefringent Polarizing Monochromator for Solar Prominences," in *Amateur Telescope Making—Book Three*, A. G. Ingalls, Ed., *Sci. Am.*, 376–384 (1953).

18

Optical Elements from Soft Crystals

A large number of optical crystals are available from optical firms such as Optovac Inc., North Brookfield, MA 01538, and Harshaw Chemical Co., 6801 Cochran Road, Colon, Ohio 44139. Optovac Bulletin No. 50 (1959) describes the techniques for working many of these crystals; for example, CaF_2, BaF_2, SrF_2, LiF, NaCl, KCl, KBr, NaF, CdF, CdF_2, PbF_2, MnF_2, KRS-5, CsI, laser crystals such as uranium doped CaF_2, and dysprosium doped CaF_2. Many of the techniques described in this bulletin are taken from it and amplified in this chapter. Exotic Materials Inc., 2968 Randolph Ave., Costa Mesa, CA. 92626, are makers of polycrystalline silicon, germanium, and gallium arsenide. Kodak Irtran infrared materials can be obtained from Eastman Kodak, 901 Elmgrove Rd., Rochester, NY 14650, which also supplies magnesium fluoride, zinc sulfide, calcium fluoride, and cadmium telluride. These are listed as Urtram 1-5, in Kodak publication U-72 and many of the optical techniques described here are taken in part from this excellent publication. To name the many entries in the ever increasing list of crystals would be beyond the scope of this book.

Further information on the optical fabrication of crystal materials can be found in D. F. Horne, *Optical Production Technology*, Crane-Russak, New York, 1972, pages 546–547. See also *Technical Manual and Application Notes*, Hacker Instruments, Inc., Northfield, NJ 07006.

1. AXIAL ROTATIONAL TESTING

Axial rotation of transmitted light in some crystals is required before working the optical surfaces. This is accomplished by using crossed polaroid sheets while the ends of the crystal are preground with 120 carborundum to which microscope cover glasses have been cemented with Canada balsam. To make the saw cuts on the crystal perpendicular to the optical axis the assembly must be mounted on a glass plate with clear lacquer and a premarked center. This assembly must be tilted on a rotating gimlet which can, if necessary, be set at the tilt of the crystal's axis for the second cut. If assembly is rotated counterclockwise and the observed fringes move inward,

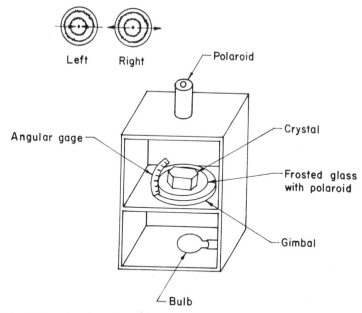

Fig. 18.1.   Testing the orientation of the principal optical axis of natural quartz, calcite, and similar optical crystals.

the crystal has a left-hand rotation; if the fringes move outward, the crystal has right-hand rotation. (See Fig. 18.1.). This assembly shown in Figure 38, Chapter 3, of *Procedures in Experimental Physics*, 5th ed., Strong, Prentice-Hall, Englewood Cliffs, NJ, 1948.

Laser crystals are costly and their axial rotation must be known before cutting. This cutting is done by a gimlet mounting which determines the axes of the required cut on the crystal. An X-ray photograph of the Laue back reflection, usually provided by the firm that supplies the crystal, reveals the axial rotation.

Natural quartz and calcite (CaF_2) crystals require this axial rotational information. Many electrooptic crystals must have their axial rotational predetermined. Gallium phosphide and U-doped-$LiNbo_3$ crystals are two of the many laser sources.

2.   HOMOGENEITY TESTS

As in all light transmission systems, the homogeneity test of crystals is important. H. Salzman [*Applied Optics*, **9**, 1943–44, (1940)] describes a simple interferometric procedure (Ronchi) to the homogeneity of crystalline materials. Figure 18.2. illustrates the principle of the Ronchi test as an interferometer. A well-collimated beam of light is transmitted through the lens and divided into a number of laterally sheared fringes. When a high frequency grating is used, the diverging light cones intersect. The Ronchigrams are observed on a viewing screen with a spatially coherent source. If the fringes are equidistance and straight, the homogeneity is excellent. The lateral sheared fringes in most cases are within one-half a fringe and may have some astigmatic slopes. Natural quartz and calcite often come from the supplier in irregular pieces and cannot be cut or ground. The natural crystals must be checked in an immersing tank, such as the glass aquarium, of 1-gal capacity. The glass walls of the tank should be tested first between cross polaroids for any strain.

The crystals are placed in a hot, concentrated solution of oxalite acid to remove oxidized iron from the crystal. After this treatment the crystals are washed with water, dried, and fully immersed in white mineral oil or a solution of 20% xylol and 80% ethyl cinnamate (by volume). It is good practice to place the crystals on a coarse wire screen with holding wires to prevent breakage.

A bright light source such as an American Optical microscope illuminator can be focused on the immersed quartz crystal and observations made from above as the crystal is slowly rotated. Striae, bubbles, and other faults will be evident. If it reveals no larger masses of striae or bubbles, cut the windows at each end of the accepted crystals. Strong and Paul provide specific details of the orientation of quartz crystals. Paul's detailed method

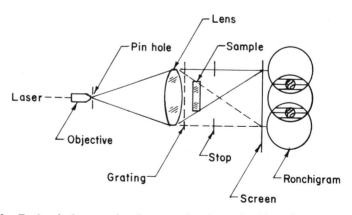

Fig. 18.2. Testing the homogeneity of test samples of crystals with a microscope cover glass wet-wrung with a matching index of refraction oil to a semiflat ground surface. Note that this is a Ronchi test setup.

illustrates many improvisations that small optical shops may be able to use. (See Bibliography in Chapter 17.)

In this section several methods have been discussed by which the homogeneity of optical crystals can be tested. The immersion test offers a means of examining the crystals without saw cuts. The Ronchi is a simplified homogeneity test with 220 carborundum grit ground surfaces and balsum cemented microscope slide windows. Also on the market are matching oils for refractive indexes below 2.0, and for others many liquids can be found to act as matching fluids.

3. GRINDING AND POLISHING CRYSTALS

The techniques of grinding and polishing natural and artificially grown crystals can be divided into three phases. Because of their extreme hardness diamond crystals must be worked with diamond grits on a rapidly rotating cast-iron plate. Sapphire (silicon dioxide) and many laser crystals fall into this category, in which metal polishing laps are required. Quartz of any variety can be treated with conventional glass-working techniques. Calcite or CaF_2 is representative of the soft third group, among which many crystals or semicrystals are sawed in rough form.

Before grinding or polishing can be done preshape the optical component by one of three methods.

3.1. Cleaving

Many cubic crystals such as diamond, salt (NaCl), and KCl, can be cleaved to form prisms, rectangles, or squares. There is a considerable art to cleaving, but like optical contacting it can be learned. Once mastered, sections as thin as 0.020 in. (0.5 mm) can be cleaved from a larger block. The block of salt to be worked is positioned on rubberized felt or several sheets of paper to prevent fracturing the bottom of the salt cube. A Gem-type razor blade is placed carefully on the block at the point at which the section is to be split and parallel to a section already cleaved. The cutting tool, generally a razor blade, is struck sharply with a small hammer. If it does not cleave after several blows, the cutting tool should be moved and realigned in a new area for another try. Smooth surfaces with rounded humps are harder to split.

A guillotine type cleaver has been designed by L. B. Harris. (See Bibliography in Chapter 17.) This instrument has a lever frame that carries a 30° cutting blade and a fulcrum that can be raised to eight different heights in the pillory blocks. The salt blocks are placed against a metal stop while resting on sheets of rubber at the base of the section. This allows the section to fall away from the larger block. A rubber bumper mounted on the heavy brass frame stops the frame from striking the base of the cleaver. This type of cleaver offers almost 100% reliability, even when used by workers unfamiliar with it, and flexibility for use with various sizes of salt block.

3.2. Wet-String Sawing

Harris also discusses wet-string cutting of water-soluble crystals. This method yields strain-free crystals. Crystals cut by other methods produce case-hardened surfaces and are difficult to remove by fine grinding.

The wet-string saw is based on the band-saw principle. The string is a Nichrome wire which has a diameter of 0.01 in. (0.25 cm) and is resistant to corrosion by the crystals. The two silver-soldered end overlap by 0.1 in. (2.5 mm). No direct heat should be applied. Superheat the silver on a stainless-steel wire and apply it to the fluxed ends of the Nichrome wire. The length

Fig. 18.3. The No. 2006 wire saw made by Laser Technology Inc., North Hollywood, CA., for slitting crystals with a running diamond pregnated wire. Coolants must be carefully chosen.

of the wire is usually 9 ft (3 m). The crystals are mounted on a sliding table that feeds the ingot into the moving wet-wire with a free-hanging weight.

The wet-string method is used for the softer crystals such as ADP, KDP, and KDDP (ammonium and potassium phosphates) and KCl, KCL, and CsI (salts).

A number of wire saws are available for soft and hard crystalline materials if the wet slurry, which is carried by the moving wire, consists of carborundum or diamond impregnated with diamond dust. One saw made by Laser Technology, Inc., North Hollywood, CA. 91604, can be obtained with a variable speed drive for harder crystals. These crystals should be cut very slowly to avoid strain patterns which are often released during grinding and polishing (the Twyman-effect). This condition causes twisting of the polished surfaces after the final surface is finished.

Very thin slices of crystal such as silicon are often required in considerable number. These slides are gang-cut with multiple diamond saw blades running at 200 to 300 rpm. Many of these crystals are fine-ground and polished on both sides by mounting them on a planoparallel cast-iron tool. Microelectronic circuits make use of the wafers and solar energy cells. (See Fig. 18.3.)

4. GRINDING AND POLISHING SOFT CRYSTALS

Several "kits" are on the market for grinding and polishing preshaped crystals with the techniques used by metallurgists. These kits contain various grades of sand-grit paper or cloth for roughing to the desired shape. Very fine sandpaper is used for fine-grinding. The lubricant is generally kerosene and the grinding is done by hand with the sandpapers anchored to the flat tool with a rubberized cement that replaces other graded grit papers with the older varieties.

The polishing of these soft crystals is accomplished by stretching a thin felt cloth over a flat tool held by a metal ring. The polishing compound is generally Linde B in rubbing alcohol. This slurry is spread over the cloth with a flat brush. As the polishing progresses less and less alcohol is used until the lap is almost dry. With this final, nearly dry polishing action a high luster is obtained.

5. PRECISION FABRICATION OF SOFT CRYSTALS

Crystals softer than rock salt crystal, such as the phosphates, potassium chloride, potassium bromide, and even potassium iodide require some special handling.

The preshaped prisms, lens, and other optical elements should have edges and bevels semipolished before final dimensional tolerances are fine-ground with 95 aloxide grit. Beveling is required because most crystals have strained particles that fall onto the fine-grinding tool or polisher and cause most of the sleeks. The beveling of round elements should be done with a deeper radius glass tool. Because the crystal element is generally mounted on a toolholder for polishing the peripheral edges and bevels, the glass tool is used with 95 grit slurry and kerosene as a lubricant. The bevels and peripheral edges can be buffed with a felt or cloth with Linde A in rubbing alcohol. Only a semipolish is required.

Prism edges are beveled on a flat glass tool with 95 grit. Larger bevels are required for crystals because they add strength and prevent breakage. These bevels must be carefully done because uneven bevels detract from the beauty of the optical element. The ground edges are polished on a flat tool over which a Pellon cloth is stretched and held by a ring. Of course, all polishing and grinding is done on a slow rotating spindle (100 rpm). Care is essential when placing the beveling tool on the crystal element and the crystal on the flat grinding tool. Many opticians do not realize that they cannot play "castanets" with crystals as they sometimes can with crown glasses.

6. GRINDING

Because of the corrosive action of salt compounds on cast iron, it is good practice to use glass tools made of commercially available glasses. For flat grinding or optical surfaces two flat tools should be available. The diameters are generally 4:1, the glass tools being the larger in relation to the diameter of the optics. One is used for rough grinding; 320 carborundum grit is the coarsest and the second is slightly convex. This special tool is used for final grinding of the crystal optic before polishing. It is better to have a slight concave surface on the optic because it is easier to keep the edges from turning down.

The finishing grit contains kerosene as a lubricant in a slurry mixture with a 145 or 95 aloxide grit grind; 145 grit is used before the 95 grit final grinding.

Much confusion is encountered in grit sizes and no standards of grading have been established. Each company that manufactures grits has its own method of grading; 95 grit is equal to 9.5-μ and 303 1/2 grit is 10-μ size. (See Appendix 5.)

The grinding mixture is washed off the crystals with a detergent solution similar to that used for dishwashing. It should be dabbed dry and placed in a dessicator with a drying agent such as calcium chloride to absorb water vapor.

Care should be taken to apply a fresh slurry of griding compound or the crystal optic will stick to the rotating tool.

The usual four grits are 320 carborundum, 225 aloxide, and 145 and 95 aloxide. The 320 carborundum and 225 aloxide are used on the roughing glass tool and 145 and 95, on the finishing glass tool.

During any flat grinding keep in mind that the outer edge of the rotating tool is moving more rapidly than the inner area and that more grinding will take place there. Thus the optic will quickly grind to a wedge into the crystal element because one side receives more grinding than the other. It is good practice to use an elliptical or eight-figure stroke across the entire area of the tool, turning the optic while grinding. This method keeps the tool in shape and prevents a wedge in the optic.

If a wedge in the planoparallel or window is determined by careful measurements with a differential micrometer, apply pressure at the high area while grinding. For small wedges it is better to use a nonrotating tool because soft crystals grind rapidly.

The fabrication methods described are for flat surface working, but for prisms or cubes of crystals the pyramidal or side angle must be kept in control. Some crystal prisms have reflecting angles of ±30 sec and others are as high as ±30 min. For the 30-min orientation a precision protector and tool-maker's square will suffice. For 30-sec tolerances use the optical setup described for prism angle testing.

7. POLISHING

Most soft crystal prisms are semipolished to relieve the strains that are setup during their fabrication. Cloth polishing described in Section 5 is followed. Fine-grinding the sides of the prisms is identical to the description given above.

Pitch polishing the flat surfaces of these soft crystal optics is done one surface at a time and no mass production blocking is attempted. Lithium fluoride and calcium fluoride windows which are a little harder, have been successfully worked when waxed to a parallel tool with lens paper.

Pitch polishing of softer crystals was practiced by the late Don Hendrix of Mount Wilson and is discussed by Strong.

The pitch lap used by Hendrix for hand polishing contained 5 lb of water-white rosin and 5 oz of pine tar. The general ratio of lap diameter to the work was 5:1. Square facets of small-finger width were cut into the pitch half way through its thickness. After the facet cutting the pitch lap was netted with smaller rippled facets by nylong or plastic screen door-netting. This is done by wetting the netting cloth with Barnsite polishing compound and placing it on the pitch lap. A cast-iron flat tool is warmed in hot water and placed on the wet netting. Make sure that only one-third of the net threads are pressed into the pitch or the pitch will pull away from the lap when the net is removed. This can be checked by sliding the warm cast-iron tool over the edge of the lap and observing the impressed depth by raising the netting a little above the pitch.

To coat the lap with beeswax Hendrix used a medium-sized paintbrush and brushed the lap with smoking liquid wax only once and in only one direction. He did not overlap the coating even if an area as large as 0.125 in. (3 mm) remained uncovered. I prefer to use Treswax or a similar paste wax, which is lightly smeared over the whole area with a small cloth into which the wax has been worked. The paste wax is then lightly worked into the pitch with slight pressure by three fingers and allowed to dry for 15 min.

The contouring of the pitch lap to a slight convex surface is accomplished by concave glass pressing plates. (See Chapter 3.)

Because a polisher with a soft wax coating and a rippled surface contours with ease, it can be hastened by placing it under a heavy weight. The contoured lap is then ready to use. A similar lap is shown in Fig. 14.2.

The fine-ground prism is grasped by both hands in a comfortable holding position and the prism is worked zigzag across the polished area in a series of W-like strokes. Some general strokes can be classified: a long stroke over the outer peripheral area of the lap tends to produce concavity; a short stroke over the center produces convexity. The polishing compound is one part Linde A to four parts isopropyl alcohol or ethane diol. Hendrix used a saturated salt-water liquid and 40-hr ground cerium. When the lap is nearly dry, the last stroke is carried directly onto a soft polishing cloth stretched over a circular wooden lap. This removes the slurry buildup on the crystal and dries it at the same time.

To achieve a one-quarter fringe polished surface is difficult because of the latent heat absorbed by the prism. This wafer windows must be mounted on a piece of lucite, with cork or rubberized felt sandwiched between the crystal and plastic and held together by masking tape.

The finishing surfaces of any optic component must be protected by spraying with lacquer or other plastic sprays that can be removed with acetone. All semipolished surfaces on prisms and other optical components must be protected by spraying or the moisture from the worker's fingers will etch them.

8. BUFFING

A buffing operation is required to renew the polish on a fogged soft crystal optic. A piece of double-nap, bonded cotton cloth with adhesive backing, made by Pellon (New York) or Microcloth (Buchler, Chicago), can be used for semidry buffing. The only lubricant is the moisture built up on the crystal optic by rubbing it over the back of the hand. Added moisture can be applied to the cloth lap by breathing on it. Never breath on the finished polishing salt surfaces after buffing. Place the crystal components in a drying box.

Part III
Telescope Systems

19

Off-Axis Paraboloids

Off-axis paraboloidal mirrors used in collimators and astronomical telescopes are especially valuable when it is necessary to measure the intensity
of reflective light photometrically. They also make excellent collimators for
testing lens systems because the light source is in an accessible position.

The fabrication of multiple small off-axis paraboloids requires unique
methods. Generally three or no more than four mirrors are core-drilled
from a large mirror with a low f/ratio and plastered back into the recess
holes. To remove the unstabilizing effects of expansion and strain new techniques such as grinding and polishing or acid-etching the small mirrors and
their cored recesses are used. These techniques, in addition to grinding and
polishing the mirror backs in the large mirror blank, help to offset the
Twyman-effect. Optical firms that make these products as shelf items use
the same large mirror blank and place the edged mirror blanks in the
original core-drilled holes. From a 2-in. (30.5-cm) mirror blank four 3.5-in
(89-mm) off-axis paraboloid mirrors can be fabricated. The f/ratio of these
elements is generally low—about f/14 from a f/5.5 paraboloidal mirror. To
get faster and larger off-axis mirrors a proportionate increase in the
diameter of the master mirror presents difficulties.

Experimentation with mirrors that have been stressed, then fabricated
and released, has met with some success. These mirrors are in medium
sizes, ranging in diameter from 20.3 to 30 cm at f/8. A 12-in (30.5-cm) at
f/6 was made by the late Marcus Brown and used extensively. Some of the
techniques for figuring these off-axis mirrors are described. Amateur telescope makers can follow them.

1. DESIGN AND CALCULATIONS

The design of an off-axis paraboloidal mirror depends on the pole of the
axis or the location of its aperture stop which is a sector of the master concave mirror. Figure 19.1 (lower) represents a sector section. Note especially
(Fig. 19.2) that the source is actually on the axis of the larger concave
master and not on the smaller (upper) mirror being made. Therefore a

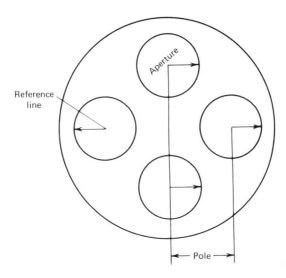

Fig. 19.1.   The layout of four small off-axis paraboloids using one large paraboloid mirror.

reference mark must be scribed on an off-axis mirror so that its axial plane is always clear, as if the large paraboloid from which the smaller paraboloids are taken were there. The upper figure is the off-axis paraboloidal mirror setup as a collimator with a pinhole source. Figure 25.8c describes the method of setting the pinhold source precisely. If used as an astronomical telescope, the eyepiece is at the location of the source.

Review the calculations for setting up the generator to produce the radius of curvature on the large mirror and its grinding tool. Assume that the shop print contains the data for off-axis paraboloidal mirrors: the large, concave paraboloid 12-in. (30.5-cm) diameter pole distance [radius from center 4 in. (10.03 cm) radius of curvature, 92.45 in. (234.84 cm)]. For good workmanship each of the four off-axis mirrors must be placed in a quadrant of the

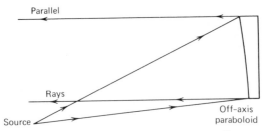

Fig. 19.2.   The design of an off-axis reflective collimator.

larger mirror. (See Fig. 19.2.) Equation (1) is a familiar formula that gives the tilt angle for the diamond cutting wheel of a known cutting edge for the radius of curvature. It is applicable to both concave and convex generation. Calculations: wheel diameter 10 in. (25.4 cm); $R = 92.45$ in. (234.84 cm).

$$\mathrm{Sin}\ A = D/2R \tag{1}$$

where A is the angle of tilt of the cutter in degrees. D is the diameter of the cutting edge; for a concave surface it is the wheel's outer edge and for a convex surface, the inner edge; R is the prescription radius. Example:

$$\sin A = 25.4/2 \times 234.84 \text{ or } 25.4/469.68 \text{ and } \sin A = 0.054079$$

From a table of sines this is a $3°6'$ angle. Note that it cannot be read closer than $2'$ on the machine's vernier scale. This is the tilt angle for generating the radius of the concave mirror with the outer edge of the diamond wheel. Calculate the angular setting for generating the convex grinding tool The inner cutting edge of the diamond wheel, for example, is approximately 24.765 cm. What is the new angle of tilt for generating the convex grinding tool? Answer: $3°1'$.

A spherometer reading is often needed is a table of sagittal values is not available. Use formula (2):

$$h = R - (R^2 - y^2)^{1/2} \tag{2}$$

where h is the sagittal depth, R, the prescription radius of curvature, and y, the radius of the sharp-pointed pins (not balls); for example, $h - 234.84 - \sqrt{234.84^2 - 15.24^2}$ or $234.84 - \sqrt{54581.145 - 232.2576} = 234.84 - \sqrt{54448.887} = 234.84 - 233.343 = 0.497$ cm or 4.97 mm. See Fig. 26.24 for the geometry of the sagittal value.

Generate the radius on the large pyrex blank and then the back surface flat. It is generally good practice to examine the glass blank for any large bubbles that may not have been removed during the generation of the radius of curvature.

The glass blank is sprayed with blue layout ink, and the centers of the small off-axis mirrors are laid out to make sure that a core drill with an inner diameter that will fit the requirement for the outer diameter of the smaller mirrors is available. Allow extra width for removing the plaster.

Before core drilling the small mirrors from the master concave mirror pitch cement a plate glass cover plate on the back to prevent the glass from chipping.

A toolmaker's wiggler is used to set up the drill press for core drilling the aperture diameters of the four paraboloids. To keep the glass blank from slipping nail two cross pieces of wood at 90° to each other on a section of exterior plywood. The glass blank is set with its circular aperture in the 90° and the wiggler is located in the premarked center by shifting the board and mirror under the wiggler. The plywood board with the mirror is clamped tightly to the drill press table.

The core drill is placed in the chuck. It can be a diamond core drill or a mild steel thin wall tubing. With the metal tubing core drill allow for the 320 carborundum grit size and the bell mounting above the small mirror. The only way this can be checked is experimentally. The core drilling is subject to bell mounting (enlargement of the top starting hole). Core drilling must be done with a pumped coolant. Too much pressure should not be applied; it is better to pull down with a constant light pressure.

After core drilling the small mirrors remove them with a rubber suction cup. Next bevel the small cutout mirrors on both edges with 145 grit.

2. POLISHING OR ACID ETCHING THE PERIPHERIES

The periphery of the large mirror is fine-ground and felt-polished with cerium oxide. A mounted brass strip about one-third of the circumference of the mirror has one end anchored and the other fastened by spring pressure. The same type of setup is used for felt polishing. Next, the small cutout mirrors are semipolished. The drilled holes are fine-ground and semipolished on a drill press with spring expansion holders. During this operation the pitch cemented cover-glass plate is not removed. This operation relieves the strains.

Perhaps the best method is to acid-etch the glass surfaces with fumes of hydrofluoric acid. This will etch all the diamond-strained surfaces, release them, and strengthen the glass element. The operation takes about 20 min and a polishlike sheen is observed when completed. It is accompanied by some danger because the fumes are poisonous and all operations must be done in waxed or vinyl containers with exhaust fans.

3. PLASTERING THE ELEMENTS

The small mirrors are plastered into the recesses of the large mirror. Place the curved side of the concave mirror on a vinyl sheet which is laid over the convex grinding tool. The small mirrors are lowered into the recesses with a rubber vacuum holder. In the next operation place wooden sticks along the peripheral edges of the small mirror and recesses to permit the plaster-of-paris to flow around the mirrors. This is the dental plaster called Hydrocal. Release the small entrapped bubbles in the plaster by using a small stick.

Let it dry for several hours or overnight. Remove the assembly and then wire brush around the recesses to a depth of 2 mm on both sides. Fill the lowered recesses with wax or paraffin heated with a soldering iron.

4. GRINDING AND POLISHING

Fine-grinding with the 250, 145 and 95 aloxide grits is common practice. Note that a check must be made from time to time with the largest pin setting of the spherometer to prevent the mirror's radius of curvature from changing from the precalculated value. If this radius is too concave, place the mirror on the bottom spindle and the convex grinding tool above the concave mirror. If the mirror's radius is too concave (low reading), the convex grinding tool is placed on the bottom and the mirror on top during grinding. For large mirrors that cannot be handled readily by one person, learn to change the radius of the grinding and maintain it by manipulating the rates of rotation of the spindles and length of the stroke. Each grind must remove enough stock of glass to be well below the preceding grinding digs. Check in particular in the recesses and small mirror areas to make sure that they have not been moved.

For the assembly with polished surfaces on the edges the back surface ought to be polished about 30%. The acid-etched surfaces need not be polished. The reference marks (four must be deeply acid-etched in the smaller mirrors. Caution! Do not use a diamond scribe because it will cause a strain in the glass. To stabilize the smaller mirrors further in the assembly, the glass cover is repitched to the bottom surface during fine-grinding.

A full-sized pitch lap is required. This is the facet polisher with smaller embossed facets described in Chapter 3. (See Fig. 3.3.)

In one interesting method of making a steep radius polisher the melted pitch is poured on several sheets of aluminum foil covering the flat surface of concave mirror. These sheets must be joined together with tape. A dam around the mirror is made with tape wide enough to contain the pitch, the thickness of which is generally 0.5 in. (1.3 cm). After this layer of pitch has been cooled it is placed on an aluminum radius holder which must be cleaned with turpentine and slightly warm to the touch. Next place the pitch tool on the concave mirror which has been wetted with polishing compound. To speed up the formation of the pitch tool lead weights may be placed on it. The channels are cut into the pitch lap and the larger facets embossed. The polishing is carried out with the mirror on the bottom and the polisher in top.

5. FIGURING

Dall's null test is used at the center of curvature of the concave mirror. (See Fig. 19.3.) The knife-edge or Ronchi grating can be used with this test. The Ronchi test is also used with a low-frequency grating (80 lines/inch). Typical Ronchigrams are shown in Fig. A14.2, the correlating focograms in Fig. A14.1. Ronchigrams for typical refracting and reflecting elements in an autocollimating optical setup are shown in Fig. A14.2 and the 45° sectional boxes must be used as shown. The slope patterns illustrate high and low areas. Observe the desired null pattern with equidistant and parallel lines along the x and y axis. What Ronchigram pattern is revealed by Dall's null test when according to the Ronchi test, the polished concave mirror is spheroid? It will be the pattern directly diagonal to the right with *pincushion* fringes, and much confusion exists among opticians in regard to the reason. A spheroid at its radius of curvature is autocollimating device when the null test is placed in the radius curvature position and the precalculated plano-convex lens has projected another type of correction (positive spherical aberration) over the surface of the spheroid. Thus to change it from the pattern of parallel lines for the spheroid to one with a pincushion Ronchigram figure the surface back to the null pattern. The other Rochigrams with slope errors in the glass surface are self-explanatory if the 45° dashed line boxes are reexamined. Focograms in Fig. 29.6 are similarly treated.

Next we consider the Dall null test for paraboloids, first published in the British Astronomical Association Journal in 1947 and revised in December 1952. A reprint appears in *Amateur Telescope Making—Book Three*, Scientific American, New York, 1953. All of these books have excellent source material and are well written. The null test is shown in Fig. 19.4.

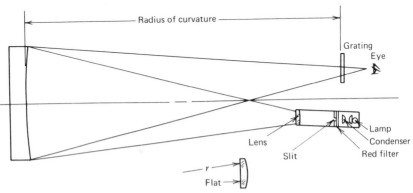

Fig. 19.3. The Dall Null test setup with a Ronchi grating at the radius of curvature of the mirror.

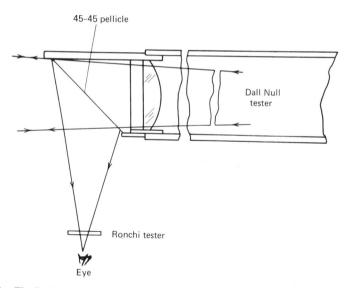

45-45 pellicle

Dall Null
tester

Ronchi tester

Eye

Fig. 19.4. The Dall Null tester makeup with a sliding tube that carries a preset planoconvex lens. Note the mylar pellicle for on-axis testing.

The null test makes use of a planoconvex lens of a certain focal length with a red filter and is set up in a sliding tube at a distance taken from a graph in either of these sources. This lens must exhibit high quality and one is placed in front of the red filter and the slit source. As the article points out, the F/f ratio should be in the middle range, where F is the focal length $(R/2)$ of the paraboloid and f is the focal length of the planoconvex lens. The apparatus is used in conjunction with a Ronchi grating. The lower part of Fig. 19.5 shows a method that uses a 45/45 pellicle to place the null tester on-axis. This is not required in the middle range F/f ratios; only when the ratio reaches 3, as shown in the graph. The convex lens used for testing a concave mirror measured 4.66 in. (11.853 cm). The calculations are given below.

Calculations. The diameter of the planoconvex lens is as follows; the $R/$ 2 or focal length is 119.42/11.852 = 10.075 value which the graph shows as 0.535. Then multiplication by the focal length of the lens, which is 11.852, equals 6.341 cm. The convex side of the planoconvex lens is then at a distance of 6.341 cm from the slit.

Most opticians prefer to parabolize a concave mirror with a starshaped polishing lap. This lap is easily formed by cutting a circle of the mirror's size from a sheet of mylar drafting paper or aluminum foil. Form a semi-

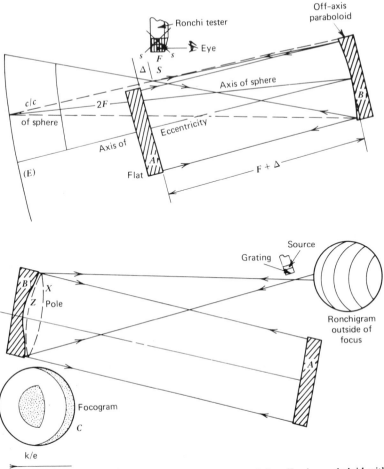

Fig. 19.5. The upper part shows the autocollimation test of the off-axis paraboloid with the Ronchi tester placed at *FS* with an optical flat. A perfect element would give parallel lines. The lower part is similar, but with this exception: it is preferable to observe the outside-of focus Ronchigram before the onset of the optical figuring.

circle by folding the paper across its diameter. Fold it again one third and bring the other section to the fold. The next step is to cut out a half rose-leaf petal on each side to form a trianglelike section. (See Fig. A14.7.) This cutoff pattern is placed on a slurry-wet polisher and the pattern is imprinted in the pitch by placing the mirror over it and by the addition of lead weights. The pitch is cut away from the six-pointed star with a razor blade. Keep the cut-out pattern for future use in any overcorrected figure.

The star-shaped lap is run on top with a one-quarter stroke, perhaps with a rpm of 15, and the bottom spindle at 100 rpm. After a 20-min run test it again to check the progress. Shorter wets or runs are made after each test, hopefully still with an overcorrection. A turned-up edge is generally observed at the end as shown by the pattern above the mull pattern on the y-axis. The star lap will not remove this edge. Therefore remake the pitch lap to full size and run it on top with very short strokes.

As pointed out earlier, a number of factors govern larger and faster off-axis paraboloids. Dall's null test limitations have an F/f ratio of about 3 and is not recommended because of the parallax of the off-axis tester. By using the pellicle 45/45 membrane, it could be placed on its axis. Perhaps it would be better to use a three-element null lens set for faster paraboloids. The faster mirrors require a larger surface which is prohibitive in cost for a one-time application, in particular because of the difficulty of figuring.

6. LARGE AND SMALL OFF-AXIS METAL MIRRORS

Making larger off-axis paraboloids from one disk of glass is a speciality item that opticians are seldom asked to attempt. Many methods have been devised. In one a crescent sector of steel plate at the 0.605 zone is anchored to a large concave mirror by epoxy cement. A second plate of steel half-sector is then anchored to the glass surface with epoxy. A third, with a thread spindle driver, is placed on the first assembly and the multiple screws are tightened to a predetermined pressure with a torque wrench. The pregenerated concave surface is then ground and polished to a figured spheroid and, on release of the screws, an off-axis paraboloidal mirror is formed. In any case, it is in a state suitable for figuring.

Another method of making off-axis mirrors uses aluminum metal mirrors which are ground and polished as spheroids. They are then nickeled and polished as spheroids. The next step is to drill and tap a second steel plate and insert push-pull screws (one assembly). The concave spheroid mirror is set up at its focus $(R/2)$ in an autocollimator with an optical flat. The push-pull screws are tightened to form a null pattern from the first observed pin-cushion pattern and Fig. A14.2. then painted with Locktite. (See Fig. A14.2.) Of course, these mirrors are not optical quality but will do for many purposes in environmental solar and thermal energy setups.

The largest off-axis mirror ever made had a diameter of 3.7 m. Its stainless steel surface was ground and polished by Ramson Physics Laboratory in the Los Angeles area. The forming was done on a large vertical lathe with a computer-developed program. The fine-grinding was done with 600 carborundum stone epoxy on 20-in. (50.1-cm) sponge rubber. The coolant was a light machine oil. The preliminary polishing was carried out with HCF (honeycomb foundation) on a flexible sponge rubber tool. The last series of polishing wets was a flexible tool with pitch facets. The polishing compound was cerium oxide and distilled water. No optical test was used.

It was noted earlier that a 12-in. (31-cm), off-axis, glass mirror was made by the late Marcus Brown. Figure 19.5 shows the test method used for an off-axis mirror with an autocollimated flat (I coedited and changed many sections to bring it up to date). The Ronchi test shown in my revised test appears in Fig. 19.5, which illustrates the unfigured spheroid in an autocollimator. The axis of the spheroid, as indicated by the dotted line that meets at the center of curvature c/c, is positioned halfway between the angle of eccentricity, as indicated at c/c. At E, the angle of eccentricity is shown; at A the plane mirror is positioned on the axis of eccentricity to permit the light rays from the spheroid B to strike at normal incidence.

The light path is marked by arrows. Although only one-half is evident, there is a return system of rays.

Some of the critical requirements in this test should be considered, but first examine the nature of the deformation to be figured optically.

When figuring a spherical surface to a paraboloid, the deformation is one of symmetry of revolution, but when making an off-axis paraboloid the surface of symmetry is destroyed and the pole of the surface shifts by an amount dependent of eccentricity or from the center of the spherical surface to the opposite side in the direction of the plane of eccentricity.

In Fig. 19.5 Z, the spheroid, is shown before correction is started. The dashed line at (X) indicates how the focogram and Ronchigram would appear before correction and at (B), the approximate shape of the finished surface. Of course, the illustration is greatly exaggerated. It does show the greatest amount of figuring is done from the pole to the outer edges.

A well-corrected surface would appear as a flat focogram and the Ronchigram would be the null pattern with parallel lines both inside and outside of focus. (See Figs. A14.1 and A14.2.)

At B the concave mirror, the focogram, and the Ronchigram (outside of focus) indicate where correction is needed. The shaded area indicates a high area if the cutoff is left-to-right and the light source is on the left side. Note that the observer is facing the patterns. Note also that the focogram extends to the right of the vertical center line; the pole of the surface has shifted because of loss of symmetry.

It is obvious that with this form of deformation the finished surface cannot be a concentric surface of revolution. Consequently it is important when setting up the mirror for tests that it be placed in the same position of orientation in the cell. This is accomplished by the permanent line etched on the back surface of the mirror before fabrication. The cell which is anchored securely to the table must have a ruled mark for repositioning the off-axis paraboloid each time it is optically figured. This is most important and cannot be neglected.

The appearance of the focogram and Ronchigram (outside of focus), as indicated in the test patterns in Figs. A14.1 and A14.2, is only representative of the deformation, for the magnitude of contrast will depend on the f/ratio of the paraboloid and the amount of eccentricity introduced. The most conspicuous feature is the pronounced high center and the pole slightly to one side of the vertical line.

The question how much material has to be removed is of secondary importance because the off-axis deformation is not a surface of revolution and must be done with fractional size tools. A triangular-shaped pitch tool contoured to the established glass surface will clean up cut up surfaces.

Many individuals have discussed a machine that would polish this contour but none is on the market. Perhaps a computer-controlled machine could be developed to keep the small polishers on the surface of the glass.

BIBLIOGRAPHY

Brown, M., The Foucault Knife-Edge or Schlieren Test (Chapter 26).

Griffith, T. and De Vany, A. S. "Off-Axis Paraboloids," in *Optical Shop Procedures and Theory*, Nortronics, Northrop Corporation, 1955.

Ransome, J. F. private communication. John H. Ransom Laboratories, 20 N. Aviador Street, Camarillo, CA.

Schulz, L. G. "Quantitative Tests for Off-Axis Parabolic Mirrors," *Appl. Opt.*, **36**, (1964).

20

Short-Focus Paraboloids

Marcus Brown and A. DeVany

The construction of short focus paraboloids, $F/2$ or less, presents unique problems in working and testing: first, because of the large departure from the nearest spherical curve, and, second, because of the difficulty in obtaining an undistorted wide angle beam of light for observation of surface quality.

Taking the problem of deforming the spherical curve to an approximation of the desired parabolic curve, the obvious procedure is to deepen the central area and form a parabolic curve tangent to the spherical curve at the periphery. The outer edge is preserved as a dependable reference.

If the focus of the parabola is critical, spherical curve that will give the correct focus without deforming the edge zone must be selected. This edge or reference zone must be preserved, regardless of other operation.

Many methods of checking the parabola curve have been tried during the grinding stage with templates, traveling indicators, pendulum indicators, and spherometers but with little success. Many of these failures, with the exception of the template, were due to an attempt to deform the sphere in the peripheral area. It is sheer luck that an $F/2$ radius sphere can be deformed to an $F/1.02$ parabola during grinding and polished without figuring to produce a high-quality paraboloidal mirror. It is possible, however, to grind the curve within reach of polishing if the proper procedure is followed.

1. OPTICAL FABRICATION METHODS

The following procedure has been used successfully and although it may not be the best way it does work if done intelligently.

In the first step select the proper radius sphere to give the desired F value of the parabola.

In the second step grind and polish the sphere to a good surface of revolution, free of astigatism, particularly in the peripheral area, say a narrow zone 0.25 in. (6.35-mm) in diameter, with a radius that will produce the correct focus. This is important.

Taking as an example a 6-in. (15.24-cm) $F/1.0$ parabola, calculate the difference between the curve of the parabola for a 1-in. (2.54-cm) zone at the periphery to show that the maximum difference from the edge is 0.000066 in. (0.001676 mm). This is not measurable on the average spherometer. There is, however, a Strasbaugh electronic spherometer on the market with this capability.

Next choose a pin setting for the spherometer that will place the pins in this area, and for a diameter of 6 in. (15.24-cm) this would be a 2.5-in. (6.35-cm) pin setting.

Now find the spherometer reading or (h) value at the center for a 2.5-in. (6.35-cm) pin setting, which is 0.2571 in. (6.53034 mm). Then find the difference in h between the spherical and parabolic curves. This is 0.0026 in. (0.06604 mm). Now add this difference to the spherometer reading, which equals 0.2597 in. (6.59638 mm). This will give the spherometer a close approximation to the required depth of the parabola.

Of course, to make these measurements the spherometer must be positioned in the exact center of the work.

Because a parabolic curve is constantly changing diametrically, the only sphericity is in the extremely narrow zones of revolution, except for a small area in the center which appears spherical when the parabolic curve is fully developed. As a result subdiameter tools which have considereable flexibility are needed. Grinding tools built up by cementing tile facets to sponge rubber have proved satisfactory because a large subdiameter tool can be used if the tools are hard. An important point to remember is that for best results these subdiamter tools should not rotate, for even though they are flexible rotation will generate a spherical surface.

The sponge rubber should be of good quality. There are many grades of inferior quality on the market that give poor results. A one-third tool for roughing emeries and a one-half diameter for smoothing are a good choice.

It is important that the number of tile facets fixed to the sponge rubber be randomly placed on the subdiameter tools without particular patterns. They must be offset by at least-one half the length of their diameters. When small-diameter tools are used, they must be somewhat smaller and a large assortment is required. Again it is customary to face the tools with tile or glass. Because the facets are quite small and widely spaced, by reducing the area in relation to the area being worked the contour of the tool will change rapidly and more nearly conform to the shape of the parabolic curve.

Do the grinding symmetrically by starting the action in the central area and as the center is ground away slowly extend the grinding by short runs out to the inner edge of the reference edge. A guide to the amount of work that should be done at the inner edge of the reference zone is as follows: when the spherometer reading indicates that the curve is deep enough at the center, the polished surface at the inner edge of the reference zone should not be a fully ground surface. Of course, the condition of the area between the center and edge zone will not be known, but with practice a parabolic curve can be produced within reach of the polishers.

The expression "within reach of the polishers" cannot be taken literally, however, because of the amount of polishing and figuring required for a short paraboloid is many times the amount required for a parabola that must be deformed by only a few wavelengths.

Here, again, the tools must not rotate, and once formed to the surface being ground they must be used in the same position of rotation. If the tools are allowed to turn, they will cut grooves and the result will be like the bottom of a wash tub.

Other types of tool with which some workers have had questionable success are those formed in star, four-leaf clover, and poinsettia patterns. These tools are also tile facets mounted on sponge rubber. Their principle drawback, however, is that their diametrical movement is so restricted that a considerable tendency to cut zones is evident, and in a short focus parabola small zones are impossible to remove.

2. WAX-GROUND OPTICAL SURFACES

R. Moreau and H. Hopkins have discussed the application of wax to fine-ground surfaces to simulate polish on reflection and transmission surfaces. My experiments with wax-ground polished surfaces indicated that a 145 or 95 grit ground surface can be treated with a product rich in Brazilian carnauba wax. It is lightly smeared on with a small pad, allowed to dry, and strongly polished with a nylon ball until it has an even, bright sheen. Of course, smeared random areas of uneven polished surfaces will remain but will cause no difficulty in optical testing. It is possible to use a Ronchi testing setup; Ronchigrams are easily observed on this wax-ground surface. Knife-edge testing is not practical for these waxed surfaces. Figure 20.1 shows a tile optical setup with illuminated sources.

If flexible-backed tools are used for polishing the same rules as in grinding such diameters and nonrotation will apply. The pitch facing should not be too hard nor the facets too large or too thick. Thick facets on sponge rubber tend to tilt and cut zones. A reciprocating hydraulic arm that carries a small flexible pitch polisher with Linde polishing compounds is used in preliminary applications to make the correction. This setup allows the dwell

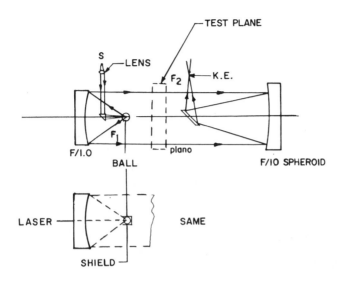

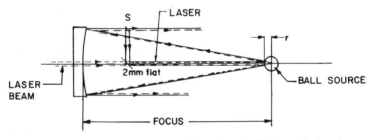

Fig. 20.1. The upper part shows the optical collimation of a $F/1.0$ parabolic mirror using an $F/10$ (not larger than 12-in. diameter). The test optical plane (optical flat) is not required but it is helpful to set up the $F/10$ spheroid to find the focal plane of the Newtonian setup. The middle shows that an illuminated (laser) ball source can be used; the back of the mirror, if fine ground, can be made semitransparent with clear plastic tape. The lower part shows an enlargement of the middle figure, but with this difference: a small 2-mm flat directing the laser beam to the ball-bearing source of 3-mm diameter.

of the polisher to be maintained at any position required for correcting the surface. It is good practice to have two other types of polisher to iron-out small irregularities between zones. One is a polisher of the diameter of the reference zone and is a solid soft pitch type without precut facets. The second polisher has the same diameter as the first but differs in that it is a facet-pitch on sponge rubber. These polishers are required mainly in the final polishing phases of correction with cerium oxide or Barnsite fine-polishing compounds. They are nonrotating and heavily weighted and run for only short periods of time before testing. To blend in various zones a slight eccentric (2 mm) can be used with caution.

3. OPTICAL TESTING SETUPS

As soon as the polish permits testing, it is important that the method chosen give a true picture of the surface conditions. A number of setups will do the job but most have some faults. All systems that make use of objective collimators will introduce chromatic problems in the knife-edge test. This can be augmented by using narrow band filters, but perhaps a yellow wavelength produces the best contrast.

The center of curvature test is not too well suited to high-quality work because the wide angles of the diverging and converging beams make it difficult to measure the intersections of zonal cones of light accurately.

One method by which many high-quality $F/1.0$ parabolas have been made is not of recent origin, for it was used by Draper and Ritchey and undoubtedly others of that time, and currently by many who are familiar with it. The instrumentation is not one that can be thrown together and it must be precisely made to achieve the results required.

Figure 20.1 is a diagram of the Newtonian setup. The various parts are indicated: SP, is the short parabola, LP, the Newtonian telescope, S, the source, and L, the lens for focusing light on the small sphere. The right-angle prism directs the light to the small sphere S which is greatly enlarged, d is the diagonal, and (f_1 and f_2), the respective foci of the mirrors.

The first consideration concerns the cells or mountings for the mirrors, which should be ruggedly constructed with fine adjustments for alignment laterally and in collimation elevation. Proper collimation is extremely important. The diagonal mount is not too critical, but the light source and lens should be mounted on a cross slide with focusing and vertical adjustments. The small sphere which is made by taking a 0.125- or 0.187-in. (3.17- or 4.76-mm) glass rod and heating it in the middle to draw it out centrally to form two sharp pencillike points. Break apart and then reheat the thinner end until a small drop of glass about 1/32nd in. (1 mm) high is formed. Remove the strain in the rod and sphere by slowly drawing the rod upward in the flame. This ball source can be silvered or aluminized.

3.1. Alignment

To align this system the first step establishes a center line on the optical rail if a guide rail or bench is not available. The center of the Newtonian must be on the center line with the short focus parabola and this is done by dropping a plumb bob at the aperture opening of the Newtonian telescope with cross hairs of thread waxed across the aperture. The short-focus mirror is aligned in the same manner. All elements must have the same elevation along the axial plane. It is good practice to move the diagonal mirror elliptical and of small diameter so that most of the central area of the parabola is not obstructed.

An aluminized master autocollimating flat, indicated as test plane in Fig. 20.1, is set into position and properly collimated. Setup the knife-edge and adjust the autocollimating flat (not the Newtonian observing telescope). Observe the knife-edge cutoff shadowgraph of the autocollimation of the Newtonian parabolidal mirror. When this is done, it becomes the referencing collimation with which the short-focus parabola will be tested. Once this is found there must be no axial change of the knife-edge. This is important. The knife-edge must remain at the focus of this parabola of the Newtonian telescope.

Use carefully positioned cross hairs and plumb bobs for this adjustment and be sure that the intersections of the cross hairs on the aluminized mirror have the same elevation as the cross hairs on the aperture opening of the Newtonian telescope and those on the cell that holds the mirror being figured. All reflection sets of cross hairs must be coincident. This alignment cannot be too carefully done. Under no circumstances should this part of the system be disturbed.

After proper alignment has been reached the master plane can be removed and the short-focus parabola can be set into place with the light source, prism, and ball source.

If the grinding of the parabola has been started, it is advisable to aluminize the outer reference zone to help with alignment. If any polishing has been done before this alignment, a mask can be placed over the central area to leave only the reference edge zone exposed. Cross hairs and a plumb bob will be needed for this mirror adjustment. Center the mirror over the center line. Adjust for elevation; the poles of both mirrors must be positioned as exactly as possible and in the same place of elevation. Center all elements with a height gauge.

For preliminary optical alignment a large light source can be used in place of the ball source. The larger light source must be set approximately at the focus of the short parabola and at the proper elevation. This is for the illumination of the cross-hair sets by which the preliminary alignment is made.

To adjust the short-focus parabola the viewing must be done from the knife-edge position; therefore it will be necessary to move the knife-edge blade down slightly. Do not move it *axially*. The small amount of movement required to view the light source reflected by the optical elements will not alter the axial position if the knife-edge vertical slide is in position.

The short-focus parabola must be adjusted until the cross hairs on the short-focus parabola is visible only in the region of the aluminized reference zone. When this adjustment if completed, the large source can be removed and the small ball source, set in place.

This ball source can be positioned first by scale measurements for elevation and with a plumb bob on the center line. Then, with an inside micrometer, set as required to position the ball source for the focal setting.

A trick to adjusting the ball source is often overlooked; that is, the surface of the ball source is not the point of measurement, but it is the center of curvature or one-half its diameter. Consequently one-half the diameter of the ball source must be subtracted from the focal length before it is positioned for focusing with the inside micrometer.

When the ball source has been positioned as close as possible by measurements, the prism can be set in place and within 1 in. (25 mm) of the short-focus parabola. If the prism is placed too close to the ball source, it will obstruct some of the reflected light. The prism should be adjusted for elevation and on the center line. A small right-angle prism or a small aluminized diagonal will be large enough.

Now the light source and condensing lens can be positioned as shown in Fig. 20.1. The lens should have a short focal length so that the light will be in sharp focus on the surface of the aluminized ball source. A zircon arc lamp is ideal here because the size of the source is approximately 0.003 in. (.076 mm) and of high intensity. The final adjustment, a "cut and try" procedure, can now be undertaken. The use of the eyepiece on the knife-edge is best, look for the astigmatic imagery that will betray optical elements in a pinched condition. The first imagery will usually leave something to be desired but it will serve for easier adjustment.

If the long-focus parabola has the focal length it should have, say 120 in. or 350 cm, the magnification of the knife-edge would be approximately 20 times that of the unaided eye. Check the aperture of the system.

The aperture of the Newtonian mirror should have been slightly larger than the aperture of the $F/1.0$ system. If the Newtonian mirror measured 8 in. or 20.3 cm, it could be a spherical mirror of $F/10$. This ratio of the two foci (6 in. for the parabola and 120 in. for test mirror) must be large enough to permit observation of the surface correction of the $F/1.0$ mirror being figured by 20X.

The care with which the rough alignment was done and the ruggedness and precision of the various adjustments can now be appreciated, as can the large circle at the knife-edge. Where to start adjusting at this stage is difficult to say. It depends on the care taken with the Newtonian setup in the preliminary stage.

The operator will need an assistant. With the light source focused on the ball source, again with the larger source, determine whether light is reaching the knife-edge. If it is not, move the light source a small amount each way at right angles to its axis in the horizontal plane. After each movement search for the image. If this does not give results, move the source back to its original position and move the collimating adjustments slightly. Search for the image after each adjustment.

Under no circumstances disturb the long-focus mirror, or its components in this part of the system.

If light cannot be brought back through the system, it would be advisable to recheck the preliminary procedures and try again.

When the image is found, a "cut and try" method is needed again for final adjustments. The sets of cross hairs will be helpful because this is a coaxial system and all crosshairs can be superimposed. The image area will give an image free of astigmatism and the knife-edge will give a uniform cutoff on the peripherial area of the reference zone. The exact position can be determined when this position is found and must not be disturbed. The source, the ball source, the long-focus parabola, the diagonal, and the knife-edge must remain fixed and any necessary adjustments must be made on the $F/1.0$ assembly unit.

As the short-focus parabola will have to be removed and replaced in the cell many times, it is obvious that the cell, in particular, must be ruggedly constructed. The mirror must face up against rigid fixed stops and be held in place by easily released spring pressures, if not the entire setup may have to be realigned.

Some idea of the sensitivity can be realized if, for instance, the distance from the short-focus parabola to the illuminated ball source is changed by 0.001 in. (.0254 mm), the change in focus of the long-focus parabola will be 24 times. Zones would hardly be detectable at the center of curvature without this magnification. Now they will look like the Swiss Alps!

3.2. Alignment by Laser

In Fig. 20.1 the use of a laser illuminated ball source is illustrated. The laser can be made to transmit through an unpolished or fine-ground glass surface with Scotch tape, which allows the laser beam to pass readily through the glass of the mirror onto the ball surface. Of course, a small 0.25 in. (6.35 mm) aluminized diagonal flat will redirect the laser beam to the ball source. This technique is fully explained at length in Appendix 12.

A laser is a tremendous help in providing for complex optical setups. The laser can be low power; one rated at 0.3 mW will serve. Still, direct observation of the beam without safety glasses or neutral filters cannot be made. It is good practice always to observe the knife-edge cutoff on a frosted screen or on a wall. The Ronchigrams can be projected similarly.

21

Schmidt Cameras

The construction of Schmidt cameras of small and large diameters is discussed as a unit; fabrication departures are emphasized, however. It is not always possible to know what tooling or what improvisations will generate large primary mirrors, especially by amateur telescope makers. Several possible methods are offered. Corrector plates for $f/1.5$ or slower are described: first, by polishing the slope at the 0.707 zone; second, by establishing a neutral zone at 0.866 for better chromatic correction with a reference curve and fine-grinding and polishing the corrector plate.

1. DESIGN

The optician is not often concerned with basic designs—only with producing them. It is our purpose here to prove a specific design for those amateur astronomers who wish to make the attempt. Professional opticians and astronomers owe a great deal to amateur opticians and astronomers who have built their own telescopes and Schmidt cameras. The first Schmidt cameras made in America were produced by amateur telescope makers long before Don Hendrix or Marcus Brown of Mount Wilson Observatory made cameras for astronomical use.

Figure 21.1 illustrates an $f/1.5$ Schmidt camera which is within the means of the amateur telescope maker. The basic design of all Schmidt cameras is a primary spheroid mirror larger than the predetermined aperture of a thin corrector plate. The aspheric slope of this plate removes the spherical aberration when placed at the radius of curvature of the primary mirror. The curved focal plane, which is equal to one-half the radius of curvature, is positioned midway between the primary spheroid and the corrector plate. The film therefore must be compressed to this radius by use of metal curved plungers and held securely in a container. The sperture of the $f/1.5$ camera is 7.5 in. (19.05 cm); it requires a glass blank approximately 1 in. (2.54 cm) larger because it is extremely time consuming to keep the edge from turning down during polishing. The diameter of the primary mirror is 12 in. (30.48 cm); its radius of curvature is 22.281 in. (56.6 cm). The film plane is posi-

An *f* 10.75 of 6 in. aperture Schmidt camera made in 1934–1935.

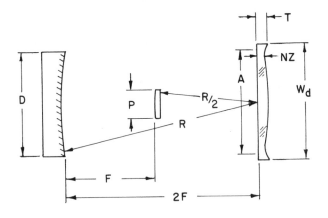

D – Diameter
A – Aperture
W_d – Working diameter
P – Plate holder diameter
T – Thickness

R/2 – Radius of film holder
F – Focal length
NZ – Neutral zone

Fig. 21.1. Data for making a Schmidt camera.

tioned at 11.45 in. (29.08 cm), and this is also the radius to which the cut
film must be curved. The diameter of the precut film is 2.75 in. (6.985 cm);
the holder assembly has a slightly larger diameter.

Schmidt cameras of $f/3.0$ as large as 52 in. (132.08 cm) have been made.
Mount Palomar has a 48-in. (122-cm) camera that has been in use for more
than 20 years and has made a survey of the star fields of a large part of the
sky. The primary mirrors must be large to accommodate the large film
plates. This camera is an $f/2.25$ ratio.

2. GENERATING PRIMARY MIRRORS

In Chapter 2 a method is discussed of generating large primary radii of curvature by a smaller diamond wheel because of the prohibitive cost of a large
wheel. Briefly, it makes use of tilting the angle for deep sagittal steps.

In another successful method the mirror is sandblasted to radius using
hardened steel shot. A radius template is made for checking and the edge of
the glass blank is protected with rubber tape cemented inside the peripheral
area.

The grinding tool is made by pitch-cementing ceramic tile to an approxi-
mate-radius aluminum tool of the mirror.

Some small optical shops that are not equipped with generators large
enough to handle medium-sized mirrors resort to turning them on a lathe.
The mirror is pitched to a flat plate and then, with a radius arm equal to the
desired radius of curvature, the mirror is attached to the tailstock with a
pivot pen and a carballoy or diamond tool cuts the glass to the desired
radius. The radius arm is fed into the glass by turning the tailstock's spindle
and the cross feed holding the radius arm in a slot completes the traversing
feed across the glass blank during hogging out. The coolant is kerosene and
the glass comes off as glass dust. The cutting time for 12 in. (30.5 cm) is
approximately 6 hr. The surface curve is stopped when it reaches 0.25 in.
(6.35 mm) of the edge of the glass blank because of the danger of chipping
the periphery. (See Fig. 21.2.)

The grinding tool is made of ceramic tiles which are oil-wrung to the
radius of the mirror. Plaster of paris is poured to 1.0 in. (2.54 cm) above the
surface of the mirror and tiles. Masking tape must be applied above the
mirror before pouring the liquid plaster. The hardened plaster is wire-
brushed away from the surface of the tiles and several coats of shellac are
painted over all the plaster-of-paris surfaces. Because large chips are caused
by lathe cutting, when viewing the results it is necessary to grind with 80
carborundum to a depth of 0.20 in. (0.5 mm) beyond the deepest chips and
to maintain the radius with the template. Fine-grinding and polishing are carried out by conventional methods.

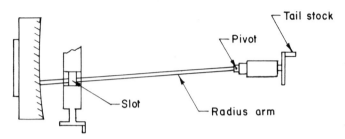

Fig. 21.2. Turning the radius of curvature on a lathe with a carboly tool bit. The cutting should be stopped short of 0.200 in. of the edge.

3. TESTING THE SPHEROID PRIMARY

Testing short spheroids with a knife-edge as described under the Foucault knife-edge or Schlieren test points out that the parallax or separation of the light source and the knife-edge will lead to a false interpretation of the shadowgraphs. Another drawback is the fact that the cone of the light source does not illuminate the full aperture of the spheroid. To overcoome this difficulty the Universal tester was developed. (See Chapter 25.) Note that the cone angle of the light can be made a more diverging source with a 20× microscope objective. A half-silver microscope cover glass is placed at a 45° angle in front of the objective to allow the knife-edge and the eye to check the spheroid on-axis. (See Fig. 21.3.) Often a 10 or 8× objective will do, but it must be determined that the peripheral area or edge of the mirror is fully illuminated.

Find a cover glass with a surface quality of one wave (two fringes), or thereabouts, which is silvered on one side. It is rather difficult to test the cover glasses on an optical flat because irregular Fizeau fringes of equal thickness and fringes of surface quality often interfere with one another. A

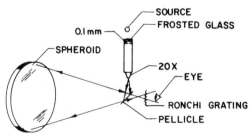

Fig. 21.3. Observing the spheroid mirror on-axis. Another methid views the Ronchigrams slightly off-axis using the setup shown in Fig. 29.3.

cover with two fringes of equal thickness (Fizeau) needs no further testing. To test the flatness of thin cover glasses place a thin sheet of lens paper on the optical flat with the cover glass on top. The lens paper is slowly pulled away from the cover glass while the glass is held with a clean eraser on a pencil. Ignore the Fizeau fringes and look for two-fringe or better surface quality. After finding a microscope cover glass with this surface quality, silver this side lightly. It can often be used without silvering if the light source (6 W) is mounted in the microscope tube with a 0.005-in. (0.13-mm) pinhold. The image of the slit is demagnified and the aerial image becomes as small as 0.00015 in.

A 45° metal tube with a lapped surface (use a dry grinding tool) and two drilled holes is mounted on the microscope objective. The flat surface of the microscope is mounted on the lapped metal holder with three dabs of Duco cement.

For a long-range instrument for on-axis testing the Universal tester should be a part of the optical shop. Interchanging tube units Ronchi gratings, or a razor-blade knife-edge can be inserted.

Amateurs can make use of a Ronchi grating over a small right-angle prism with a frosted surface extending over its end. (See Fig. 29.3.) Ronchigrams with equiparallels are the desired pattern. (See Fig. A14.2.)

4. CORRECTOR PLATE

The materials for corrector plates have ranged from synthetic calcium fluoride, to quartz, Vitaglass, and plate and optical glass. The Schmidt camera used on the moon had a small aperture of calcium fluoride for ultraviolet star survey. Vitaglass was a plate glass available in the 1940s and I used it for several Schmidt cameras at that time. It is no longer available, however. Plate glass ranging from 0.25 in. (6.35 mm) to more than 1.0 in. (2.54 cm) can be obtained. The 48-in. Schmidt-camera corrector plate is made of 1.0 in. (2.54 cm) plate glass. Plate glass suffers from green-edge coloration and like all glass used for corrector plates must be checked for homogeneity and strain. The best available optical glass of crown type is supplied by many firms both here and abroad and Schott's BK-5 for some ultraviolet coverage and K-50 are on the market. Crystallex glass is excellent if it can be located. Optical glass is free of coloration.

Because the glass is semipolished, it can be tested for strain, striae, and bubbles. To test for strain take a sample section across the aperture of the glass before edging and use two cross polaroid sheets in a diffused flood lamp. A small desk lamp with drafting paper covering the aperture and with one sheet of polaroid film attached with masking tape will serve as a strain tester. Use this sample section and with the second polaroid on top, rotate it over the lighted polaroid of the lamp. No colors or Maltese crosses should be observed. The edge is tested by cementing balsam and two microscope cover glasses diametric to the edges; one polaroid is then rotated in front of the second on the opposite side.

The glass disk can be tested for homogeneity and striae by using Murty's method with a finished spheroid primary mirror. Because the ratio of the thickness of the corrector plate is small between the R/D number of the mirror (where R equals the radius of curvature of the mirror and D is the diameter of the mirror) the glass thickness contributes only a small amount of spherical aberration during the test.

To test the quality of strain-free glass for the corrector plate, the glass disk is set in front of the finished spheroid and Foucault-test combination. The presence of striae or a variable index of refraction can be tested for with a knife-edge or Ronchi grating. Striae are veinlike streaks that may look like water running out of a faucet when large ones are massed together. Refractive index variations can give a number of dark ring patterns (like pistol targets), rather evenly spaced from the center to the edge. Glass disks with pistol patterns should be discarded if they are set off to one side. These disks can be used provided that the patterns can be edged in a central position with an aperture slightly larger than the effective aperture of the camera. If in general there is one large symmetrical pattern in the full aperture, the plate can be used. Acceptable glass usually shows a zone directly in the middle and two or three rings rather uniformly and evenly spaced toward the edge.

Glass cut from a larger sheet will sometimes exhibit tinted shadowgraphs or Ronchigrams that are not symmetrical and that tend to appear astigmatic under the test. These small refractive changes will cause no difficulty in figuring and can be ignored. For further methods of checking variations in the glass sheets see Chapter 5.

5. PRELIMINARY WORK ON THE CORRECTOR PLATE

It is necessary that all corrector plates be made larger than the required aperture of the camera. A general rule for corrector plates smaller than 6 in. (15.24 cm) requires that they be larger by 1/12 of the diameter. Larger corrector plates can be the size of the primary mirror itself, for this cuts the work time by one-half. The edges of corrector plates of lesser diameter become badly turned down in figuring.

The parallelism of the two surfaces of the corrector plate should be held to 0.0001 in. (0.0025 mm). This can be checked with a differential

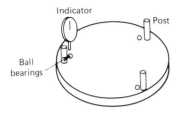

Fig. 21.4. A testing fixture to parallel both fine-ground surfaces to a tolerance of 0.001 in.

micrometer or a dial indicator on a properly designed fixture which can be improvised by using three ball bearings epoxed to a metal plate. The ball bearings are positioned at 120° intervals and slightly inside the circular diameter of the corrector plate. The dial indicator is placed over one of the ball bearings. (See Fig. 21.4.)

Two flat, tiled tools of equal diameter required for grinding are made by cementing small facets of tile to an aluminum flat tool with hard pitch. The procedure is based on the old three-disk method of grinding the corrector plate to an optical flat surface free from astigmatism. Number the tile tools 1 and 2 and the corrector plate, 3; grind 1 on 2, 2 on 3, and 3 on 1.

Because of its flexibility, the corrector plate must be removed from its aluminum holder. It is almost impossible to hand-grind variations in thickness by pressing the plate with one hand on an unmounted plate while grinding on a rotating spindle. An excellent holder is made of light aluminum with a piece of rubberized felt between it and the plate. A substitute is an aluminum holder with masking tape running radially from center to edge. For large-diameter corrector plates lead weights must be placed in the peripheral area in which the wedge occurs. The plate is held to its aluminum holder by wrapping masking tape around the periphery.

A tolerance of 0.005 in. (0.013 mm) parallelism is sufficient for large-diameter corrector plates. After the final fine-grinding with 95 aloxide grit one side is 25% polished and the second side is completely polished. This will remove any Twyman effect that causes polished surfaces to warp. Support for polishing is important and can be provided by a grinding tool which has thin masking tape placed over the tops of the tiles. Before placing the plate on the tile holder Karo syrup is smeared lightly on the masking tape and waterproof tape is tightly wrapped twice around the periphery of the corrector plate. The syrup is easily washed off the plate. The use of olive oil leaves much to be desired.

The flatness of the finished polished surface ought to be two-fringe or better for a 6-in. (15.24-cm) or a 20-in. (50.8-cm) corrector plate. This surface will be retouched with a rose-petal polisher to remove any astigmatism present.

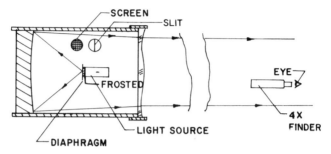

Fig. 21.5. The screen test of the Schmidt camera, which is placed at its focal plane. A perfect corrector plate projects parallel lines.

6. THE TESTING UNIT

The testing unit with its assembled components is made of thick plywood that will allow the corrector plate to be removed during figuring. The corrector plate is set at a distance equal to the radius of curvature of the aluminized primary spherical mirror. A four-legged spider holder that carries a fiberglass grating (used for boat and car dents) is preset at the focus of the primary mirror. This is accomplished with a diaphram to close the aperture down to 2 in. (50 mm) and thus allowing the paraxial image to form an image of the moon or sun on the holder. The holder at the final film diameter of 2.7 in. (7 cm) is screw-threaded and the best focus is found by turning the flat holder. The fiber glass square grating can be obtained from fiber cloth suppliers. It generally measures 16 squares/in. (2.5 cm). The screen holder should also be fitted with a lighted slit formed by separating two razor blades at a distance of 0.004 in. (0.1 cm) on a small brass diaphram with an aperture hole and this can replace the fiberglass grating over the frosted glass. (See Fig. 21.5.)

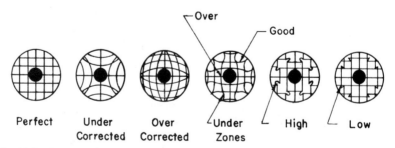

Fig. 21.6. Patterns of the wire test, showing the degree of correction of the corrector plate.

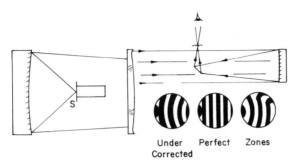

Fig. 21.7. Using a smaller telescope aperture to cover at least one-half the aperture of the Schmidt camera. A lighted slit is used with a Ronchi test.

This permits two test methods to be used for testing the figure of the corrector plate.

1.   If the frosted glass and the fiberglass grating are lighted and the magnified projected image of the grating is inspected, the contour of the grating over the entire aperture can be observed at one time. Figure 21.6 illustrates a number of screen patterns of the figure of the corrector plate. Note their similarity to inside-of-focus Ronchigrams discussed in several chapters. (See Fig. A14.1.) In both cases these patterns are similar to the inside-of-focus patterns observed in autocollimation or parallel light.

2.   A lighted slit at the focal plane allows the use of the Lowers and other optical tests. In this category ball sources are illuminated with lasers. (See pages 529 and 530; Don Hendrix and W. H. Christie describe several tests in *Amateur Telescope Making, Book Three*.) Figure 21.7 shows one of these tests in which a sector of the correcting plate is covered by a well-corrected objective or Newtonian telescope. Note that the aperature can be 4 in. (10.2 cm) for a 10-in. (25.4-cm) aperture corrector plate. A 20-in aperture telescope system could test a 52-in. (132-cm) camera that is $f/3.0$ or faster.

7. THE PROFILE AND FABRICATION OF CORRECTING PLATES

Schmidt pointed out that the neutral zone of corrector plates for chromatic differences of spherical aberration should be placed in the 0.866 zone. When the neutral zone is in the 0.707 zone, the slope thickness of the edge is equal to the central zone. The following data will illustrate the fabrication that will successfully produce corrector plates with neutral zones.

The basic biquadratic parabola formula given by Hendrix and Christie for the slope or profile is

$$x^4 - kr^2x^2/[4(n-1)R^3]   \qquad (1)$$

where x is the radius of the zone, k is a chosen constant, r is the semi-diameter of the corrector plate, n is the refractive index of the glass, and R is the radius of curvature of the spheroid.

With a modern pocket electronic computor these calculations are more easily made by letting $4(n-1)R^3 = 1/K$ and factoring x^2:

$$x^2(x^2 - Kkr^2)   \qquad (2)$$

which is in a convenient form. Caution! Any R^3 value above R of 200 cannot be carried out on an eight-digit calculator, in which case (1) is used.

The neutral zone radius of the correcting plate is calculated by using one of two formulas:

$$x = kr^2/2 \quad \text{or} \quad \begin{array}{l} x = 0.707d/2 \\ x = 9.866d/2 \end{array}   \qquad (3)$$

where x is the radius of the neutral zone and k is equal to 1 for the neutral zone at 0.707 and 1.5 for the neutral zone at 1.5. This gives the radius of the neutral zone from the center, and to find the depth of the profile slope formula (4) is used:

$$k^2x^4/4K   \qquad (4)$$

where x is the neutral zone value from formula (3).

With this information calculate a number of Schmidt cameras and compare the best fabrication procedures for a typical Schmidt camera of any aperture.

In Inches

(a) 7.5 corrector plate (working diameter 10). NZ (.707)
10.0 spherical primary (radius of curvature 22.281)
1.75 diameter curved film holder (radius 11.140)

From (3) $x = 1 \times 3.75^2/2 = 14.0625/2 = 7.375 = 2.652$ or $0.707 \times 7.5/2 = 5.3025/2 = 2.651$.

In a calculation of $1/K = 4(n-1) R^3$ is $1/K = 4(1.5 - 1) (22.281)^3$ or $1/K = 4 \times 0.5 \times 11061.245 = 2 \times 11061.245 = 1/K = 22122.49$ and finally $1/22122.49 = 0.0000452 = K$ as a reciprocal multiplier in formula (4).

From (4) $= 1^2 \times 3.75^4/4 \times 0.0000452 = 1 \times 197.7539/4$
$= 49.4385 \times 0.0000452$
$= 0.00223$ inch or $0.00223/0.000011$
$= 202.7$ fringes from center to neutral zone.

The same parametric values for the diameters and radii; the neutral zone is now 0.866 or k is 1.5.

In Inches

From (4) $= 1.5^2 \times 3.75^4/4 \times 0.0000452$
$= 2.25 \times 1977359/4 = 444.94627/4$
$= 111.23656 \times 0.0000452 = 0.0050$ in.
$0.005/0.000011 = 454.5$ fringes from edge to neutral zone.

If the two Schmidt cameras are compared, one with a 0.707 zone and the second with a 0.866 zone, it is apparent that for the 0.707 zone the delta slope value is 0.00223 in.; for the 0.866 zone the delta value is 0.0050 in.

A polishing technique I developed in making an $f/1.5$ Schmidt camera with the neutral zone at 0.866 uses Cox's method which has a reference convex surface reaching from the center to the edge of the corrector plate. The final correction is drag polished to remove the same amount of glass as for the $f/1.5$ with its neutral zone at 0.707.

In Graph A the reference polished sphere (solid lines) shows this aspheric slope only 0.0023 in. below the reference sphere. This reference sphere is a polished convex surface with a radius of approximately 807 in. (4590 cm). To calculate it use formula (5) to arrive at the upsweep of the aspheric curve at the edge of the 7.5-in. (19.05-cm) aperture. Here one has the corrector plate at approximately 10. (25.4 cm). This does not enter into the calculation, for if a spherical surface occurs at the edge of the corrector plate it will have the same radius everywhere. (See Graph A.)

To calculate the radius of the upsweep of the aspheric curve from the 0.866 zone use formula 2. Substituting

$= 3.5^2 (3.5^2 - 1.5 \times 3.5^2) 0.0000452$
$12.25 (12.25 - 1.5 \times 12.25) 0.0000452$
$12.25 (12.25 - 18.375) 0.0000452$
$= 12.25 (6.125) 0.0000452 = 0.00339$ in.

Because the spherical reference curve in Graph A requires figuring in areas

Graph A

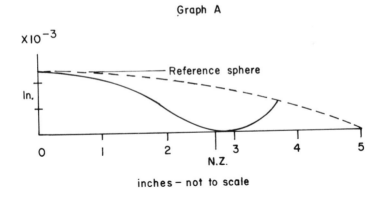

inches – not to scale

of the upsweep portion of the aspheric curve, use spherometer formula (5):

$$R = r^2 + h^2/2h \qquad (5)$$

where R is the radius of the sphere, r is the radius of the area, and h is the depth. Substituting $h = 0.00339$,

$$R = 3.5^2 + 0.00339^2/2 \times 0.00339$$
$$R = 12.25 + 0.00001/0.00678$$
$$R = 12.25001/0.00678 = 1806.7 \text{ in. (4589.2 cm)}$$

It is apparent that this is a shallow reference curve and gives the optician a symmetrical surface at the onset of figuring. These values are shown in Graph A.

Now make a choice of the $f/1.5$ Schmidt camera to be built. A corrector plate with a 0.707 zone is perhaps easier to make than one with an 0.866 zone. Some large Schmidt cameras for astronomical applications are slower, with a ratio of $f/2.25$ and 0.707 zones that can be polished to full correction without fine-grinding if drag polishing is practiced. Cameras with smaller apertures and this low ratio can be made by testing the figuring of the corrector plate on an optical flat.* With R, y, and x and formula (2) a full slope profile of the corrector plate with appropriate (n) values for the refractive index of the glass and the monochromatic light source used can be developed. The number of fringes per area can be counted over one-half the area of the corrector plate and in Professor Cooke's corrector there were only 14.5 fringes of sodium light illumination. I devised this test in 1938.

* See *Sky and Telescope*, "Figuring and Testing a Schmidt Correcting plate," F. Cooke, 560–563 (October 1956).

First to be discussed is a Schmidt camera and corrector plate with its neutral zone at 0.707 radius. In an earlier section the corrector plate was an excellent planoparallel with one side reasonably flat and the second side, a semipolished surface. Before doing any further work on this surface it would be well to place it in its position in the test unit. (See Fig. 21.5.) Walk back a short distance and observe a projected, magnified image of several squares of the grating with pincushion contours which shows undercorrection. Now remove the glass disk and the pattern remains identical. The projected pattern becomes enlarged. A well-corrected Schmidt camera will project magnified, parallel lines. As an exercise, determine the magnification of the projected image of the grating if one square is encircled by the aperture. The number of squares per inch is 10. The answer is approximately 200×.

Two polishers are made up on a rubberized canvas conveyor belt which has been cemented to a light aluminum 7.5-in. (19.05-cm) holder. Sponge rubber has often been the choice but it is too flexible for drag polishing. Its use is discussed in another section (see p. 55). The canvas belting has but a slight yielding base and the polishing pitch and honeycomb foundation adhere strongly to it.

Pour medium hard hot pitch (two-second thumbnail test) over one pitch toolholder and to keep the pitch from running off wrap masking tape around the periphery. A rubber hex-mat larger than the poured pitch lap is wetted with the slurry of polishing compound and placed on the semicooled pitch. When wet corrector plate is set on the rubber mat, the pitch oozes up into the hex-openings. When cool, the pitch is removed from each facet with a Gem razor blade. The second polisher will have the honeycomb foundation with precut rose-leaf configurations waxed to a slightly waxed canvas belting.

The configuration of the precut polisher is generally a five rose-leaf pattern and the major polishing action takes place in the neutral zone. To make the precut pattern in the polishers shown in Fig. 21.8 fashion a template of heavy mylar drafting paper and develop a formula to determine where the active areas may be. Their widths are approximately

$$z_x = y_x \times r_x \times \text{a constant} \tag{7}$$

The equation is not discussed. Figure 21.8(1) illustrates this configuration for the rose-leaf pattern which has a series of squares for the layout for another profiling polisher. To do this enlarge or foreshorten the squares to the ratio desired. Actually it has the same slope as Graph B.

This template is placed on the facet polisher and the pitch around the active polishing area was removed by cutting the pitch away with a Gem-type razor blade. All facets must be separated without other pitch

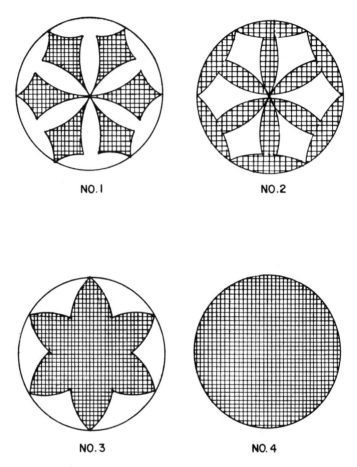

NO.1   NO.2

NO.3   NO.4

Fig. 21.8.   Various profiling pitch polishers used in correcting the corrector plate.

interference. The honeycomb foundation (HCF) is similarly precut, using the template. The HCF pattern should be slightly larger because it is attached to the lightly waxed canvas belting with a small soldering iron to melt the wax to the precut pattern. To hold the wax pattern on the holder the soldering iron is touched on several central areas of the wax foundation.

As discussed in an earlier section, use should be made of the tile grinding tool as a holder for the corrector plate during polishing with double-coated tape. This holder has been used for large corrector plates that range from 20 in. (50.8 cm) to 52 in. (132.08 cm). For smaller corrector plates other holders can be selected if thicker plates are being worked; otherwise the tile grinding tool is the best. Other types include a flat aluminum holder with thin rubberized felt inserted between corrector and the metal and a model on which a honeycomb foundation is covered with a sheet of mylar drafting paper; the glass corrector plate rests on the HCF and paper and is held to the holder with masking tape around the periphery.

Drag polishing requires one of several polishing actions. For a large corrector plate the polishing action is provided with by a hydraulic rotating mechanism for driving the polisher on top of the corrector plate. The driving pin has a square facet on the end that fits into a round hole in the polishing tool. The polishing tool rotates in the direction opposite to that of the corrector plate on its rotating holder. A Draper-type polisher* which has an encompassing ring or an articulating alligator that actuates the holder and the corrector plate is sometimes used. For the smaller correcting plates a standard polishing machine is prevented from rotating while translating across the corrector plate with a wire tied to one of the three tapped bolts placed at 120° intervals around the periphery of the holder. In all cases the eccentricity of the top stroke is always small. Lead weights similar to spoke wheels added where the heavier mass lies over the neutral zone will speed up the polishing action.

The speed of the rotating spindle and the eccentric stroke depends on the diameter of the corrector plate. For the $f/1.5$ camera the spindle rotation is around 100 rpm and the 0.25-in. (9-mm) eccentric stroke translating across the corrector plate is approximately 25 rpm. The lead weights around the neutral zone can be as heavy as 8 lb (3.63 kg). The polishing compound is cerium or Barnsite water slurry. Caution! Keep the slurry of the polishing compound between the polisher and the glass. If it is allowed to dry, the pitch polisher's contours especially the honeycomb foundation type, will soon be ruined. A can with pumping action can be used to jet water between the glass and polisher. Another area to watch is the masking tape which often becomes too wet to adhere to the glass and holder. Plastic holders around the periphery are sometimes required and by removing only one the corrector plate can be taken off for testing.

After a prolonged period of polishing, say about 3 hr, the corrector plate is tested by using the fiberglass screen and viewing the magnified projection at a short distance, perhaps at 20 times the aperture. No great change should be apparent at this time and the rate of polishing action can be checked with a small optical flat and by observing the number of concave fringes in the neutral zone. Generally five fringes an hour over 1 in. (2.3 cm) in the neutral zone shows that drag polishing is progressing. Here it is important to maintain the contour of the rose-leaf patterns by cutting away

---

* This tool is described by J. H. Hindle in *Amateur Telescope Making*, Book One, 1953, pp. 234–240.

pitch that has spread into other areas with a razor blade. Check with the plastic template. The wax lap shows little spreading of the honeycomb foundation, which can be kept in shape by melting away the wax with a small soldering iron.

If the polishing speed begins to slow down at the neutral zone, speed it up by removing some of the pitch or foundation. [See Fig. 21.8(1).] It will take approximately 24 hr before the corrector plate noticeably change the contour of the undercorrection pattern during optical testing. It is good practice to observe the straightness of the screen grating line with a low-power finder telescope that has a ruled reticle in the field of the eyepiece and to check considerably off-axis of the central zone near the film holder. If the unaided eye is used, false interpretations will be made because of the power of the aspheric slope at the neutral zone.

The Cox method of fine grinding with tile facets in the configuring pattern for removing glass in the neutral zone is discussed in Section 21.6.

Making a corrector plate of large diameter by drag polishing will take many hours to finish and is generally free of sharp zones.

As the correction nears completion and the projected lines of the screen test become more parallel examine the figure with other optical tests. One of the simplest, which gives an amplification ratio for the optical figure, uses a telescope with a longer focal-length and a well-corrected lens system. The aperture of the telescope can be less than one-half the aperture of the corrector plate. The focal length should be at least twice the focal length of the camera.

A simple Newtonian reflecting telescope will serve and need not have full aperture to cover the corrector plate. Because all zonal slopes are symmetrical the sector of the corrector plate of the Schmidt camera covered by a portion of the test telescope system is true for other sectors. (See Fig. 21.7.) In this test it is necessary to replace the fiber screen with a slit source(s) at the infinity focus. This is done with a test telescope that has been checked by observing the moon or a star with the eyepiece. The telescope is placed at a comfortable distance from the testing unit and does not interfere with the removal of the corrector plate. The slit is in the focal plane when it is well defined by observation with the eyepiece at the far field focus of the telescope. Adjust only the slit source and not the focusing eyepiece. Check to see if the slit is vertical to the line of sight by rotating the slit unit. If the aperture of the telescope is smaller than the aperture of the correcting plate, set the telescope to one side, cover just a small portion of the focusing film holder, and make sure that the extreme edge is well covered. (See Fig. 21.7.) After the slit is located in the eyepiece anchor the telescope to its cradle holder and secure the camera test unit to prevent it from moving while the corrector plate is being replaced. Determine whether the slit is still in the field of the eyepiece before removing its tube. Observe the light source spread over the portion of the aperture covered. The Ronchi grating is inserted in the eyepiece tube and the Ronchigrams are interpreted as shown in Fig. A14.2.

Some of the most difficult interpretations concern the raised or grooved zones often called rings that are polished in the corrector plate. Figure 21.7 shows the contour pattern of these ring zones. Note the sharp edge of the pattern in Fig. 21.6; the barb ends point away from one another like arrows, the area is a grooved or low zone, whereas in the raised zone the barbs point toward one another. Recall that these are inside-of-focus Ronchigrams (see Figs. 14.4–14.6). The outside-of-focus Ronchigrams are reversed. The screen test shows the same contour patterns and these sharp zones are always found when grinding occurs. (See Fig. A14.2.) The drag polishing has done most of the correction and only retouching remains. This will often take considerable time.

The solid polisher used to correct zonal areas or any astigmatic surfaces is now slightly warmed with hot water and placed on top of the corrector plate which has been wetted with polishing compound; the pitch is then allowed to compress to the polished contour. [See Fig. 21.8(4).] The addition of lead weights hastens the process which should take approximately 1 hr. This solid polisher with raised facets of pitch is placed on the corrector plate and drag-polished for 15 to 20 min. Retest the plate in the test unit and observe how some of the scrambled irregularities and zones have been ironed out. This process also removes astigmatic surfaces that may have been produced. This polishing procedure can be continued if the correction improves, but always cold-press the pitch polisher before drag polishing.

The final correction depends upon what the test reveals. Of course, the higher zones are always worked down, leaving the low zones untouched. Do not make use of drag polishing here, but allow the polisher to ride free without weights on the corrector plate. The polishing configuration of the polisher is controlled by cutout patterns. (See Appendix 14.)

Before finishing the full correction on this one side it is good practice to go back to the first surface which was polished completely and observe for sleeks. Drag-polish it for a short period to remove any astigmatic areas. This polisher must be cold-pressed to the polished surface before work is done on it. Caution! Do not attempt localized figuring because sharp zones are generally the result. A thin wet slurry of polishing compound works better than a heavy one.

These figuring techniques work well for medium-diameter corrector plates as large as 20 in. (50.8 cm), beyond which the polisher becomes too large to be handled by one person. Plates as large as 52 in. (132.08 cm) require other types of handling. It is interesting to compare the neutral-zone departures of the largest Schmidt $f/3.0$ camera made—the 52 in. For a neutral zone at 0.866 or 22.516 in. (57.19 cm) the slope departure is approximately 0.00423 in. (0.000452 mm) or 385 fringes (mercury source). For the 0.707 neutral zone at 18.314 in. (46.520 cm) the slope departure is only 0.00188 in. (0.04775 mm) or 171 fringes. This could be drag-polished without grinding as needed in the neutral zone at 0.866 without a reference sherical surface. If a reference sphere is polished on the corrector plate with 0.866 zone (best color correction), approximately the same amount of glass will have to be removed.

8. COX'S METHOD OF FINE GRINDING THE CORRECTOR PLATE

Cox's method of fine-grinding was mentioned briefly in an earlier section. This method mades use of preground reference sphere that touches the center and the up-sweep curve. (See Graph A.) Next a grinding lap is made by cementing small tiles that can be obtained from any builder's supplys shop. Choose the smallest available; a random size will do. Using the precut template, cut out five sets from a large square section of tile. Becuase these tiles are fastened to a netting, some care must be taken not to break them from the netting when shaping them to the pattern.

These tiles are cemented to a fine-grain sponge rubber by a contact cement or heated in a pan with hot pitch. The tiles must be hot to the touch. The sponge mat is cut to the diameter of the corrector plate and laid out with a marking pen to the 0.866 zone and the position of each rose-leaf pattern. The template form is identical to the one developed for the polishers already described. The widest portion of the tile tool must be over the 0.866 zone. (See Graph B.)

The sponge-rubber tool can be placed on an aluminum holder by securing it with contact cement or by encircling the periphery or by encircling the periphery with waterproof masking tape. The emery is 145 aloxide grit, followed by 95 grit. The corrector plate can be ground on top or bottom. There should be only a small eccentric throw on the translating spindle and if the rotation of the spindle is about 75 rpm at about one-third the rate of rotation of the spindle. It is good practice to have a spherometer with adjustable radius pins (sharp edges) to check the neutral zone and two intermediate zones as a rough test during grinding. Otherwise, use the oil test.

The oil test makes use of an illuminated fiber-glass screen as described. The fine-ground corrector plate can be tested before grinding the preground reference surface to decide what must be done. The projected illuminated

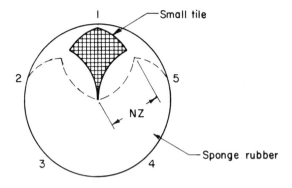

Graph B

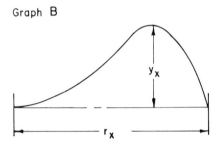

Graph B. The design of a flexible tile tool used to fine-grind the correction in the corrector plate with a reference convex radius.

screen can be seen if kerosene or paint thinner with some oil mixed in it is smeared lightly over the ground surface. Wipe away the surplus kerosene or mixture with a small rag. The projection and magnification of all but a few squares of the screen depend on the distance at which the observer views the projectd image. Ten times the focal length is a good distance at the start of the test. The reference spherical surface causes no appreciable change in the encircling pincushion patterns. The grinding is continued with 145 grit until the slope of the grating begins to straighten out and becomes parallel. The 95 grit is used when the correction passes the neutral zone aperture. The fine 95-grit grind produces a better illuminated projected screen test. Figure 21.6 shows a number of possible contours of the correction ground in the corrector plate.

The center ofter develops a barb zone as shown in Fig. 21.6. Two things can be tried immediately:

1. The eccentric throw should be shortened.

2. Remove some of the tile grinding facets at the center and shorten up the eccentric stroke. Also try to remove any lead weights on the top corrector plate.

A high zone, also shown in Fig. 21.6, is easily removed by increasing the eccentric throw of the translating stroke. Caution! Watch the correction past the neutral zone because it might revert to undercorrection.

A state of overall correction is not likely to occur during grinding but if it does a new type of tile facet tool must be made. This tool is similar to the one shown in Fig. 21.8(2) (back-up tool). It is directly opposite the previous grinding tool.

High ridges or low valleys are not likely to occur if random-sized facet tiles have been used. These zones are caused during grinding by unequal lead-weight distribution. Remove the weights and grind slowly. Drag grinding will iron out the zones. The correct interpretation of high ridges or low valleys in intermediate zones is essential. See Fig. A14.2 for Ronchigrams (inside-of-focus refraction) which are similar to screen one-half patterns in Fig. 21.7.

Note that a raised ridge has spurlike projections, whereas low valleys have a Z-like spurs that enclose a larger area. Only high areas can be ground down and this can be done with microscope cover glasses or thin brass sheeting with 95 grit. For low valleys tile is removed from the zones and by drag grinding the low zones can soon be brought back to a common slope.

Two-facet polishers are made on canvas rubberized belt. (See Section 9.) One is made with the rose-leaf petal configuration and the other is a solid-facet type. The rose-leaf is used first with the drag polishing technique. The second polisher is applied in the final figuring phase according to the profiling pitch tool technique. (See Appendix 14 and Chapter 22.)

9. OPTICAL FIGURING OF THE CORRECTOR PLATE

Drag polishing helps to approach the correct form and conventional polishing with other polishers is used for final correction. Keep a notebook to record progress and avoid repetition of errors. Profile polishers can be used to advantage. (See Appendix 14.)

Before completing the optical figuring it is good practice to compress the rose-leaf polisher for a period of time by removing its slopes and placing it, slightly warmed with hot water, on the semiflat side. Recut the configuration of the rose-leaf pattern with a razor blade and check it with the template. The flat-surface corrector plate is drag-polished for two wets, each 15 min long, and followed by a 20-min run with the polisher operating on top with medium spindle rotation and a small eccentric stroke. After every 10 strokes of the eccentric arm hold the pitch lap for several seconds to stop its rotation. This removes any astigmatic surface that may have been built up by the Twyman effect or other holding method. Because the corrector plate has a large slope departure, hold the corrector plate to the holder. A rubberized felt or slightly dampened newspapers will suffice. Do not attempt to make the final correction on this surface because zoning will develop and there will be two surfaces to figure.

10. OTHER METHODS OF MAKING CORRECTOR PLATES

Several vacuum deformation methods are used to make corrector plates. Schmidt first adopted one for the $f/1.75$ camera and figured it on both sides to avoid possible breakage if one side were slumped. This is called the dioptric elastic method. Gerard Lemaitre describes it at length in *Applied Optics*, Vol. 11, July 1972 and a French patent, ANVAR 70, 19, 261, has been awarded. Lemaitre's method deforms the oversized disk supported at the neutral zone on a narrow metal ring. Air is partially evacuated from the inner ring and a primary vacuum is formed below the outer peripheral area. The elastically deformed area is fine-ground and polished flat, and on removing the corrector plate a smooth profile is obtained. Corrector plates for $f/1.0$ cameras cannot be made by this method.

Another vacuum method is described by Edgar Everhart in *Applied Optics*, Vol. 5, May 1966. It is pointed out here that there is no danger of breakage of the corrector in $f/2.5$ and slower Schmidts. It is recommended that both sides of the corrector plate be figured. The article deals primarily with building an $f/5$ Wright camera. Further information in regard to mathematical solutions is provided by Robert Cox in his article "The Vacuum Method of Making Corrector plates," published in *Sky and Telescope*, June 1972. This article deals at length with vacuum holding of the corrector plate and grinding and polishing the deformated glass blank with a controlled test-plate spherical surface. Also examine other information in the same category published by Frank Cooke of North Brookfield, MA, in the January 1972 issue of *Applied Optics*.

Finally, the patent method of Thomas Jackson of Celestron, in which the enlarged, thin glass blanks are held by a vacuum to a prefigured opposite glass foundation (test-plate configuration). This foundation has sawed radial lines from the center running to a drilled aperture close to the periphery of the thin corrector plate. Vasoline is rubbed on the periphery, the vacuum is pulled, and the surface is fine-ground and polished. The secondary mirror of the Celestron, a Schmidt-Cassegrain, is figured null to a parallel light setup. The primary is spheriod.

These vacuum methods produce corrector plates that are reasonably free of astigmatic surfaces and no sharp zonal departures due to loose grit grinding on facet tools, but they are rather expensive because they need vacuum equipment and metal holders for the corrector plates.

In 1973 I made a Wright $f/2.95$ telescope with a 12 in. (30.5 cm) corrector plate and only the drag-polishing method. Correction was done on both sides and was completed in less than 40 hr. It is obvious that the correction required for a Wright corrector plate that is close to the focal plane is greater than that for a similar f-ratio Schmidt camera that has a corrector plate at the radius of curvature of the primary mirror.

BIBLIOGRAPHY

Cooke, F. "Making a Schmidt Plate by Vacuum Deformation," *Appl. Opt.*, **11**, 222–225 (1972).

Cooke, F. "Figuring and Testing a Schmidt Correcting Plate," *Sky and Telescope* 560–563, (October 1956).

Cox, H. W. and Cox, L. A. "The Construction of A Schmidt Camera," *J. Br. Astron. Assoc.*, **38**, 308–313 (1938).

Cox, H. W. and Cox, L. A. "Further Notes on Schmidt Cameras," *J. Br. Astron. Assoc.* **50**, 61–68, (1939); reprinted in *Amateur Telescope Making, Book Three*, 345–354 (1953).

De Vany, A. S. "A Rapid Method of Making a Schmidt Correcting Lens, *Pop. Astron.* **47** 197–200 (1939).

De Vany, A. S. "Four Tests for Schmidt Cameras," *Sky Magazine* (forerunner of the *Sky and Telescope*), April issue 1944.

Everhart, E. "Making Corrector Plates by Schmidt's Vacuum Method," *Appl. Opt.*, **5** 713–718 (1966).

Hendrix, D. O. and Christie, W. H. "Some Applications of the Schmidt Principle In Optical Design," in *Amateur Telescope Making, Book Three*, A. G. Ingalls, Ed. *Sci. Am.*, 354–365 (1953).

Hodges, P. C. "Bernhard Schmidt and His Reflector Camera," in *Amateur Telescope Making, Book Three*, A. G. Ingalls, Ed., *Sci. Am.*, 365–373 (1953).

Johnson, T. J., Celestron International, Torrance, California, private communication, 1965.

Lemaitre, G. "New Procedures for Making Schmidt Corrector Plates," *Appl. Opt.*, **11**, 2264 (1972).

Lemaitre, G. "New Procedures for Making Schmidt Corrector Plates, Errata," *Appl. Opt*, **11** 2264 (1972).

Paul, H. E. "Schmidt Camera Notes," in Amateur Telescope Making, Book Three, A. G. Ingalls, Ed., *Sci. Am.* 323–340 (1953).

Schmidt, B. "A Coma-Free System," *Amateur Telescope Making*, Book Three, A. G. Ingalls, Ed., Sci. Am., 373–375 (1953).

Smiley, C. H. "II. Chromatic Aberration and Other Factors Governing Design," *Pop. Aston.* **48**, 43 (1940).

Wright, F. B. "Theory and Design of Aplanatic Reflectors Employing a Correcting Lens," *Amateur Telescope Making*, Advance, A. G. Ingalls, Ed., *Sci. Am.* 401–409 (1959).

22

A Baker-Type On-Axis and Off-Axis f/0.35 Solid Schmidt Camera

A mounted Baker-Schmidt off-axis camera with spring loaded plunger to contour the precut film sector.

The conventional Schmidt camera which consists of a corrector plate at the radius of curvature of its primary mirror is necessarily limited by two factors; first, the steepness of the reflecting angle from the spherical mirror and, second, the chromatic aberration of the corrector plate at low f-ratios. A rather simple formula for the diameter of the image in a given color is

$$\delta = [\Delta n/4(n - 1)]\,(h^2/R^2) \tag{1}$$

where Δn is the difference in index of refraction between the chosen wavelength and that of perfect focus, h is the diameter of the clear aperture, and R is the radius of curvature of the spherical mirror; for example, in an $f/0.5$ h equals 52 mm and R equals 59 mm. The diameter of the image in red light (C-line 6563) for a mirror of excellent focus in blue light (F-line 4861) is 0.07 mm for quartz glass. Because the higher order terms for δ must be considered, δ will be about 0.1 mm. Nothing can be done to a single correcting plate of glass to reduce the color error. It is already minimized. The

305

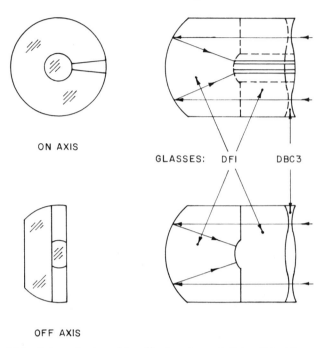

ON AXIS

GLASSES: DFI   DBC3

OFF AXIS

Fig. 22.1. Schematic view of the off-axis and on-axis Baker-Schmidt camera.

only approach would be to use an achromatic corrector plate of crown and flint. Someone might suggest switching to the solid-glass type for which $f/0.5$ is easy and for which the chromatic errors are $1/n$th of those of the conventional Schmidt of the same physical size. For the same focal length and focal ratio (and therefore the same scale of sky on the photographic plate) the chromatic error of the solid Schmidt is $1/3n^3$ or about one-third that of the usual Schmidt. This is still too large, as Baker has pointed out, and the solid Schmidt camera of two types (on-axis and off-axis) can be made. These solid Schmidt cameras have three deep aspheric surfaces of the form used on correcting plates of ultra systems. For the cemented surface mating positive and negative aspheric correcting surfaces are needed. The depths at the 0.88 neutral zone are 1.6734 and 2.6622-mm, respectively on a DBC-3 corrector plate. The difficulties of working on small apertures, combined with large aspheric departures, are well known.

1. FABRICATION OF THE ON-AXIS CAMERA

In building the off-axis Schmidt it was decided initially to start with a conventional on-axis Schmidt because the optical working techniques would be easier to handle than those of the laminated, solid off-axis Schmidt. (See Fig. 22.1.) Later on this proved to be unnecessary, the techniques are essentially the same for on- and off-axis Schmidt cameras. The steps needed in the construction of conventional on-axis and special off-axis Schmidts can be summarized as follows: the main body is composed of two DF-1 flint glasses, Bausch & Lomb dense flint 1.60500 (ud) and 38.0 V_d, and the corrector plate which is made of DBC-3 dense barium crown 1.61100 (ud) and 57.2 (V_d). The corrector plate is cemented to the main body.

In Fig. 22.2 are two different forms of corrector surface. The first involves the surface cemented to the main flint body to be matched by the corresponding surface on the barium crown element; the second is the other side of the barium crown element, which is an air-surface and the first surface of the assembled system. The two aspheric shapes are similar in form

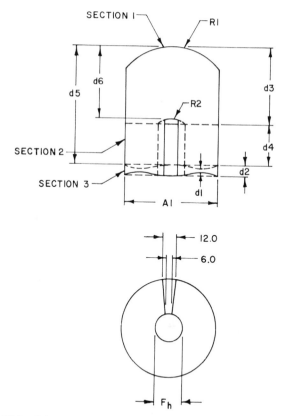

Fig. 22.2. Schematic design of the on-axis solid Baker-Schmidt camera.

TABLE 22.1. BAKER-TYPE, SOLID SCHMIDT CAMERA: ON-AXIS AND OFF
AXIS SPECIFICATIONS[a]

D_1	Working diameter	59.00
A_1	Clear aperture	58.00
A_s	(I) Corrector plate coordinates	Positive
A_s	(II) and (III) plate coordinates	Positive and negative
R_1	Radius of spherical primary	56.210 ± 0.050
R_2	Radius of spherical focal plane	31.090 ± 0.254
F_d	Diameter of focal plane	$16.00 + 0.254$
		$- 0.000$
F_h	Diameter of focal plane hole (on-axis)	17.5 ± 0.254
d_1	Thickness corrector plate (neutral zone)	5.59 ± 0.254
d_2	Thickness corrector plate (central zone)	55.88 ± 0.254
d_3	Thickness of Section I	25.043 ± 0.050
d_4	Thickness of Section II	25.679 ± 0.254
d_5	Total thickness of Sections I and II	50.722 ± 0.254
d_6	Focal plane critical distance	24.927 ± 0.002

Glasses; Bausch & Lomb, Catalog No. E 40

1. Dense flint $\qquad$ N_d 1.60500 $\qquad$ V 37.9
2. Dense barium crown $\qquad$ N_d 1.61090 $\qquad$ V 57.2
Laminated off-axis camera flint slabs
Main body: 3 pieces (one extra) $12.7 \times 63.5 \times 63.7$
Focal plane: 3 pieces (one extra) (12.6 and 6.35) are $+0.254$
-0.000
Plate glass slabs (fill-ins) off-axis 13 to 15
Vee slow cut (on-axis) in Section II and corrector 12×6

[a] Dimensions in millimeters.

but different in slope departures and were arrived at in calculations supplied at an earlier date by Baker. (See Tables 22.1 and 22.2.)

The bottom No. 1 section of the flint body with its 2.213-in. (56.21-mm) primary spherical mirror, 1.224-in. (31.09-mm) spherical focal surface, and flat surface were made and tested with test plates. All three surfaces were worked to one-eight-wave surface quality.

The second part of the flint body (see Fig. 22.3) for the three sections was made next. The 0.501-in. (14.7-mm) film hole was core-drilled within 0.5 mm of the inner face. After completing the other aspheric surface this flint section was brought to the proper thickness and concentric within 0.001 in. (0.025 mm). The plug, filled with low-melting pitch, was removed by running it under hot water which slowly freed it from the glass body. (See Table 22.1 for thickness dimensions.)

TABLE 22.2. BAKER-TYPE SOLID SCHMIDT CAMERA

Specifications: Aspheric coordinates in millimeters

Zone No.	Corrector plate first surface		Second surface of corrector plate and first surface of flint body	
	Y	X	Y same	X
1.	1.2294	0.07112		0.01140
2.	2.4562	0.02840		0.04521
3.	3.6728	0.06299		0.10033
4.	4.8819	0.11074		0.17628
5.	6.0757	0.17120		0.27229
6.	7.2517	0.24079		0.38303
7.	8.4074	0.32080		0.51286
8.	9.5402	0.40843		0.64973
9.	10.6477	0.50292		0.80010
10.	11.7297	0.60123		0.95656
11.	12.7813	0.70231		1.1174
12.	13.8024	0.80518		1.2809
13.	14.7930	0.90729		1.4423
14.	15.7480	1.0079		1.6033
15.	16.7256	1.1049		1.7569
16.	17.5616	1.1966		1.9037
17.	18.4150	1.2827		2.0406
18.	19.2380	1.3622		2.1672
19.	20.0254	1.4336		2.2807
20.	20.7797	1.4966		2.3810
21.	22.5011	1.5502		2.4663
22.	22.1894	1.5946		2.5370
23.	22.8473	1.6292		2.5918
24.	23.4747	1.6541		2.6314
25.	24.0716	1.6685		2.6546
26.	24.6405	1.6734	Neutral zone	2.6622
27.	25.1816	1.6688		2.6548
28.	25.6946	1.6546		2.6322
29.	26.1823	1.6309		2.5946
30.	26.6446	1.5984		2.5430
31.	27.0840	1.5578		2.4783
32.	27.5006	1.5095		2.4016
33.	27.8906	1.4531		2.3119
34.	28.2702	1.3894		2.2103
35.	28.6258	1.3190		2.0985
36.	28.9611	1.2426		1.9769
37.	29.2811	1.1603		1.8458
38.	29.5859	1.0719		1.7054
39.	29.8729	0.9789		1.5573
40.	30.1473	0.8798		1.3998

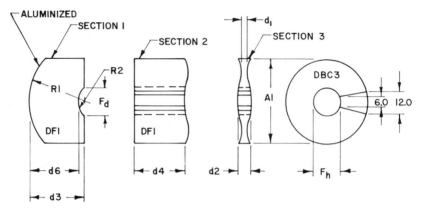

Fig. 22.3. Three glass-sections of the solid Baker-Schmidt camera (exploded view of the three glass-sections).

To match the correcting surfaces of the DF-1 body and adjacent thin, dense barium disk two sets of tools were provided. The first was made of mechanite cast iron and the second, of medium-hard brass formed on a lathe with a "cross-feed" and the "feed-in" slide. This metal blank had a 0.250-in. (6.35-mm) hole drilled and reamed to permit the location of a precise starting area. This produced a rough surface that was brought to approximate form with a scrapper and emery paper. The final values for the x- and y-coordinates of the aspheric surfaces were measured with dial indicators. The y-values were taken from an Ames indicator that read to 0.0005 in. (0.013 mm). The x-values were established by a precise Metron indicator that read to 0.0001 in. (0.0025 mm) with estimates of one-half of one ten-thousandth and a travel of 0.200 in. (5.1 mm). The cast-iron surfaces were brought within 0.001 in. (0.025 mm) of the final aspheric form. The corresponding surfaces of the brass tools were within 0.0001 in. (0.0025 mm). The indicator for the x-coordinate must have a hardened ball no larger than 0.2 mm to prevent a reading error in the measurements of the sagitta of the concave or convex surfaces.

The corrector plate was diamond-generated with a 320-grit wheel, with two convex surfaces generated on both sides to minimize the amount of glass to be removed. The first surface of dense barium crown had a 15.2-in. (335-mm) radius; the radius of the second surface was 15.5 in. (397 mm). The corrector blank was edged to 2.250 in. (57.15 mm), which is approximately 4 mm larger than the aperture required. The grinding of the corrector plates was carried out with No. 320 carborumdum and the cast-iron tools and the parallelism was constantly maintained and checked by placing them on a fixture with three ball bearings 120° apart. The Cleveland electronic gauge had the capability of measuring the parallelism within 0.0001 mm. The cast-iron tools departed by 0.025 to 0.050 mm in the rough and fine-grinding emeries through aloxide 95. The brass tools were needed to clean and shape the aspheric surfaces which in final form on the corrector plates were ground into shape in two 5-min aloxide 50 wets.

The fine-ground surfaces had several high rings when first observed. It was felt that these rings would work down and blend into the surface when fully polished. The polisher was made on a base of small-pore sponge rubber which did not spring so readily as some of the larger pore varieties. The pitch was poured over the sponge rubber which had an approximate thickness of 4 mm and was cut to a diameter of the same size as the surface to be polished. The pitch polisher was formed by heating under running water and placing it on a well-soaped corrector plate. The facets cut in the polisher were 5.0-mm squares separated by 1.0-mm channel cuts through the pitch to the base sponge rubber pad. This was all hand work; the correcting surfaces on both elements were worked on top and the polisher was rotated at approximately 100 rpm. The polisher looked worn at the end, marred by cracks and chips. It is my opinion, however, "that the world's worst looking optical tool does the world's best work." This was all handwork and little side motion was needed.

The second surface of the corrector plate of thin, dense barium crown 1.27 mm thick at the neutral zone was ground and polished, as described for the aspheric areas on the first surface and the aspheric surface on the DF-1 main body. It should be noted that in final form the first air surface of the barium crown provides the quality control of the final image.

The mating brass grinding tools in final aspheric form must be lapped together almost dry with no abrasive between them and checked in two directions at 90° intervals with dial indicators. One attempt to use a poor grade of cast iron for finishing showed hard spots and could not be brought to the final shape. The only figuring necessary after testing was a cure of the high center. This was brought down quickly with a smaller flexible polisher.

The corrector plate was core-drilled with 320 grit and a brass tube. A lipped template fitted over the diameter of the corrector with a centrally located hole was drilled in both directions.

The assembly is straightforward. The corrector plate is cemented to the matching surface of the main body with Lens Bond but is not treated with anticoating because of the danger of slumping. The core drill holes are checked for concentricity. If it is not concentric the core drill is run through both holes. Be sure that each has a 1.0-mm chamber.

Because this on-axis camera must have a vee slot cut in it to allow precut circular film to be placed on or picked off the focal plane surface, the slot must be carefully sawed through the two cemented sections with their

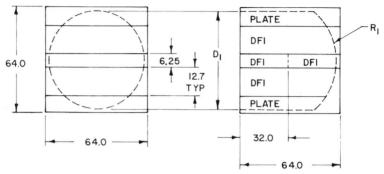

Fig. 22.4. The assembly of the laminated glass sections of the off-axis camera into a square block prior to edging and generating the primary reflector surface. Parallelism must be maintained on all surfaces.

cemented corrector plate and one section of the flint body. The vee slot is approximately 12 mm wide at the periphery and 6 mm in the central area of the camera. A well-designed fixture must be made to fit over this section of the diamond-cut slot. It is good practice to spray it with several coats of acrylic and then allow it to dry before placing the fixture around it and sewing. This saw vee slot is then beveled and ground with several grades of carborundum, starting with No. 320 and ending with No. 600. A 1.0-mm brass blade and a 150-mm diameter blade installed a drill press or lathe facilitates this grinding. This is necessary to prevent twisting the polished surfaces or to relieve the Twyman effect, and 0.5 mm of glass should be ground off.

After this operation the two polished surfaces of the flint body were cemented together with Lens Bond and allowed to dry in a vee block. It is important that the two sections have the same diameter, and they should have if the flint body is edged as one section before being diamond sawed into two for grinding and polishing. This completes the assembly of the on-axis solid Schmidt camera. It should be painted with black auto enamel.

2. FABRICATION OF THE OFF-AXIS CAMERA

The optical glass for this camera should be doubled because it is possible to build two off-axis cameras at one time. In essence, the slabs of optical glass are ground and polished on one side, combined with plate-glass slabs, cemented together, and edged as a cylinder. Thereafter the work is confined to making a full-aperture, solid-glass Schmidt. After the correcting surfaces and mirror body have been completed the camera is taken apart and assembled into two off-axis cameras.

Figure 22.4 shows the assembly of the three flint-glass and two plate-glass sections into a square block before being edged into a cylinder. The flint pieces are ground parallel and polished. Three pieces (one-extra) are 0.50 in. (12.7 mm) thick and 2.5 in. (63.5 mm) long. The three other flint pieces are 0.25 in. (6.35 mm) thick, 2.5 in. (63.5-mm) long, and 1.25 in. (32.2 mm) wide. The tolerance of all flint pieces when ground and polished was plus 0.015 in. (0.38 mm) and their parallelism was within 0.0001 in. (0.025 mm) over the 2.5-in. (63.5-mm) width. Both surfaces of the flint glass were polished to a quality of one-half wave. (See Table 22.1 and Fig. 22.5.)

The polished surfaces of flint glass are protected by lens paper pitched between each flint piece and dummy plate-glass section when the square block is assembled before the mirror surface is edged and generated. The block is edged to a working diameter of 2.25 in. (57.15 mm) and the radius of the primary mirror is 2.213 ± 0.002 in. (56.21 mm). It is generated after edging and ground and polished to a surface quality of one-eight wave, controlled by radium test plates. The block is disassembled and the flint glass pieces are cleaned and beveled. One thinner, parallel piece, 0.25 in. (6.35 mm) thick, with a mirror surface is put aside for further work. The second [0.25-in. (6.35-mm)] flint piece is reassembled into the cylinder block but placed 1 mm beyond the primary radius surface and hand-ground to match the preground radius of 2.213 in. (56.21 mm). It is then ground, polished, and aluminized. (See Fig. 22.5.) The parallelism of each assembled block during grinding is maintained by a three-ball checking fixture with a dial indicator.

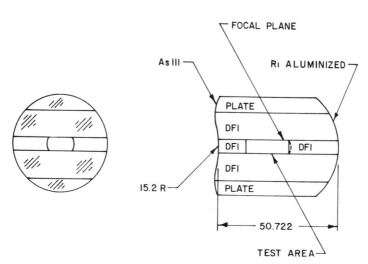

Fig. 22.5. The reassembling of the camera with its polished primary. The top aspheric surface is fine-ground and polished.

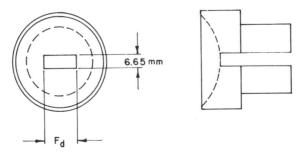

Fig. 22.6. A special grinding tool of flint glass for grinding and polishing flint sector with the critical concave focal plane.

The control of the radius focal plane and its critical distance to the radius surface of the primary mirror called for a specially designed holding tool. A cast-iron concave tool with a radius and with a slot cut through the tool was first tried, but to polish a brittle flint and cast iron at the same time makes a sleek-free and completely polished surface almost impossible to obtain. A glass toolholder in the same flint glass was designed. See Fig. 22.6, in which is shown a 2.250-in. (57.2-mm) diameter, round glass blank, 1.0 in. (25.4 mm) thick with a 0.265-in. (6.73-mm) slot cut through the 1.011-in. (31.09-mm) concave radius. The slot length is 0.625 in. (16.8 mm).

The flint-glass pieces were mounted into the slot with lens paper and pitch and the 1.011-in. (31.09-mm) radius was fine-ground and polished to the critical focal length of 0.982 in. (24.92 mm). The tolerance was plus or minus one ten-thousandth, and the parallelism between the primary radius and focal plane was controlled by the milled slot in the glass holder. The second flint-glass slab was also ground and polished, and any elongation of the 0.625-in. (16.8-mm) diameter of its focal-plane radius could be reestablished by grinding down the top of this glass piece because it does not effect the critical focal-plane distance.

The flint-glass pieces were assembled with Canada balsam as shown in Fig. 22.5. The extra flint piece was cut in two and one piece was cemented to the top of the assembled cylinder to allow the negative matching slope of the correcting plate to be fabricated. This also produced a slot near the focal plane for testing purposes.

The primary-mirror grinding tool helps to keep the hot assembly of the glass slabs in alignment as it cools. It is good practice to smooth a sheet of aluminum foil over the concave radius tool before placing the warm assembly in it. When the assembly has nearly cooled down, place it and the primary mirror on a 2.0-in. (50.4 mm) diameter brass edging tool that had been preheated. This is pitched to the brass holder in the conventional manner. *Watch* the small, central flint piece to keep it from sinking below the top. Masking tape can be used to hold it in place. The holder and assembly are slowly rotated on a horizontal spindle and must be made to run concentric with the holder by moving the whole assembly 0.005 in. as it cools. It will then have a concave radius of approximately 13.2 in. (393.7 mm) generated to a depth of 0.010 in. (0.254 mm) above top tolerance when the total length of the assembly is 1.903 in. (50.88 mm). This will give 0.020 in. (0.51 mm) of glass stock that can be removed to reach the nominal tolerance thickness of the flint body. Again, the concentric thickness should be checked constantly with the three-ball testing fixture.

Corrector-plate fabrication is identical to that of making the corrector plate for the on-axis camera.

After the corrector plate is finished it is cemented to its matching negative corrector surface on the main camera body. A temporary cement of Canada balsam was used and before its application each element was slightly heated. The cemented slab camera is shown in Fig. 22.1.

3. TESTING

Several methods of testing were used on the solid camera while it was being figured optically and when it was finished.

3.1. Lower's Test

This test was used first. (See Fig. 22.7.) A line 0.05 mm wide was ruled through a semidry lacquer paint across a plastic rod with a convex radius to fit the focal-plane concave surface. This rod had a semipolished top surface and the line was ruled on the convex end. This gave a projected line across the aperture that could be observed with a small light source in a small telescope with crosshairs in the eyepiece. The line could be displaced in any zone by moving to a different position for observation. The straightness of the projected line could be observed in any zone to examine correction of the camera. Any zonal aberration is detected by curved lines in a zone or intermediate zone. (See Fig. 22.7.) The observer must make no observations with the unaided eye because of the distubing effect of the strong aspheric surface.

3.2. De Vany's Test

This test in many respects is the same as Lower's but is a refinement in that it projects a series of ruled squares that can be observed through a telescope or projected on a screen some distance away. (See Fig. 22.7.) To do this a small section of fiberglass netting (12 × 6 mm) of 150 squares per inch was used. A small section of white plastic tape was placed over it to hold the

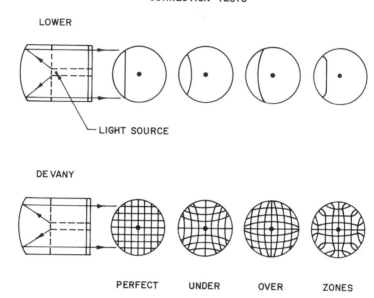

CORRECTION TESTS

LOWER

LIGHT SOURCE

DE VANY

PERFECT   UNDER   OVER   ZONES

Fig. 22.7. The nature of the Lower and De Vany tests.

The De Vany optical test using a mesh screen at the focal plane. Note magnification of a single 1-mm square. The camera was not on axis causing some upper and bottom distortion.

316

netting in place and act as a light diffuser. Note that it is placed midway between the opening in Fig. 22.5. The focal plane is an elongated spherical surface at this central area of the slab-assembled camera. Make a small convex plastic radius fitting block with ruled lines for the focal plane. A microscope illuminator lamp is placed over the central area. For the laminated build-up camera a small sector of the screen or a small plastic block with a convex surface and its ruled line can be positioned easily by going through the open space left in the assembled cylinder. The microscope illuminator lamp can be used by simply directing it through a small right-angle prism waxed to the center of the corrector plate. The image projected was 1 m in diameter. A multiple series of squares make it possible to follow the contour of the figuring. (See Fig. 22.7.)

4. OPTICAL FIGURING

In optical figuring of deep aspheric surfaces of small diameters flexible polishers are used. One technique was developed in which one polisher can be changed to another of a different configuration by impressing or compressing the Zonal soft pitch with folded cutouts.

The mylar cutouts were made in the following steps: (See also Appendix 14.)

1. Cut a circle of sheeting slightly larger than the diameter of the polisher.
2. Fold across one diameter to make a semicircle.
3. Fold over one end of the semicircle into two equal 60° sectors.
4. From this 60° sector of six folds the desired profiles can be cut with a pair of scissors; for example, if a high zone occurred in the central zone of the correcting plate, the profile pattern could be a six-pointed star with curved areas. The main central area of the profile must cover about three-quarters of the maximum area of the high central zone. The purpose of the other six points of the star's profile is to minimize any sharp zone that may develop during the polishing action. The reason that three-quarters of the area must be covered is that the sweep of the stroke of this polisher must be considered during the polishing action as the assembled camera body is rotated.

A general procedure for impressing or compressing the flexible polisher and using a cutout profile polisher is follows. The Mylar drafting sheet is laid on the slightly warmed pitch polisher and the optical element which has been wetted with polishing compound is rested on the spindle. A lead weight can be placed on the polisher to speed up the process. The polisher must be recut with a razor blade to remove any pitch overflows in the precut channels.

The wide variety of profiles that can be made cannot be discussed at length here. (See Appendix 14.) There are times however, when it is necessary to lower the pitch in some areas of the polisher and raise it in others. Thin runners of sheeting are directed to several areas to maintain the profile cutout pattern as a unit. It would be well to save all cutouts for possible reuse in figuring.

The amount of stroke used over the corrector plate as it rotates on the machine is important. The polisher is held in both hands and the sweep of the stroke should be no more than 2 to 5 mm, depending on the area figured. The amount of pressure is also a function of the speed with which the glass is being removed and the rotation of the spindle holding the camera body.

Often in the case of extreme up edges a solid ring polisher must be used because the flexible polisher is not working hard enough in this area. I have often had to resort to rubber erasers (pencil type) to ride down the extreme edges. Rippled optical surfaces that have developed during this harsh action can be removed with the flexible polisher. When figuring optical surfaces with smaller zonal polishers, these areas tend to become astigmatic, a condition that is overcome with a full-sized solid polisher compressed to the corrector plate's surface. The spindle is turned at its maximum rotation and the polisher is held stationary for several minutes at a time. The rpms of the spindle are cut to one-half and the polishing period is repeated. Because the polisher has no preset facets it has a few radial razor-blade scrapes into which the pitch can flow. It is obvious that several things can take place. First, if cylinderlike surfaces (astigmatism) are polished with a small polisher, the solid polisher works harder on the higher zonal areas and has a limited polishing action on the lower. Second, this removes the smaller and higher ripplelike scrambled zonal areas. The solid polisher tends to make the corrector plate to move toward an overcorrect figure therefore the preceding correction must have been very close.

It is noteworthy that in the making of the two Baker-type solid Schmidt cameras the only zonal correction needed to complete their construction is to figure a central high zone of overcorrection. It is important to emphasize that in figuring deep aspheric surfaces of small diameters no drilled holes or vee cuts be made to handicap the working of the flexible pitch polisher. These operations must be done after the completion of the figuring, preferably with fine carborundum emery and not with diamond grit saws or drills. As discussed before, the first diamond sawing of the vee slot into the aperture of the on-axis camera is followed by the removal of 1 mm of glass from the sides of the vee slot to suppress the Twyman effect.

5. ASSEMBLING THE TWO OFF-AXIS SOLID SCHMIDT CAMERA

Assembling the two solid off-axis cameras is rather straightforward. The glass cylinder is reheated in a electric controlled oven with temperature increases of 5°C every five minutes until the balsam melts and the glasses can be separated. The cooled flint-glass pieces are cleaned and examined for stain and all the glass pieces are retouched on a solid polisher, but preceded by fine-gounding with 145 aloxide grit of one side surface of each piece. This means one surface on the small focal-plane flint glass and one surface on a larger flint piece. The primary mirror surface must be *properly orientated* before grinding is started. Figure 22.8 is the assembled unit.

The flint-glass pieces are retouched on the polisher and the stain removed. These repolished flint pieces were rechecked for flatness with a flatproof plane and were flat to one-half wave. It must be pointed out that flint-glass pieces can depart from their original flatness but would not have effect on the cemented surfaces to harm the final imagery.

The corrector plate must be carefully diamond sawed into two pieces because of the thinness of the glass at the neutral zone. The width of each sector of the corrector plate is left with 0.5-mm of surplus glass on each side in order to fine-grind the assembly on each side after cementing it with Lens Bond in its final configuration. The cement can be mixed to harden within 30 min. The three elements are cemented together. First the soft aluminized coating used during the fabrication of the two flint bodies is removed and a new aluminized coating with an protective overcoating is applied. These two

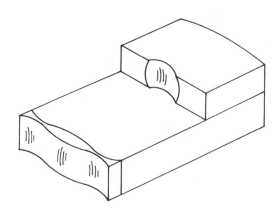

OFF-AXIS CAMERA

Fig. 22.8. View of the finished off-axis Baker-Schmidt camera:

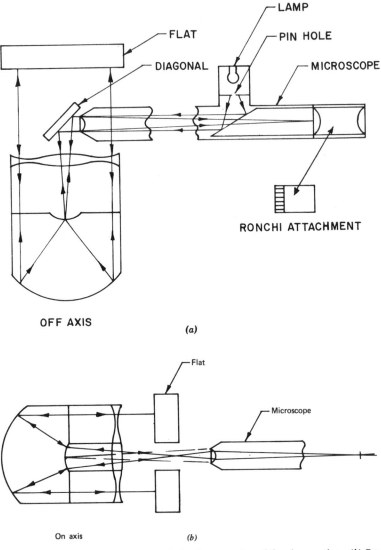

Fig. 22.9. (*a*) The testing of the camera during its cementing of the glass sections. (*b*) Same as Fig. 22.8*a* but with flat with a hole.

pieces of the flint body are cemented and allowed to dry firmly in a holding fixture. Second, the section of the correcting plate coarse ground on both sides within 1 mm total is cemented to the previously cemented flint pieces of the main body. This had to be done with much care and an autocollimated setup was devised. [See Figs 22.9 (*a*)–(*b*).] Here the optical arrangement is an aluminized autocollimating flat. The off-axis flat redirects the light diverging from a microscope objective and fills the curved focal plane of the camera. The resulting parallel light from the off-axis camera is retrodirected into the microscope and the image of the light pinhole source is magnified in the eyepiece. The camera body and the cemented (not hardened) section of the corrector plate rests on the rigidly held main body of the camera.

The general purpose is to watch for elliptical or oval imagery as the cement hardens and corrector plate is firmly oriented. No deterioration of the imagery was observed while the corrector plate was held by adhesive forces. It was purposely moved and elliptical imagery was observed, but it became round and symmetrical as the corrector floated back onto the flint body. After the cement had hardened (36 hr) the surplus glass was ground off the corrector plates and made even with the main body. The fine grinding of the flint body, on which the focal plane is located, must be done on a flat brass strip and the aluminized and polished surface must be protected by treating with an acrylic spray. All the fine-ground surfaces are given two coats of black enamel after the spray has been removed with acetone. Final testing of the imagery was accomplished by checking the autocollimation setup in Figs. 22.9 (a)–(b) or by using a long focal length collimator bench (Askania type).

6. CONCLUSIONS

The following conclusions can be drawn:

1. The optical working techniques of an on-axis solid Schmidt camera and an off-axis Schmidt system are the same.

2. The fabrication of deep aspheric surfaces depends entirely on the final form of the brass lapping plates. The two cast-iron and brass aspheric laps had 0.2-in. glass plugs inserted in their reamed holes and were lapped together in sequential order. Therefore no high zone was ground on the corrector-plate aspheric surfaces.

3. The abrasive used in the last grinding on the brass aspheric laps must be broken down to prevent film from forming between the lapping tool and correcting plate.

4. It is good practice to bevel all sharp edges on all the glass elements to be assembled.

EXERCISES

1. Using formula (1), calculate the wavelength difference of the (C-line) and (F-line) for UBK-7 glass for a sample system. In an $f/0.6$ h equals

41 mm, and R equals 52 mm. The C-line equals 6563 Å and the F-line equals 4861 Å.

2. Describe how aspheric grinding tools are made. Which surface of the aspheric is the most critical and why?

3. What three optical tests can be used to test the assembled camera?

4. Describe the outside-of-focus Ronchigram for an autocollimated optical system which has a turned-down edge and an overcorrected central area.

5. Describe the making of the pitch polisher for polishing the aspheric surfaces.

BIBLIOGRAPHY

Baker, J. G. *Proc. Am. Phil. Soc.*, **82**, 322–338 (1940).

Cox , H. W. and Cox, L. A. "Further Notes on Schmidt Cameras," in *Amateur Telescope Making, Book Three*, A. G. Ingalls, Ed., *Sci. Am.* 349 (1953).

De Vany, A. S. "A Rapid Method of Making a Corrector Plate," *Pop. Astron.* **47**, 197–202 (1939).

De Vany, A. S. "Two New Telescope Design," *Appl. Opt.* **2**, 201–204 (1963).

Lower, H. "Notes on the Construction of an $f/1.0$ Schmidt Camera," in *Amateur Telescope Making, Book Two or Advance*, A. G. Ingals, Ed. *Sci. Am.* 410–416 (1959).

23

A Schmidt-Cassegrain Optical System
with a Flat Field

A Schmidt-Cassegrain telescope with a flat field, made in 1970 with a 25 cm aperture.

Reflecting telescopes of the usual Newtonian or Cassegrain form have drawbacks that often dull the enthusiasm of eager observers. Among the deficiencies are the difficulty of maintaining alignment, air currents in an open tube that spoil the images, and aluminized surfaces that become sleeked by efforts to remove dust and moisture condensation. A short-focus Newtonian may not provide sufficient magnification for planetary observations, whereas a long-focus Cassegrain is at a serious disadvantage when a large field is desired. (See Fig. 23.1.)

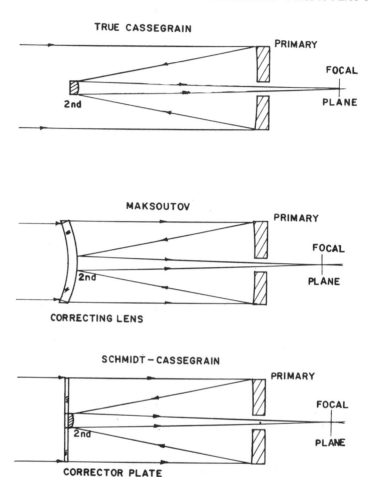

Fig. 23.1. A number of telescope systems used by amateur and professional astronomers.

Therefore many amateurs have turned to catadioptric optical systems such as the Schmidt-Cassegrain and Maksutov in which a correcting lens balances the aberrations of the primary mirrors. In the original Schmidt camera of 1930 the corrector was placed at the mirror's center of curvature. The result was a telescope twice as long as its actual focal length. Thus the Schmidt camera was made a short-focus, wide-field camera because its strongly curved focal surface was quite difficult to reach for visual work.

In 1940, however, James G. Baker described a family of flat-field cameras reduced in length, yet equivalent in performance to the Schmidt; later Schmidt-Cassegrains for visual work were designed with the same overall dimensions as the more popular Maksutovs; for example, in *Sky and Telescope* for April 1962 (p. 191 and 226), Ronald R. Willey, Jr., described a telescope in which the secondary (an ellipsoid) can be mounted on a corrector plate to eliminate the diffraction effects of a spider support. With a spherical primary this system has less curvature than a true Cassegrain, yet the focal surface is behind the primary mirror, convenient for visual observation with a zenith diagonal. Furthermore, because the corrector can be as large as the primary mirror, the effective aperture is greater than that of an equivalent Maksutov-Cassegrain. The off-axis performance of Mr. Willey's Schmidt-Cassegrain is satisfactory according to his comprehensive ray-trace computer calculations.

1. POSSIBLE DESIGNS

Various combinations of spherical and conic surfaces, such as a paraboloidal primary and spherical secondary, an elliptical primary with a spherical secondary (the Dall-Kirham system which must be over- or undercorrected for use with a corrector), an oblate spheroid primary with an oblate secondary (Mariner's telescope for Mars) with equal radii on the primary and secondary mirrors, are possible. It is apparent that any degree of intermediate correction (under or over) can be achieved, according to how turned-up the primary and how turned-down the secondary are toward the solution free of spherical aberration, coma, and astigmatism. The telescope system shown in Fig. 23.2, however, is the easiest of all compound Schmidt-Cassegrain systems to fabricate because both the primary and secondary can be spherical and have the same radius of curvature for System I. System II differs slightly; both the primary and secondary are of equal radius of curvature and the secondary is a weak ellipsoidal conic surface. System I is described first.

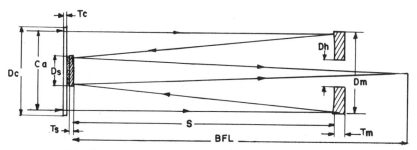

Fig. 23.2. A Schmidt-Cassegrain system with equiradii on the secondary and primary mirrors which give a flat field.

2. SYSTEM I: PRIMARY AND SECONDARY MIRRORS SPHERICAL

These equal radii satisfy the Petzval condition for zero field curvature and the field is flat. The scaling factor for comparing astigmatism of the standard Cassegrain and this designed system is three times smaller than the standard Cassegrain because of its low 1.71 amplication factor and at $f/11$ the coma would not be noticed.

At the same time the tube is closed against air currents and the corrector supports the secondary mirror. This compact optical system fits a tube length about that of an $f/5$ Newtonian reflector, yet its effective focal ratio of $f/11$ provides considerable versatility. For the four sizes tabulated here the focal lengths range from 66″ for the 6-in. and 132″ for the 12-in. Thus relatively high powers can be achieved with short-focus oculars and moderately wide fields with low-power eyepieces. The focal plane is satisfactory for many kinds of celestial photography. Table 23.1 list the parameters of the design.

3. MAKING THE MIRRORS

Because both mirrors have the same radius of curvature, they can be made together, one disk serving as the mirror, the other as the grinding tool. Oscar Knab, who has done similar work on his 4.5-in. schiefspiegler (*Sky and Telescope*, October 1961, p. 232) recommends that the curve be checked constantly during grinding. He used a homemade spherometer, but an ATN can adopt any coventional method of working a spherical surface of about $f/6$, which is only slightly deeper than the conventional $F/8$ Newtonian. The radius-of-curvature tolerance is plus and minus 1 in. because allowance for the final value can be made when the optics are assembled for testing. As an alternative a radius template made of brass or thin aluminum sheeting can be used to check the radius of curvature of the concave mirror.

Before proceeding with the 320 carborundum grinding the central holes in both mirrors are core-drilled to the correct aperture diameter listed in Table 23.1. Because the secondary made from the grinding tool blank and the back hole in the primary mirror are of near-equal diameter they can be core-drilled with the same cylinder core drill.

To do this accurately a circular template of plywood equal to the primary mirror's diameter, with a central hole equal to the secondary mirror's diameter, is made up. The inside diameter of the core drill governs the diameter of the secondary mirror and its outside governs the hole size in the mirror. Also the 80 carborumdum grit size enters into both diameters and it is good practice to make the diameter of the secondary mirror $\pm\frac{1}{32}$ in. and

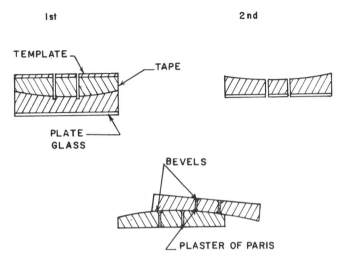

Fig. 23.3. Core drilling setup for drilling the hole in the primary mirror and the secondary mirror from the convex grinding mirror.

let the mirror hole be governed by the wall size of the core tubing and grit size. This provides a wider field for photography. (See Fig. 23.3.)

Brass or steel thin-wall tubing must fit the central hole of the plywood template snugly. The tubing must be fitted with a central stem for mounting in a drill press. The core drilling is done by placing the template on the concave surface of the preground mirror. Pour a slurry of 80 carborundum and water into the central hole of the template and wrap several layers of masking tape around the edges of the mirror and template to hold them securely in place. After drilling about ⅛ in. remove the tape and the template. This permits carborundum to be added. Drill with a steady pressure and then ease up because the bottom surface can chip as the core drill breaks through. Place the template on the other side and drill partly into the bottom surface; then finish the core drilling to meet the section already drilled.

The next step is to core-drill the secondary mirror by placing the concave mirror on the grinding tool. Use only 320 carborundum during the starting recess drilled into the convex tool. The glass blanks must be centered over one another before the drilling is started and masking tape must be wrapped around both edges. After this recess circular ring has been drilled it can be finished like the hole in the concave mirror.

Each section drilled from the glass blanks must be strongly beveled ⅟₃₂-in. with 145 grit. The drilled holes must also be strongly beveled. In the next step the cores are replaced in the glass blanks with melted paraffin. Be sure to space each core uniformly with three match sticks.

The mirror's surfaces are fine-ground with each finer grit, and the radius of curvature is checked with a spherometer or radius template. The mirrors can be kept to the desired radii by fine-grinding alternately, one on top of the other. If the radius of the concave mirror is too steep, it is placed on the bottom; conversely, if the concave surface is shallow, it is worked on top. Note also that both back surfaces are fine-ground for surface polishing. It is assumed that a beginner will have several front-surface mirrors with spherical or parabolic surfaces.

The primary mirror is polished first and must be a spheroid verified by a knife-edge or Ronchi test at its radius of curvature. (See Figs. 14.1 and 14.2 for the test patterns of focograms and Ronchigrams.) The back surface of the concave mirror should be given a semipolish to keep all surfaces normalized to the surrounding temperature. Determine whether a change in the pattern of the Ronchigram of focogram has been caused by the Twyman effect. If so, refigure the spherical surface.

Next polish the secondary mirror's back and front surfaces. A 50% semipolished flat surface on the secondary mirror is all that is required to make sure that a spherical surface is maintained. As an added check test the convex surface by using it as a reflector to transmit light rays through the flat surface. The test setup is foreshortened over the radius of curvature test. A Ronchi test with a yellow or red cellophane filter is the best choice. A slight change in the secondary mirror's radius does no harm; examine for the null pattern (spheroid).

The glass plugs are removed from both glass blanks by running boiling water over the paraffin surrounding the glass cores. Clean both surfaces with xylene and wash with a detergent-water solution and dry. Both surfaces can now be aluminized and overcoated.

4. SELECTION CORRECTOR-PLATE GLASS

The correcting plate is made of planoparallel glass plate devoid of striae. Corrector plates for Schmidt-Cassegrain telescopes are made of BSC2 or BK7 crown glass. High-grade plate glasses, such as Water-Wite, and twin-ground and polished plate or float (selected) can be used. The plate glasses can be obtained from Semon Bache & Co., Morton and Greenwich Streets, New York, NY 10014. For ultraviolet transmission Schott's UBK 5 or calcium fluoride are essential. Plan to make the corrector at least 1 in. (2.5 cm) larger than the design aperture to facilitate optical processing. It is easy enough to fit a lens holder the larger size corrector. One-quarter-inch (6.4 mm) is good for a 6-in. (152.4-mm) instrument, 3/8 to 1/2 in. (9.6-mm to 12.5 mm) for the 12-in. (304.8-mm) sizes; but for a first instrument it might be well to use the thicker glass to avoid possible flexure of the corrector that would introduce astigmatism. There will be no noticeable difference in the figuring process or telescope performance.

At least two circular pieces cut from a sheet of plate glass are needed. If the optical glasses are obtained from Schott, Corning, Bausch & Lomb, or their suppliers, good quality can generally be assured. Test the quality by setting each piece up against the finished aluminized spherical primary mirror and doing a Foucault test of the combination. The presence of striae or a variable index of refraction can be detected by using a knife-edge or Ronchi screen. Striae are veinlike streaks that may look like water running out of a faucet when the large ones are massed together. Refractive index variations can produce dark ring patterns (like pistol targets) when seen in this test. Glass or optical quartz with pistol targets off to one side should be discarded. If, however, one large pattern is symmetrically centered in the full aperture, the plate can be used.

Acceptable glass generally has a zone directly in the middle and two or three rings rather evenly spaced to the edge. Glass cut from a larger sheet will sometimes show color shadows that are not symmetrical and tend to appear astigmatic under the test, but these small refractive changes will cause no difficulty in figuring and can be ignored. For further methods of checking small variations in glass disks or sheets see Chapter 5.

5. PRELIMINARY WORK ON THE CORRECTOR

Because the correcting plate is a weak aspherical lens with about six wavelengths maximum deviation from a purely planoparallel figure, it could be polished and figured from a perfect piece of plate glass without grinding. In practice, however, a number of steps will probably be required before figuring can begin.

First the two surfaces must be parallel with the wedge or variation in thickness amounting to no more than 0.0002 in. (0.005 mm), checked at many places around the edge. If the sides are not true, one or both must be ground with 225 aloxide grit within the prescribed tolerance. First, however, chamfer the edges with a fine carborundum all around the periphery. Optical work is often sleeked and scratched because there is too little chamfer or because the worker allows an excess of polishing compound and pitch to collect around the edges; this material works loose and drops on the pitch polisher.

The glass disk for the corrector plate with a wedge of 0.0002 in. or less is mounted on a flat aluminum holder on which a sheet of mylar (drafting-paper) or rubberized felt of equal diameter has been placed. Masking tape is then wrapped twice around the disk and holder. Corrector plates as thin as 0.25 in. of 6.5-in. diameter have been made successfully.

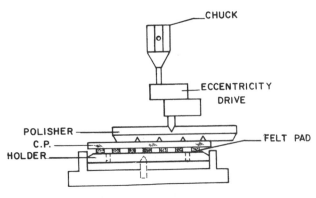

Fig. 23.4. The drill press method for grinding and polishing with an adjustable slide for an eccentric throw.

6. POLISHING MACHINES

Optical processing of the corrector plate may be done by a variety of methods, especially by experienced workers. With the grinding and polishing machine shown in Fig. 23.4 the fabrication is carried out by conventional practices. Amateur telescope makers may have to improvise by turning a drill press into a polishing machine. J. H. Hindle describes machine methods in *Amateur Telescope Making, Book One*, pp. 218–220 and 234–240. To slow the drill press below 100 rpm add another cone pulley or secure an adapter. Even the primary mirror can be polished and figured by constructing a small "crocodile" mechanism (p. 236) and setting it up as illustrated on p. 239 (*left*). Some workers however, may find it difficult to adapt the rotating base or cell diagrammed on p. 219. This is not part of the usual drill press. To get around this problem drill and ream a 0.5-in. hole in the center of the drill-press plate for a freely turning brass plug; its other end will fit snugly into a center hole of the corrector-plate or pitch-lap holder. It will be found that in the normal working of the polisher this base assembly will rotate by itself. This action can be helped by oiling the drill-press plate surface or by inserting a ball bearing in the bottom of the hole under the plug.

The sketch in Fig. 23.4 shows a simple adjustable eccentric arm that will swing the piece off-center over the polisher to permit work on different zones of the corrector-plate surface. I believe it is well worth the time and effort to adapt a drill press in this manner. Of course, a milling machine can be similarly set up.

7. RONCHI TESTING

There are three methods of testing the telescope system with a parallel light setup:

1. The use of an autocollimating flat (one-tenth fringe surface quality).
2. A reflecting oil pool.
3. A distance sodium or mercury street lamp.

The three-quarter flat is setup to cover the central and edge areas. Because all Ronchigrams are symmetrical, the same interpretations can be made from the patterns. Roger W. Tuthill and others have used oil reflecting pools of heavy oil. This is illustrated in *Sky and Telescope*, March 1964, p. 184. It is worthwhile to review the diagrams on p. 343 for the oil-pool setup and also the inside-of-focus Ronchigram (undercorrection) on p. 341 of that publication. Note the diagram for testing a Schmidt camera with an objective lens. It is obvious that a large Newtonian telescope could be set up to form a parallel light arrangement. Testing with a distant monochromatic street lamp has been done by Jeff Schroder (see Appendix 15) and others as a source of parallel light. Of course, weather conditions are a disturbing factor.

The primary mirror is aluminized and set approximately at the distance from the corrector plate given in Table 23.1. The secondary mirror is also aluminized and mounted on the corrector plate with double-coated tape; several 0.25-in. squares around the rim are sufficient. Permanent marks can be scribed with a carboly-tip scribe at the centers of the corrector plate and secondary mirror when these elements have been optically aligned in the autocollimation test set. It is then easy to place them in the same relative position while the plate is being figured. (See Fig. 23.5.)

The test pattern is observed through the hole in the primary mirror. When the light source fully illuminates the secondary, it and the corrector are moved toward or away from the primary until the proper distance behind the primary is obtained. Generally, for zenith observation with a star diagonal this distance is approximately equal to $\frac{9}{10}$ of the aperture of the primary.

When an oil flat or distance street lamp is used, it is easier to interpret the test patterns of the Ronchi screen than to apply the Foucault knife-edge, but for checking the final surface smoothness the knife-edge can be applied (a strong light is needed).

With parallel light the Ronchi test yields nine basic patterns of the overall surface figure, shown in Fig. A14.2 for tests inside and outside of focus for refractive (corrector plates) and reflective elements. No account is taken of

TABLE 23.1. SCHMIDT-CASSEGRAIN TELESCOPE (IN INCHES)

CA	Clear aperature of corrector	6	8	10	12
D_c	Diameter of correcting lens	6.5	8.5	10.5	12.5
T_c	Thickness of correcting lens	0.25	0.375	0.5	0.5
D_m	Diameter of primary mirror	6.0	8.0	10.0	12.0
T_m	Central thickness of primary	1.0	1.5	1.75	2.1
D_h	Diameter of hole in primary	2.5	3.5	4.375	5.125
D_s	Diameter of secondary mirror	2.25	3.5	4.375	5.125
T_s	Thickness of secondary mirror	0.375	0.4	0.5	0.625
R_1	Radius of curvature of primary	76.52	102.204	127.754	153.06
R_2	Radius of curvature of secondary	76.52	102.204	127.754	153.06
S	Separation of mirrors	22.133	29.515	36.891	44.273
BFL	Back focal length	27.872	37.168	46.456	55.752
EFL	Effective Cassegrain focal length	66.136	88.193	110.233	132.290

Dimensions in Millimeters

CA	Clear aperture of corrector	152.4	203.2	254.0	304.8
D_c	Diameter of correcting lens	165.1	251.9	266.7	37.9
T_c	Thickness of correcting lens	6.35	9.52	12.7	15.88
D_m	Diameter of primary mirror	152.4	203.2	254.0	304.8
T_m	Central thickness of primary	25.4	34.92	44.45	53.34
D_h	Diameter of hole in primary[a]	50.8	63.5	76.2	88.9
D_s	Diameter of secondary mirror	66.68	88.9	111.13	133.35
T_s	Thickness of secondary mirror	9.53	10.16	12.7	15.24
R_1	Radius of curvature of primary	1943.61	2595.98	3244.95	3887.72
R_2	Radius of curvature of secondary	1943.61	2595.98	3244.95	3887.72
S	Separation of mirrors[b]	562.18	750.44	937.03	1123.44
BFL	Back focal length	707.95	944.07	1179.98	1416.10
EFL	Effective Cassegrain focal length	1679.85	2240.10	2799.92	3360.17

[a] For photographic use the hole can be made 25.4 larger if desired.
[b] This distance is variable, depending on the mounting thickness of the cork supporting the secondary and on the secondary mirror's thickness; variation of 25.4 to 50.8 mm is tolerable.

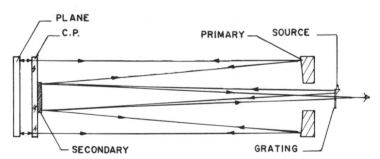

Fig. 23.5. The autocollimation setup of the Schmidt-Cassegrain system used during testing. An optical flat is shown (see text for other test methods).

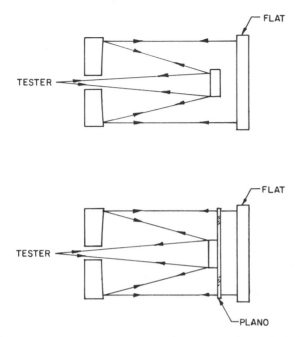

Fig. 23.6. An autocollimation mode of two equiradii primary and secondary mirrors set at a parametric distance.

the high or low areas (step zones), intermediate scrambled zones, or other localized variations (nonhomogeneneity in the glass) often observed while figuring is in progress.

It is important to realize that these Ronchigrams are completely reversed when reflective mirror elements are tested, as Fig. A14.2 illustrates. An exception is a perfect refractive-lens system tested in an autocollimation mode or a concave mirror tested at its radius of curvature (self-autocollimation mode). This shows a null pattern of equiparallel lines both inside and outside of focus. Fig. A14.1 illustrates the focograms. Examine Fig. A14.2, for inside-of-focus Ronchigrams, parallel sections (box) and refracting surfaces.

8. FIGURING THE CORRECTOR

Consider the optical system made with equiradii on primary and secondary mirrors separated at a parametric distance for an 1.72 amplification. Establish the two coated reflectors of the system in an autocollimation setup with a flat shown in Fig. 23.6. Two optical tests can be used if a slit source

STEP ZONE

Fig. 23.7.   Similar to the mode shown in Fig. 23.6, but differing in that the secondary is mounted to the planoparallel component.

is added to the knife-edge apparatus when a Ronchi grating is placed over the lighted slit with some extension for eye viewing. (See Fig. 28.3.) The oblate spheroid pattern of undercorrection (pincushion) is shown in Fig. A14.2. Because this involves a *reflective* setup of the two mirrors, rotate Fig. A14.2, 180° to read the inside-of-focus Ronchigrams. Of course, the outside-of-focus Ronchigrams could be read just as well with Fig. A14.2 in its upright position; this would give the inverse Ronchigram (barrel) of undercorrection. Note in any case that the 45° cross section show reflective optical testing; refractive optical testing as described by the parallel lines.

Question.   What test pattern is observed in Fig. 23.7 if the convex secondary is mounted on or near the planoparallel element?

Answer.   It would still be an oblate spheroid of undercorrection.

This is not the contradiction described for refractive elements. It is not until one of the flat surfaces is four or more waves aspheric that the *refractive* test patterns are observed. Therefore a planoparallel element does not change the incoming parallel rays from the autocollimating mode or setup.

The first Ronchigram observed in the autocollimation test with the unfigured planoparallel is an oblate spheroid figure. At this stage the corrector sides will have been rendered parallel and polished smooth. It is well to have one surface convex at the start by 6 to 10 circular fringes because it is easier to begin the figuring "downhill."

It is also advisable to plan some correction on both sides of the corrector to eliminate the astigmatism worked into the flat surfaces. If one side is convex, as in case of planoparallel elements, the second side will generally be concave and near zero power. Even if the corrector has some power it would cause no harm. Using the concave side, plan 15 to 20 min work on this side to smooth it and eliminate any astigmatic surface or irregular sharp zones. The polisher is the the rose-leaf shown in Fig. 23.8. Then mark this side to identify it and never work it again. Be sure the bottom holder has been rotated with the corrector 5 to 10 times a minute by hand; this will remove the astigmatism on that side. Using a polishing machine of the conventional type shown in Fig. 2.27, set the eccentric arm to about 0.5-in. rotating at near 10 rpm and the bottom spindle perhaps at 125 rpm. The polisher must be held stationary for more than one-half the allotted time. A 0.25 eccentric throw on the drill-press mechanism is sufficient for this operation.

A certain amount of astigmatism should be expected to appear in the early stages of the final figuring of the second side. Small amounts of astigmatism are difficult to detect by Ronchi or knife-edge testing. Astigmatic surfaces are caused by lack of or insufficient rotation of the corrector-plate holder on the drill press or on a piece of glass that is too thin.

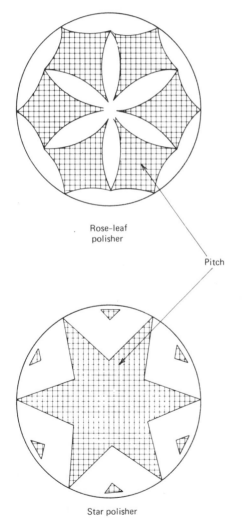

Rose-leaf polisher

Pitch

Star polisher

Fig. 23.8. Profiling polishers developed for optical figuring the corrector plate or secondary mirror.

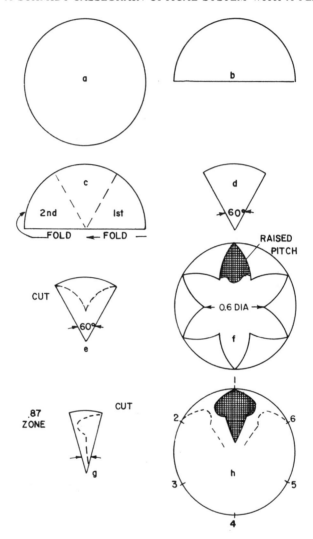

Fig. 23.9. Directions for how to fold and cut out the configurations of the patterns for forming the pitch polishers.

In the optical system under discussion the neutral zone of the corrector (through which the light rays pass undeviated) should be about the 0.86 zone (in a 10-in., the 4.3 zone). Because the greatest amount of glass will be removed at the neutral zone, the pitch lap should have the largest area of pitch at that location. This is the case for the rose-leaf pattern of polisher No. 1 shown in Figs. 23.8 and 9. All polishers are made of soft pitch similar to Zobal soft. Cut the rose-leaf pattern in a piece of cardboard and place it over a smooth disk of pitch; use a shop razor to cut the pitch away from the white areas of the illustrations. This deepening of the 86% zone is generally the same kind of correction adopted in the primary mirror correction of an $f/15$ Gregory-Maksutov. The rose-leaf pitch lap is used as outlined by Warren Fillmore in his booklet, "Construction of a Maksutov Telescope." Polishers Nos. 2 and 3 are similarly made up, but No. 3 is opposite No. 1 in its design and working action. All polishers must be embossed with smaller facets with plastic screening.

To begin the figuring place polisher No. 1 on the corrector which is wet with polishing compound (cerium or Barnsite) and adjust the eccentric arm as shown in Fig. 23.4 to a small distance, perhaps 0.25 in. At this stage the drill press can be used without the auxiliary pulley to slow it down; use the slowest spindle speed. Hang a 2-lb weight on the pull-down handle of the drill press and be sure to grease the rotating stud in the drill-press plate and keep the polisher wet. If the glass holder is not rotating under the pitch polisher about once each minute, it should be turned by hand.

It is obvious that these settings are easily carried out with a conventional polisher.

At the start do not work more than 20 min before checking the autocollimation test setup. After several runs the Ronchigrams may still show the pattern in Fig. A14.2 in quadrant at $+1$, the old pincushion pattern of undercorrection. It could become any one of the many configurations with step zones or badly turned edges. However, the edges have not been touched with the polisher! It is evident, on second thought, that the full correction has not reached the edge. Therefore, the work is continued until the Ronchigram shows the null central pattern with parallel lines and a turned down edge. This is called a step zone and can be found in any figured element. Set the eccentric arm at almost zero displacement and continue; determine also that the embossed facets are still there. If not, renew them. The polishing must be continued even to the point at which some slight overcorrection (barrel) is observed at the central and peripheral zones. Then No. 3, the backup polisher, is used to bring the figure to a null pattern that shows full correction.

Work especially on the high areas and adjust the eccentric arm for the longer or shorter strokes required to control the figure. Slow rotation of the corrector holder will offset the development of sharp defined zones.

When making corrector plates "the work follows the polisher." Therefore it is essential to keep the tool is proper shape by constant trimming of the pitch laps in Nos. 1 and 3 and keeping the small embossed areas for better contacting. Again the polisher works better with a watery rather than a heavy polishing slurry because better contact is made. Intermediate sharp zones can be smoothed out with a soft, full-sized polisher with no precut channels. A small 0.5-in. hole should be dug out in the center of the polisher to allow the pitch to flow into it.

When the correction has neared completion begin to use the technique of profiling the polisher with cutout patterns to contour it. (See Appendix 14.)

To learn from experience keep a detailed log. It would take much too long to recount all the figuring conditions that can develop and the means of correcting them. An up-edge is a difficult problem because the rose-petal polisher tends to stay away from the edges. It is quite possible, even for an experienced optical worker, to work the gamut of configurations before final success is achieved.

9. NOTES ON TESTING

If a fine Ronchi screen is used, the Ronchi test will produce imagery as satisfactory as the knife-edge test: the screen lines should be only 0.003 in. wide. Of course, an oil flat limits the fineness of the grating, as does the 5-mile sodium lamp.

Small irregularities of about one-eight the width of a pattern will not effect the quality of the image.

Star testing with the Ronchi grating is a good final check. Use the finest grating that observing conditions will allow on a bright star. The Ronchigrams are exactly the same as shown here and their interpretation is also the same. It is possible to cut the full aperture with a one-line Ronchi grating to obtain a Foucault reading if the atmosphere permits it.

Incidently, nine Ronchigrams in central areas occur in parallel light testing of objectives, parabolodial mirrors, or any telescope system, whether the light is provided by an autocollimation mode, a collimator, or a star. Figures A14.1 and A14.2 must be rotated for the reflective and refractive elements being figured. There is a difference.

10. MOUNTING THE CORRECTOR AND SECONDARY

As illustrated in Fig. 28.2, the secondary is attached to the corrector by an expoxy-cemented cork buffer. This arrangement, of course, requires that the corrector plate have an adjustable mounting to permit the secondary mirror to be properly aligned in the optical system. This adjustment of the corrector does not affect the star images if the tilt is not too great. Correlate with a two-amplification change in the back focus increase or decrease as the adjustment is made. .

Some designers may prefer to mount the secondary in a hole drilled in the corrector with a carborundum-fed cookie cutter. Be sure to protect the finished surfaces of the corrector by pitching window glass at each side. The hole should be at least two-thirds of the diameter of the mirror holder and its adjusting screws. Insert two "O" rubber rings where the assembly touches the corrector plate.

11. SCHMIDT-CASSEGRAIN WITH ASPHERIC MIRRORS

The aberration called astigmatism, in which off-axis images seldom lie exactly in a true plane when there is primary astigmatism in a lens system, is now dealt with. This aberration can be eliminated by making one or both mirrors a slight ellipsoid or oblate spheroid. In any case the small differences in the two conic surfaces are insignificant in an $f/12$ system.

As mentioned at the beginning of Section 2 several conic surfaces can be used to eliminate most of the astigmatism and improve the off-axis imagery.

One solution is to overcorrect the corrector plate slightly by one fringe pattern to give the test the appearance of the Ronchigram in Fig. A14.2 (barrel-shaped). Then refigure the secondary mirror with a small rose-leaf polisher to make the system null in its autocollimating setup. This means that an aspheric surface (ellipsoid) must be figured on the secondary mirror.

In another solution the primary and secondary mirrors are figured as oblate spheroids, after which the primary is figured to an oblate figure at its center of curvature test to 1 in Fig. A14.2. Next give the corrector plate a slight overcorrection and then correct the secondary mirror. Of course, to do this properly during autocollimation testing the mirrors must be silvered. (See Appendix 7.)

12. CONCLUSIONS

In summary the optical figuring of a system always depends on the optimum design; for example, the optician would generally aspherize the central area of a paraboloid mirror during polishing. For a Schmidt-Cassegrain corrector plate the 0.86 zone would be lowered first with a rose-leaf polisher to a slight overcorrection; the secondary mirror would then be retouched. Again it is interesting to note that the apparent focal plane can be estimated by optical testing, as projected backward or forward to any major optical component surface, and that surface can be figured. Perhaps it is well to consider a smaller surface—one with a high amplification factor or eccentricity value as in the secondary or tertiary mirrors in Baker's tertiary system. In a Maksutov compound-telescope system less glass is removed by optical figuring of the primary mirror; figuring the corrector lens requires that four times the amount of glass be removed. For optimum performance of a well-designed lens system the optician must be told what surface or surfaces to retouch to achieve a well-corrected system.

BIBLIOGRAPHY

De Vany, A. S. "A Schmidt-Cassegrain Telescope with a Flat Field, *Appl. Opt.*, **4**, 1353 (1965).

De Vany, A. S. "A Schmidt-Cassegrain Optical System with a Flat Field," *Sky and Telescope*, **29**, May and June 1965.

De Vany, A. S. "A Schmidt-Cassegrain Telescope with a Flat Field: II. *Appl. Opt.*, **6**, 976 (1967).

De Vany, A. S. "Optically Figuring a Schmidt-Cassegrain System," *Appl. Opt.*, **18**, (1979).

Slevoght, H. *Z. Instrumentenkd.*, **62**, 312–327 (1942).

Willey, R. R. *Sky and Telescope*, **23**, 191 (1962).

24

A Planointerferometer with an Aspheric Lens

The planointerferometer patterned after the Fizeau interferometer offers many methods of testing flat surfaces, checking parallelism, homogeneity, radius test plates, and reflecting flat surfaces mounted in assemblies. Polished metal surfaces can be tested if the reference flat surface is coated with aluminum or zinc sulfide. Because flat optical surfaces can be set approximately 1 cm from the surface of the reference flat, sleeks and digs do not appear on prisms or in other areas.

A typical design contains a planoconvex lens in which the planosurface serves as the reference flat and the convex becomes a hyperbolic curve. The making and figuring of the aspheric convex surface are discussed later in this chapter. The light source for the pinhole must have the wavelength of the intended source of the planointerferometer. A mercury source (reed-tube type) has an approximate 1.5-cm separation of the flat optical surfaces without losing the contrast of the observed fringes. It is possible to use a low-power laser (0.3 mW) that offers methods of testing widely separated flat surfaces. Generally, the flat surfaces can be 3 in. (75 mm) apart and the fringes can be projected on a frosted screen. The lens material for any size of $f/6.0$ planointerferometer can be ULE quartz or preselected Cer-Vit and Pyrex. (See Fig. 24.1.)

One other design has a doublet lens that is color-corrected for the intended monochromatic source (generally mercury) and a wedge element (10 min of arc) with one excellent flat reference surface of one-sixteenth fringe or better. The wedge angle is used to offset any reflection of the illuminated pinhole from the second surface of the reference flat element.

In either of the two designs chosen for the planointerferometer fold the light path by interposing one or two aluminized folding flats to ensure a compact design. A swinging eyepiece with a large field is used to help locate the two images that form the interferogram when they are superimposed. Neutral filters of sufficient density must be located at the exit opening for the laser-illuminated pinhole source. This is an assembled unit that must have the

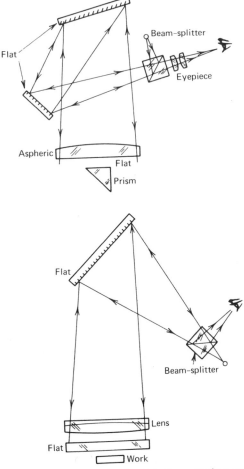

Fig. 24.1. Two designs of a planointerferometer in a compact unit.

capability of vertical movement on a lathe-type bed. A tilting table must be a separate unit of kinematic design with fine adjustments to set the contrast of the interferograms.

1. USING THE PLANOINTERFEROMETER

This device has many uses in the optical shop and can be cheaply made. It can be used to test parallelism of windows, wedges, and planoparallel surface quality of flat surfaces of prisms and flats, homogeneity of glass, using the Perry and Twyman method, the sphericity of test plates (see Chapter

15), and checking the assembled reflecting flats and windows in optical instruments. Here the contrast of the coated windows and flat surfaces can be tested if one-half diameter of the testing flat is coated with titanium oxide. Each half can then be used to the best advantage.

Laser illumination may have to be used for flat elements in deep-seated assemblies but for most testing a mercury source performs well.

Many 45° reflecting diagonals can be checked for pinched assembly with metal tap holders or contact cement. Many a costly Newtonian telescope has been ruined by these practices.

One of the better ways of mounting flats is to recess three small ball bearings in the metal holder and then apply PRV or a pliable cement for mounting the balls on the glass. Dab each recess hole and place the flat on the balls. It is well known that equal mass distribution of a circular diameter is in the 0.707 zone. Never attempt to cement the whole element to a metal holder with PRV or other pliable cements because these materials cause wrappage of the optical flat.

2. MAKING THE ASPHERIC PLANOCONVEX LENS

Preselected low-expansion materials used to make the aspheric lens can be reannealed pyrex, ULE quartz, or Cer-Vit. They should be carefully selected for the absence of heavy striae, cords, dirt, and bubbles. Faint striae and cords cause no harm and can be tolerated. The focus of the $f/6$ lens can be calculated with the thin-lens formula. (See Fig. 24.2) For a 4-in. (102-mm) diameter the focal length is 24 in. (35.6 cm) and its radius of curvature is approximately 12 in. (30.5 cm). We should allow sufficient thickness at the peripheral edge for easier working. If the edge is too thin, difficulty will be

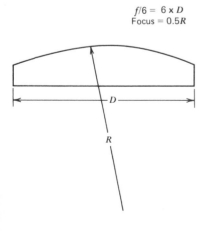

$f/6 = 6 \times D$
Focus $= 0.5R$

Fig. 24.2. Design of the aspheric element for a planointerferometer.

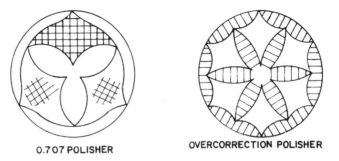

0.707 POLISHER   OVERCORRECTION POLISHER

Fig. 24.3. Configurations of the pitch polishers used for figuring the convex surface of the planoconvex lens into an aspheric.

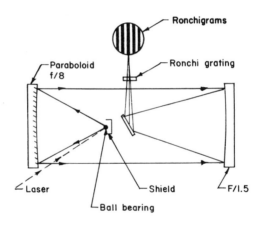

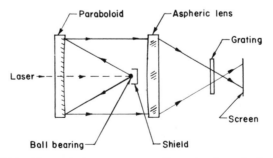

Fig. 24.4. Two optical geometry setups for testing the aspheric element.

encountered during optical figuring of the aspheric convex surface. (See Fig. 24.2.)

The surfaces of the planoconvex lens is generated in the conventional manner. Next the surfaces are fine-ground with 320 carborundum, followed by 225, 145, and 95 aloxide grits. Both surfaces are given a preliminary 70% (approx.) polishing. The flat surface is then figured to one-sixteenth fringe or better.

Pettit has proposed a formula for removing the spherical aberration at the 0.707 zone by polishing the planoconvex lens. The formula is

$$\Delta x = 0.0123 \, d/R^3 \tag{1}$$

where d is the diameter of the lens and R is the radius of curvature.

Substituting for the lens, $0.0123 \times 4/1728 = 0.000028$ in. or 1.3 wave.

Two types of pitch polisher are illustrated in Fig. 24.3. The 0.707 polisher is used to polish the aspheric curve in the convex surface of the lens. This precut polisher is run on the convex surface with a drag-type polishing action. The polisher is prevented from rotating by tying a wire on one of three hooks mounted at 120° intervals. A small eccentric 0.25-in. (6.3-cm) throw is set for the translating stroke. The pin must be weighted down or .the air pressure set for 2 lb (1 kg). After several hours time the lens is setup in one of the two optical processes shown in Fig. 24.4. The upper figure is similar to Norman's test. The Ronchi tester has a mercury light-reed source

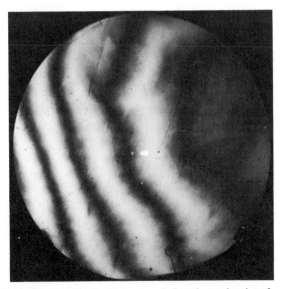

Fig. 24.5. Photograph of a 25.4-cm made for a large planointerferometer.

and the filter is for the mercury green line. The lower diagram shows a red laser ball source in use for testing the figured planoconvex lens with a paraboloid mirror for the parallel light source. The planoconvex lens is intended for a laser-illuminated planointerferometer. Figure 24.5 is an interferogram of this lens.

It is important that the precut configuration in the polisher be maintained, especially during drag polishing. Any overflow of the pitch into the precut areas should be removed at once with a shop razor blade. Drag-polishing is continued until a definite correction is observed. The polishers are then run on the convex surface in a conventional manner. Different profiling polishers can be made with the techniques outlined in Appendix 14 to meet the requirements of a null figure. This is shown in Fig. A14.2 for Ronchi testing and Fig. A14.1 for focograms or knife-edge testing.

3. CONCLUSIONS

The planointerferometer will pay for itself in time because the optical flat surfaces of flats and prisms are free of scratches and digs caused by proof planes. One other use not yet mentioned is that of making prisms by optically contacting master prisms to a copy being made and testing it interferometrically as described in Chapter 9.

BIBLIOGRAPHY

De Vany, A. S. "Reduplication of a Prism Angle Using Master Angle Prisms and A Plano-Interferometer," *Appl. Opt.*, **10**, 1371 (1970).

Pettit, E. "The Interference Polarizing Momochromator," in *Amateur Telescope Making, Book Three*, A. G. Ingalls, Ed., *Sci. Am.* 413–428 (1953).

Part IV
Optical Testing

25

A Universal Tester

A universal tester can be made cheaply or improvised with a conventional microscope tube (160-mm). It must be remembered that the length of the microscope governs the optimum design of its objectives and eyepieces. Other tube lengths cannot be substituted if the correction of spherical and other aberrations is to be maintained. A similar instrument was described by B. K. Johnson, author of *Optics and Optical Instruments* Dover, New York, but it was not given extensive use because of the multiple imagery caused by the thin-glass beam splitter; but since the advent of excellent thin plastic members with 45/45 (R/T) coatings this instrument has many capabilities for on-axis imagery testing.

1. MAKING THE INSTRUMENT

For any one wishing to build a tester the membrane pellicle is mounted on a special insert tube holder. The tube carries a small thick-walled insert on which a 45° diagonal is cut. A flat surface is lapped on the 45° cut which must be within one wave. This is easily provided by a charged cast-iron grinding tool on a dry charged surface. There must be a 1-mm bevel all around the 45° cut to anchor the membrane to the tube. (See Fig. 25.1.) The tubing is blackened nickel and the coating is removed by relapping. To keep the plastic pellicle from being damaged in transit, a second small sliding tube with a top cover is made to fit over the tube with the membrane. A small 0.625 in. hole near the top must be drilled to prevent the membrane from being broken by air-compression of the sliding assembly. Because this pellicle holder must be adjusted during assembly, it must be a sliding fit for adjustment in the microscope tube for axial rotation of the 45° coated pellicle and the proper distance of the pellicle from the pinhole source. This pellicle mount and its shipping protector is illustrated in Fig. 25.1.

The instrument make-up with its four interchangeable units is shown in Fig. 25.2. One sliding unit (A) consists of a 0.05-mm pinhole and is strongly illuminated with a small, fast condenser system and zirconium arc source. This unit fits in the tube (b) of the 160-mm microscope tube (a). The second

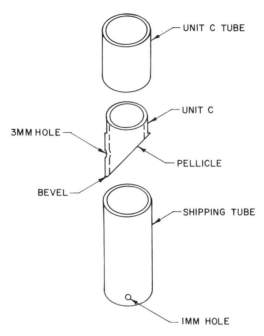

Fig. 25.1. A pellicle Unit A for holding the 45/45 (R/T) plastic membrane for the test instrument.

unit (B) consists of a sliding tube (c) which carries the 45/45 aluminized pellicle. This unit consists of a measuring bifilar with interchangeable eyepieces. These three units are placed in the main microscope tube at optimum foci of the instrument. A rotary holder with four microscope objectives, a leveling, height and squaring adjustment on accessories, completes the instrument. Figure 25.3 shows these adjustable units. It is good practice to have a 1× microscope objective and a series of three other objectives up to 10×. The low power permits the location of the reflective images by the test setups.

A number of adjustments must be made before the instrument can be used:

1. Look directly into the side tube to determine whether the coated pellicle is reflecting objects from the surrounding area (remove the eyepiece and objective). Rotate unit A for best direct-object viewing.

2. Illumination of the pinhole is done by observing with the 1× (80-mm) objective and adjusting the sliding unit (A) which is carrying the light source

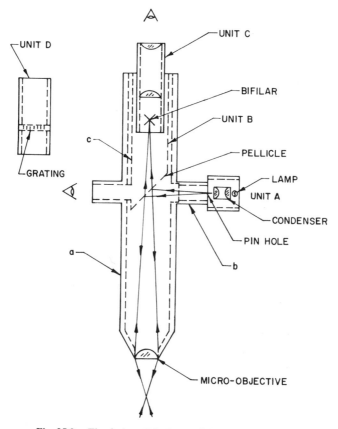

Fig. 25.2. The design of the four-unit Universal Tester.

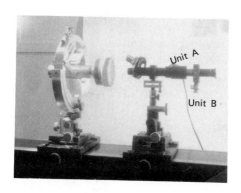

Fig. 25.3. The Universal Tester with several axial adjustments.

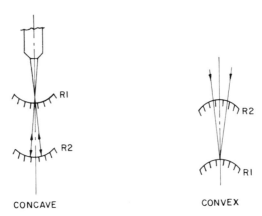

CONCAVE   CONVEX

Fig. 25.4. A reflective imagery setup for measuring and testing concave and convex surfaces.

until the brightest illumination appears. Unit A has been adjusted for the condenser system to focus the arc on the pinhole.

3. The two foci of the instrument must be in the same plane. Observe a ruled reticle with the instrument, using a moderate power 5× objective and a 10× eyepiece. The image of the projected pinhole and reticle rulings must be in the same focal plane; if not, slide unit A back and forth until the rulings and pinhole correspond. Record the diameter of the projected star image by comparing it with the ruled reticle. This should be done for each microscope objective on the rotary holder.

4. The star image must be located in the center of the field of the eyepiece and on the stationary cross hairs of the bifilar. This adjustment is made by sliding or slightly rotating unit B. Now lock the set screws of units A and B.

5. A number of methods will fix the squaring and leveling of the instrument to the optical rail, the best being an ellipsoidal concave reflector of two foci. One focus could measure 125-mm and the other (2000-mm); the leveling and azimuth rotation are checked in turn, at each focus. Do not change the height which has already been adjusted centrally at the working distance of the microscope objective. Observe the surface of the ellipsoidal mirror with an ink mark at its center. The working distance of a microscope objective, R_2, is shown in Fig. 25.4.

2. TESTING CONCAVE AND CONVEX SURFACES

Concave test plates can be measured by picking up the front surface (see left diagram in Fig. 25.4) image and finding it at the radius of curvature R_2.

Long radii of curvature on concave tests are difficult to locate even with a 1× objective. A low-power (0.3 mW) laser can be directed through the microscope tube from which the eyepiece and 1× objective have been removed and then to the reflector. Finally the laser beam is reflected back on itself. *Hint:* Cut a small hole (6-mm diameter) in a small piece of paper and place the paper on a stand in front of the laser. Allow the laser beam to pass through the hole and through the microscope tube onto the concave test plate. Adjust the holder of the test plate to reflect off the paper. This method does not require a second person to help the observer. Caution! Wear safety glass when working with a laser beam in this manner.

The observer should establish a procedure of measurement. Always have a 1× objective on the rotary holder to locate the reflective image at the center of curvature and then proceed to the higher power objectives. A 10× objective will serve for most measurements but be certain that the entire aperture of the reflector is illuminated.

Convex radii can be measured as shown in Fig. 25.4. Generally only short radii can be measured because of the limitation imposed by the working distance of the objectives. Some representative working distances are 80-mm (1×), 145-mm; 40-mm, 57-mm; 25-mm, 12.5-mm; and 16-mm, 5.5-mm. Longer working-distance reflecting objectives are supplied by the Ealing Corporation, 22 Pleasant Street, South Natick, MA 01760. It is also possible to obtain all the necessary units to make this tester.

The convex radii can be measured in two steps. First, the projected image is located at the working distance of the microscope objective. Second, the image formed by the convex surface is located. The radius of curvature of the convex surface is the difference between the readings of the first setting and the second, R_1 and R_2 (See Fig. 25.4.) There is one important adjustment. The first image must be expanded and made circular, especially for convex radius testing. [See Fig. 25.5(*A*) and (*B*).] If it is similar to Fig. 25.5(*A*), the holder is adjusted in the vertical plane; if it is similar to Fig. 25.5(*B*), the holder is adjusted in the horizontal plane. After adjustment the convex surfaces can be measured and tested by Ronchigrams, as can the concave reflectors. [See Appendix 14, Fig. A14.2 (reflective).]

A   B

Fig. 25.5. Off-axis image patterns due to misalignment of the optical element.

3. TESTING LENS SYSTEMS WITH STAR TEST

This instrument has the capability of testing image-forming systems of mirrors and lens by autocollimation with an aluminized flat. Because the artificial star image is demagnified proprotionally to the power of the objective used, it can give images as small as 0.005 mm. W. T. Welford has shown that in star image tests, when applied to image-forming systems, the sensitivity is such that it can indicate aberrations of $\lambda/60$ for sharp and $\lambda/20$ for slow variations. Welford is the author of Chapter 11 in *Optical Shop Testing*, edited by Daniel Malacara, Wiley, New York, 1978. This is an excellent book on optical shop testing for the optical engineer and optician and also the amateur telescope makers who can find in it areas of use. Welford illustrates many of the well-known optical setups for testing image-forming systems. This universal tester can project on-axis star imagery in autocollimation setups or in null testing configurations, such as the Hindle test for hyperbolic surfaces and the Ritchey-Common test for flat surfaces.

The chief versatility of this instrument for on-axis-forming systems is its ability to form Ronchigrams. This is unit D in Fig. 25.1, which consists of Ronchi gratings (2 to 10 lines per millimeter) in various tubes. It is interesting that an $f/0.7$ radius dome can be tested with a $40\times$ microscope objective (mount a microscope cover plate in front of the objective if required) and its surface quality found. Note that the interpretation of the Ronchigrams, is difficult, especially when higher frequency gratings are used. Always observe Ronchigrams first with the unaided eye to achieve better correlation. Again it is better to examine the outside-of-focus rather than the inside-of-focus Ronchigram. A Ronchigram with equal distance and parallel lines is a pattern that indicates excellent sphericity of a concave or convex surface and a well-corrected image-forming system of mirrors and lenses. This pattern does not change when the instrument is defocused on each side of the focal plans. The low and high areas, along with the degree of correction in the image plane of image-forming systems, are well covered in Appendix 14. This instrument does the work of the more complicated sphericity interferometer.

Aspherics being optically figured with this instrument are examined by the Holleran immersion null test. This novel method makes use of an immersion oil or liquid whose refractive index is equal to the eccentricity of the aspheric surface. The index is given by a simple formula:

$$n = (1 - e^2)^{1/2} \qquad (1)$$

where n is the index of refraction of the liquid that fills the full aperture of the aspheric mirror.

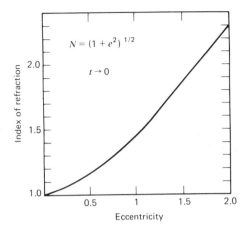

A typical optical test setup is shown in Fig. 25.6. A container that holds
the mirror and immersion liquid just covers the top of the mirror. A leveling
table helps to adjust the diverging and converging rays into the test instrument, where a Ronchigram interpretation is made of the progress of the
figuring. Because this is a reflective autocollimation setup, rotate the
Ronchigram chart as discussed in Appendix 14. When the defocused

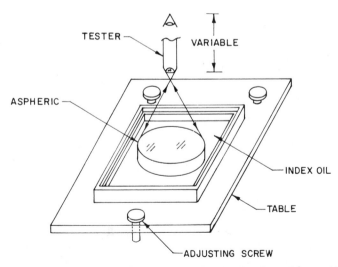

Fig. 25.6. The optical test setup of an aspheric mirror (concave) that uses a matching refractive index oil that equals the eccentricity value of the system involved. This is a null test.

Ronchigrams are straight and parallel (null), the aspheric figure is 0.1 wave or better.

The graph of the refractive index liquids to be used for the eccentricity values was devised by Robert T. Holleran and appears in the December 1963, issue of *Applied Optics*, pp. 1336–1337. Various indexes of refraction liquids can be purchased.

Another method is Norman's test for making secondary mirrors for Cassegrain systems. The eccentricity of the hyperboloid to fit the Cassegrain system is easily calculated by the following formula:

$$e = (p' + p)/(p' - p) \tag{2}$$

where p is one focus of the hyperboloid and p' is the second.; p is calculated from the formula

$$p = b + f_m/A + 1 \tag{3}$$

where b is the focal point back of the vertex of the paraboloid mirror, f_m is the focal length of the primary mirror, and A is the amplification ratio

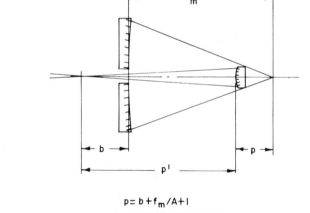

$$p = b + f_m/A + 1$$
A = AMPLIFICATION RATIO

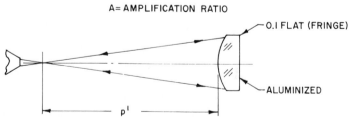

Fig. 25.7.   Norman's test for null testing a Cassegrain secondary hyperbolic mirror.

desired (generally from 3 to 5). This is shown in the upper diagram of Fig. 25.7.

Select from Schott's or O'Hara's glass catalog the nearest refractive index for the n_e value (the mercury line 546.6 nm). The tester requires a narrow band filter and a zirconium arc that has several lines near this wavelength.

The test is carried out by setting the secondary mirror at the distance p'. This mirror has the prescribed calculated radius of curvature, calculated by the following formula:

$$R_{cv} = (e - 1)p' \qquad (4)$$

where R_{cv} equals the radius of curvature of the secondary mirror, e is the eccentricity value of the system from Eq. (2), and p' is the back focal length measured from the secondary mirror's surface to the final focal plane. The back of the secondary mirror should be a semipolished flat with a 0.1 wave surface. This mirror is a true hyperboloid if the test is a null pattern. (See Appendix 15.) Norman's test was described in *Sky and Telescope* in the November 1957 issue, pp. 38–39.

4. SETTING THE OPTIMUM FOCUS OF PINHOLE OR RETICLE

This instrument has the further capability of setting the optimum focus of a pinhole or reticle of a collimator lens and of determining its axial chromatic aberration.

In Fig. 25.8(a)–(c) observe the testing instrument that projects the diverging rays through the pinhole or the reticle, which can be precisely set by referencing it to the autocollimating rays that reconverge on the reticle or pinhole. The peripheral edge of the pinhole, or one ruled line of the reticle, is observed simultaneously at the working distance of the microscope objective, which is coincident with the star image formed in a properly adjusted instrument.

It is extremely difficult to locate the returning autocollimating star image with a 10× or higher power microscope objective. As mentioned before, start with the lowest power 1× objective. For measuring axial chromatic aberration the higher power objectives ought to be apochromats or reflective objectives for the best correction of spherochromatic aberration. The lowest power objective locates the autocollimated image and the higher powers are rotated in turn. Because this artificial star image is demagnified in direct proportion to the power of the objective used, it can produce images as small as 0.005 mm.

It is possible, therefore, to find the pinhole source or the reticle of the collimator lens. Many optical benches have a series of various diameters on

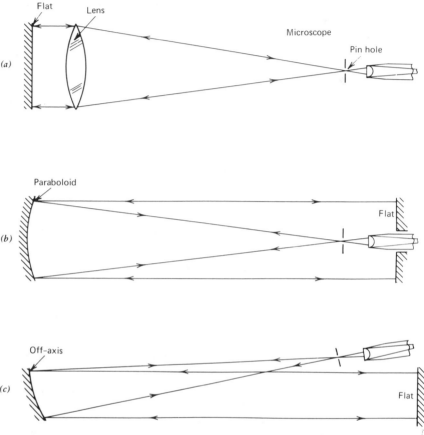

Fig. 25.8. Determining the optimum distance of the collimator lens projecting a star image through the pinhole or recticle. The autocollimated image is autocollimated back for viewing.

a circular disk that is mechanically centered. Each of them can be set by the procedures described. With this instrument it is also possible to rotate a series of different diameter pinholes electronically to determine their optimum diameter and electronically or by direct observation to determine the axial chromatic aberration (Δ_{chr}) for each wavelength by interposing narrow band filters and changing the light source. The Δ change is the repositioning of the pinhole source from a preselected wavelength.

Care must be exercised, when using an autocollimating flat, to set up off-axis and on-axis parabolic mirrors. Do not mistake the radius of curvature image for the autocollimating image. In the off-axis mirror it is possible to place the autocollimating flat well out of range. [See Fig. 25.8(c).]

The positioning of the on-axis parabolic mirror is more difficult because of the near coincidence of the autocollimated image and the radius of the curvature image. These images can be moved farther apart by inserting the test instrument in the hole in the flat, well past its aluminized surface [see (b)]. If a flat with a hole is lacking use one without it. Here the flat is positioned to cover the 0.707 zone and the outer edge; the instrument then picks up the autocollimated image. For another check place a Ronchi grating in the test instrument and observe the outside-of-focus Ronchigram. If the Ronchigram has equidistant and parallel lines, the autocollimated imagery will be seen. Any curved lines should be viewed with suspicion because they are the radius of curvature patterns if the optical figuring is well corrected. The use of the smaller flat, which scans the 0.707 zone and the edge, causes a foreshortened effect and not enough area to determine whether the correct image is being observed. Scan across the whole area of the collimator reflector to verify that the correct image is being viewed.

BIBLIOGRAPHY

De Vany, A. S. "Making a Schmidt-Cassegrain Optical System with a Flat Field," *Sky and Telescope*, June 1965.

De Vany, A. S. "A Universal Tester," *Appl. Opt.*, **5**, 867–888 (1966).

Hindle, J. H. "The Compound Telescope—Cassegrainian and Gregorian Types," in *Amateur Telescope Making, Book One*, A. G. Ingalls, Ed., *Sci. Am.* 226–228 (1953).

Holleran, B. A. "Immersion Null Test for Aspherics," *Appl. Opt.*, **2**, 1336 (1963).

Johnson, B. K. "Short Radius and Small Diameter Surfaces," Dover, New York, 1960, pp 179–180.

Norman, B. A. "A New Test for Cassegrain Secondaries," *Sky and Telescope*, November 1957.

26

The Foucault Knife-Edge or Schlieren Test

Marus Brown

Among the many methods of testing optics no single one will fit the demand for all requirements. Each has its limitations and qualifications.

Although the Foucault knife-edge was the first laboratory method invented for testing front surface mirrors, its use was confined to scientific institutions until recent years when it became widespread among amateur telescope makers.

It is certain that Clark's, Brasher, Fitz, and others in the United States used the test after 1856 when Foucault published it. Other famous European opticians such as Fraunhofer and Cauchoix in France and Tully in England were making achromatic refractors of large aperture before 1856, and before the knife-edge test, but their methods are lost in the limbo of secrecy. It is thought that they used Sharp's method by defocusing the eyepiece which gives star images of equal diameter and intensity in the expanded ring system of an objective free of spherical aberration.

Little is known of the first reflecting telescope made by Newton (1668) and others and their testing methods other than an examination of the star image formed by the mirror at night.

Herschel (1738–1822) wisely built his mirrors of moderate relative f/ratio ($f/10$ to $f/20$). His 48-in. mirror was a $f/10$ ratio that would require only a slight degree of asphericizing from a reference sphere. Most of the telescopes he sold were of smaller aperture (6 to 14 in.), $f/10$ or greater, and a spherical mirror that would meet the most critical needs.

These methods were used by Newton until about 1856 when Foucault invented the knife-edge which made the Schlieren effect visible to the naked eye.

Foucault's invention used an artificial star at the center of curvature and to one side of the axis of a spheroid mirror, which formed an image by reflection of the artificial star at the center of curvature but to one side of the axis of the spheroid mirror. This exposed the image, and the image and light rays could be intercepted with the knife-edge at the center of curvature. A spherical mirror with surface inequalities, when placed in front of this setup, would reveal the Schlieren effect to the otherwise unaided eye.

Under many conditions the knife-edge is sensitive to 0.1×10^{-7} or $1/10,000,000$ in. which is sensitive to the most critical requirements and has no size limitations.

Today most large telescopes could not be made without the knife-edge test. Although the knife-edge is indispensible to large optics, its application to small optics is limited because considerable experience is necessary to interpret the observed inequalitities.

Since 1970 Kent's single wire test has been used extensively for large telescope mirrors by Loomis at the Optical Center of Arizona University. (See Chapter 28.) Don Hendrix applied the star test for the final correction of the 200- and 120-in. mirrors at Mt. Palomar and Lick. See Appendix 16.)

An important feature of the knife-edge test, regardless of size in most cases, is that the entire area under test can be seen at a single glance. Consequently surface irregularities, such as astigmatism or an aspheric condition, can be readily detected.

Another important factor in the Schlieren test is that the object can be supported in one of its most stable conditions. Deformation resulting from improper support will ruin or deform the most precise optical surface and make it impossible to determine surface quality.

To overcome the deformation caused by the cell mounting, most mirrors since 1939 have been married to their cells and have worked with them in place with vertical towers built for testing. Parson-Grubb and the Optical Center at Arizona University use test towers, and Don Hendrix optically figured the 200- and 120-in. mirrors in the observatory dome.

The knife-edge test can be of great value in testing front surface optical elements that do not involve refraction because it is free of achromatism and white light can be used to give a black and white contrast. If, however, the system involves refractive elements, it can be readily adapted to the use of monochromatic narrow band filters and becomes valuable as a collimated source of parallel light.

Many optical elements can be tested with the knife-edge: spheres and plane, parabolic, ellipsoidal, and hyperbolic conic sections, among others. It can survey homogeneity, striae, bubbles, seeds by the Schlieren effect, Ronchi tests, and polarization.

It finds widespread use in the study of aerodynamics and window tunnels.

1. REFLECTION FROM A PLANE OF CURVED SURFACES

The principle of the knife-edge involves light rays as a lever arm that magnifies surface irregularies or inequalities; the surface tested functions as the fulcrum of the system.

Although it is desirable to use a lever arm as long as possible, there are some limitations, such as vibration, homogeneity of the air in the test path, and the ability of the eye to resolve the surface inequalities. The diameter of the work to a great extent governs the length of the lever arm. As a rule the larger the work, the longer the lever arm. This is due to the multiplication of the lever arms of the light rays: the longer the lever arm, the greater the sensitivity.

The principle of the knife-edge or Schlieren test is a simple one: the light ray appears incident to a plane surface at the same angle but on the opposite side of the mirror, where S equals the origin of source, F is a plane surface, I_1 is the image, i the angle incidence, and r is the angle of reflection. (See Fig. 26.1.) The light rays are indicated by the lines with arrows.

For practical purposes when discussing the behavior of light rays, assume that they travel in straight lines unless otherwise specified. Therefore say that sin i = sin r. It is important to remember this rule (Snell's law) because it is important to the sensitivity of the Schlieren test, as the reflected ray is deviated by twice the surface error.

The behavior of a light ray reflected from a plane surface applies to a curved surface. This is because a light wave is extremely small, that is, of the order 21×10^{-6} in. or approximately the wavelength of the green line in the visible spectrum. The human eye is more sensitive to the green than any other color. Therefore think of a curved surface composed of a multitude of tiny plane facet surfaces with areas of $21/1,000,000$ in. lying of a curve with normals of each of these facet planes facing the center of curvature.

The analogy of small planes on a curved line can be applied to surfaces of revolution which are the primary interest here. Because it would be impossi-

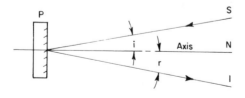

Fig. 26.1.   Reflection at a plane mirror.

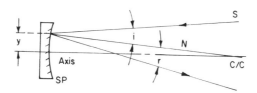

Fig. 26.2. Reflection at a curved surface.

ble to make light-ray tracings on a 1-to-1 scale, all tracings will be greatly exaggerated. It is necessary, however, to learn to think in terms of wavelengths. A fine line 0.001 in. would be 476 wavelengths for a mercury green line.

Figure 26.2 illustrates the behavior of a single ray reflected from a curved line: the curved line at SP, the normal N to SP at point y, s the origin, I_1 the image termination, i equal to the angle of incidence of, and r the angle of reflection. The light rays are indicated by lines with arrows. Here a single ray is reflected from a curved surface that is identical to the reflection from a plane surface for most practical purposes.

2. A SPHERICAL SURFACE OF REVOLUTION

In Fig. 26.3 there is a large circle D that represents a sphere or hollow ball; R is its radius. Assume that the inside of this sphere is a perfect spherical surface of revolution and that the point c/c (center of curvature) is the exact center of the sphere. Now suppose that a small spherical source of light is placed at the center of curvature, as indicated at S, and could project light rays in all directions through 360°; all rays would strike inside the sphere at normal incidence and would be reflected back along the same path to the point of origin. This would occur because the inner surface of the sphere is a spherical surface of revolution, indicated by the shaded area.

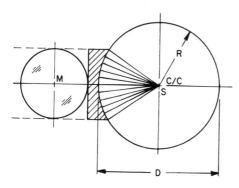

Fig. 26.3. Reflection on a sphere with the source at the center.

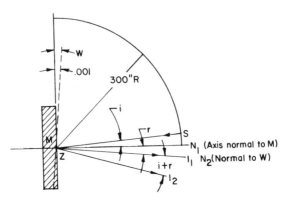

Fig. 26.4. The effect of slope errors on reflection.

The shaded area in Fig. 26.3 is a side elevation section of a zone sector with the apex at s and is typical of the conditions that will exist in the knife-edge to test spherical surfaces of revolution.

The reason that surface errors are observed in any spherical or spheroid mirror is due to slope errors magnified by the use of light rays. (See Fig. 26.4.)

Here a ray of light with its origin at S is incident on a curved surface z and then reflected to I_1; angle r = angle i. The included angle $I_1 = r + i$. Now introduce a change in slope at z, as indicated by the lines with arrows; call it W and trace N_2 which is perpendicular to the change in slope.

Because S remains stationary the image will move from I_1 to I_2. For the ray from S to Z, I_1 = angle i + angle r, and I_2 = 2(angle i) + angle r, where the change in slope at w = angle i. Because angle i + angle r are constant, the shift in the image from $I_1 = 2(\sin WR)$.

The angle I_1 formed by Sz represents the normal condition when testing with a knife-edge formed by I_1, z. I_2 represents the multiplication of the change in slope w.

To determine the size of this magnification consider an example. Let the distance from S to s equal R (300 in.) and let the change of slope at z be 0.001 in. as indicated at w. The magnification in inches is $2wR$ or $2 \times .001$; 300 = 0.6 inch for the shift from $I_2 = I_1$.

The problem just outlined was used for illustration only, but now use the same problem to consider the change in terms of wavelength and let the change in slope at $z = \frac{1}{2}$ or $\frac{1}{4}$ wavelength or approximately 0.000,005 in. Then $2(wR)$ would equal $2 \times 0.000,005 \times 300 = 0.003$ in. or 1363.6 wavelengths. This illustrates how, in this instance, a small change in slope can be magnified by optical levers to give a quantity visible to the unaided eye.

Most methods of testing optics magnify small inequalities to quantities that the eye can resolve.

3. THE KNIFE-EDGE SCHEMATICALLY

To consider the function of the knife-edge see Fig. 26.5, again greatly exaggerated, the source S, a pair of condenser lenses, and a small aluminized flat or right-angle prism near the axis of the system. The second lens from the source focuses the pinholes of several wavelength diameters just in front of the diagonal or prism which directs the cone of light toward the mirror M. Note the diverging and converging light paths in the setup.

It is important to remember that the included angle of the incident divergent light must be as great or greater than the included angle formed by the reflecting surface M and its center of curvature at c. The reason for this is that if the included angle of the incident light were smaller than the included angle of surface M being examined surface M would not be entirely filled with light and only the illuminated part would be seen.

The long dashed lines in Fig. 26.5 are the normals that have their origin at the center of curvature and touch the surface M at the point that intercepts the light rays from the source. The incident and reflected rays are in solid lines with arrows.

The knife-edge (Fig. 26.5) is shown in three axial positions. Consider the position c/c; if the knife-edge is moved *in* (indicated by arrow) to intercept the reflected rays, the surface M would start to shade uniformly because all the light from M is concentrated in a small area at c/c. This, of course,

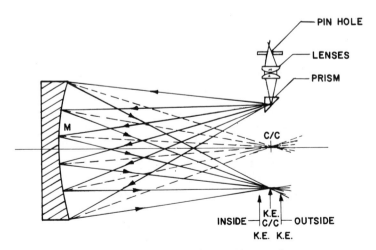

Fig. 26.5. The knife-edge test. Not to scale.

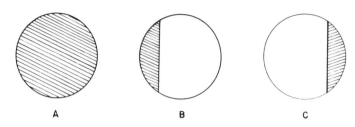

Fig. 26.6. The observing surface M for the knife-edge test.

assumes that M is a perfect spherical surface of revolution and the area at c/c is 0.005 in. in diameter. At a partial cutoff the surface will assume a steel gray appearance; then as more rays are intercepted it will look black. [See Fig. 26.6.] Now, if the knife-edge is moved to the outside of the c/c in Fig. 26.5, a study of the figure will show that rays from the opposite side would be intercepted and the area would begin to shade over. The balance of surface would be light. [See Fig. 26.6(B).]

If the knife-edge were moved out to a position *outside* c/c (see Fig. 25.5), the shadow would travel *in* from the opposite side, as at (C) in Fig. 26.6. It can be pointed out here that this is the way to determine when the knife-edge is at the center of curvature of the surface M. In making the cutoff with the knife-edge, the shadow appears on the left in the horizontal plane and the knife-edge is inside the center of curvature. If it appears from the right, in from the horizontal plane, the knife-edge must be moved in or out until the surface shades evenly and simultaneously; it will then be at the center of curvature. [See Fig. 26.6(A).]

4. ALIGNMENT OF THE KNIFE-EDGE

To align the knife-edge there are rules that the observer must follow to eliminate errors in the test. (See Fig. 26.7.) Note the dashed line in the figure designated as a "line of sight" in all planes through 360° rotation. [See (*b*), (*c*), and (*d*).] The knife-edge is double-bladed so that cutoffs can be made in either direction for all diameters although it swings only through 180°; also note Z–Z in Fig. 26.7(A). If the knife-edge intercepted the rays at one of these angles, finding the focus or c/c would be difficult and the correct interpretation of the shadows, impossible. The shadows would be accentuated from the direction of the cutoff and travel in toward the center of the mirror. The knife-edge should be adjusted to cut the line exactly at right-angles to Z–Z.

Included in this set up is a swinging eyepiece holder and a 10× eyepiece of good design. The eyepiece is set in place between the eye of the observer and the knife-edge. To adjust the image in the eyepiece set one blade of the knife-edge mount in and out each side of the c/c in the horizontal and vertical plane. Good adjustment is found when the image of the slit or pinhole is centrally located in the eyepiece. The in-and-out images should be equal in diameter (expanded star image for pinhole and diffraction lines in case of a slit) for a truly spherical mirror.

A desirable knife-edge mount includes a swivel base, a longitudinal slide, a right-angle slide, and a vertical slide. All but the swivel base should be screw-operated. The knife-edge should be operated with a fine thread and a half-nut. The longitudinal screw should be geared to a dial counter to read in one-thousandths of an inch. The screw is held in the half-nut with a blade spring which permits a rapid interchange of positions.

The sketch (Fig. 26.8) illustrates the principle components of a satisfactory mount for the knife-edge. All slides move freely with a minimum of play and lost motion in the screws. The large circle on which the knife-edge rotates is desirable because it reduces the trouble encountered in finding the image. It should have a 3-in. aperture. A small disk is attached to the nose of the eyepiece.

One circular disk should be made up with three pinholes and an extended light source of larger diameter. The pinholes measure 0.002, 0.004, and 0.008 in.; a $1/32$-in. slot with a frosted screen, which permits interchangeable viewing of the Ronchigrams thorugh the Ronchi grating is included. This interchange from focograms to Ronchigrams permits the correlation of findings. (See Appendix 14.) These four sizes of pinhole will fullfill the requirements for most purposes.

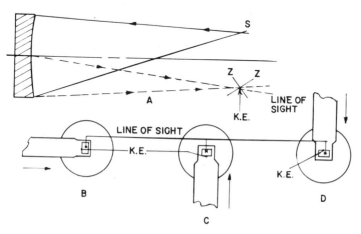

Fig. 26.7.   Alignment of the knife-edge test.

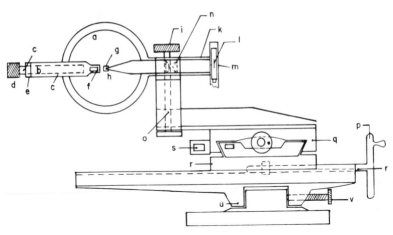

Fig. 26.8. The knife-edge mount.

A 108W General Electric movie projector lamp is satisfactory because it provides a large source and ample light for most work. A large light source is necessary when a wide-angle optical system, perhaps an $f/1$, must be observed. The eyepiece is not shown, but it should be hinged from the lens housing support and free to be swung out of way. For general use a $10\times$ Ransden eyepiece is also adequate, but in some cases a higher magnification is required.

Much confusion is caused by determining on which side the light source is located and in what direction the knife-edge is moved horizontally. Brown, Porter, and Texerau set the knife-edge and light source differently. Brown moves it and its attached light source left-to-right; the light source is on the right. (See Fig. 26.8.) Porter moves the knife-edge right-to-left and his *stationary* source is on the right. Brown and Porter state that the slopes observed should be illuminated on the *opposite* side of the cutoff of the knife-edge blades. Texerau makes a rule to have the stationary light *always* on the left-hand side facing the mirror and inserts the knife-edge blade right-to-left. His method of interpreting follows Brown and Porter: the rule is to assume that the light comes from the opposite side of the blade's cutoff when testing any reflective combination of optical elements; for example, in reflective optics for (1) figuring a secondary mirror of a Cassegrain system with its parabolic mirror in an autocollimation setup and using a flat; (2) figuring a primary mirror of a Maksutov system with its corrector lens and aluminized stop fixed in an autocollimation set with an optical flat; (3) a concave reflector setup for testing at its radius of curvature.

Texerau's *rule*: always assume that the light is illuminating the slopes from the opposite side of the knife-edge (for testing an objective lens the rule is

reversed when the lens is set up in parallel). All three optical setups for reflective optics testing follows his rule. Even Porter's setup has the knife-edge on
the left-hand side and the stationary light source on the right. Brown's setup
can lead to some misunderstanding because he has a slot blade in his knife-
edge that can be translated in both directions, thus leading to cutoffs in the
wrong direction in relation to the light source and the blade of the knife-edge.
Always use the right side of the slotted knife-edge blade.

For transmitting glass optics, objectives, corrector plates for Schmidts, correcting lens for Maksutovs, aspheric lens, and so on, think of the light as
behind the knife-edge blade as it makes its translating cutoff; in other words,
the shadows are cast forward.

5. TESTING FRONT SURFACE SPHERICAL MIRRORS

When testing with the knife-edge, consideration must be given to the size of
the pinhole or slit source. Theoretically, a 1×10^{-4} or a 5×10^{-4} in. is
necessary for the most minute details. Pinholes of this size, however, are not
practical for general shop use and are required only for metrology.

For general use a pinhole source of 0.004 to 0.005 in. will satisfy most
requirements for detail and surface quality. A source of this size gives sufficient light for testing single or multiple surfaces without excessive eyestrain,
and with experience accuracy of 0.1×10^{-7} in. is possible.

Instead of a circular source, some opticians prefer a slit source that will
increase sensitivity and give more light.

There is one disadvantage, however. When testing for astigmatism with the
eyepiece, the ovallike, out-of-focus ring patterns with the defocusing of the
eyepiece are not obtainable.

6. DIFFRACTION

With the knife-edge diffraction is always present and visible as a series of
narrow, bright, circular rings around the periphery of the work under test.
Diffraction is of no particular importance in testing with the knife-edge;
however, its presence may confuse the beginner. As a rule, if a reflecting
surface is viewed with full aperture, a series of narrow, bright, encircling
rings of light will appear around the periphery. The innermost ring will be
the brightest, with the outer rings gradually diminishing in intensity. This is
diffraction; it is caused by the sharp rounding of the extreme edge or an
edge that is turned. This causes the light rays to diverge abruptly. The
intensity of the first ring originates near the crest of the slope and its area is
the greatest; the other areas diminish in size and the rings, in intensity. This

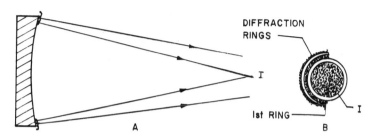

Fig. 26.9. Diffraction effect in the knife-edge test.

sudden divergence of the edge rays causes them to separate from the light in the image. [See (*I*) of Fig. 26.9.]

It is possible to discern diffraction rings by using the iris of the eye as an aperture stop. Move the eye around while viewing the illuminated surface. In certain positions diffraction rings will be recognized as a number of faint circles of light on one side of the mirror and just outside the bright area. (See (*B*) in Fig. 26.9.)

These straight-line diffraction fringes are also visible when the cutoff is made with the knife-edge and always appear on the side opposite the direction of the cutoff.

These shadow fringes can be seen with the knife-edge inside the focus following the blade and are valuable in determining up-or-down-turned edges. If their ends tend to circle to the right-hand side, the edges are turned down; if they tend to circle the periphery to the left, the edges are turned up.

The observer soon learns to disregard the diffraction rings in the knife-edge. For those who may wish to study the theory of diffraction most textbooks on optics contain adequate discussions.

7. PARALLAX

Parallax is another inherent characteristic of the Foucault test. Many attempts have been made to develop an on-axial system that would eliminate parallax but without much success. Consequently the observer must learn to live with it. It is possible, however, to offset the effects of parallax in locating the crest of the zones. This subject is discussed under the treatment of the zones in which it is detrimental.

Parallax is the result of separation of the source and image that produces a small angle between the incident and reflected beams of light.

Since the advent of strong membrane pellicles with their 45/45 reflecting and transmitting coatings and halogen light sources, on-axis knife-edges can be designed. Mirrors with low f/No. and an R/D of 3 or lower (where R is the radius of curvature and D, the diameter) require this on-axis tester.

8. ZONES

In the discussion of spherical mirror testing various inequalities and irregularities are considered individually to avoid as much confusion as possible. As a rule at least two and often more irregularities are found in a system. A zonal condition frequently encountered is astigmatism which has a number of forms and many causes. These are treated under zone testing. The zonal condition that appears most frequently is the turned-down edge; therefore it is discussed first and for simplicity it is assumed that the balance of the spherical surface is a perfect sphere, free of astigmatism.

Again it is necessary to exaggerate the zones and defects in the illustrations.

It might be well at this stage to explain in more detail why surface irregularities are visible with the knife-edge. One of the simplest examples is looking down from an airplane on a range of hills with the sunlight striking at an angle. (See Fig. 26.10.) All slopes receiving direct sunlight would appear shaded. The contrast would make the hills stand out in pronounced relief and accentuate the change in slopes; this contrast is not very sensitive to elevations, although elevations can be calculated.

The explanation is that the eye is influenced by the amount of light it receives from a given area, and as a result a long slope that, at its highest point, may not depart from the true sphere by more than a small fraction of a wavelength will appear as an extremely high zone. On the other hand, a narrow slope at its highest or lowest point will be negligible. [See (L) and

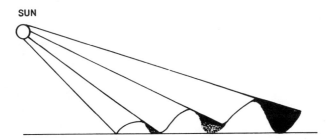

Fig. 26.10. The shading of a series of hills illustrating the effect of slope differences under oblique illumination.

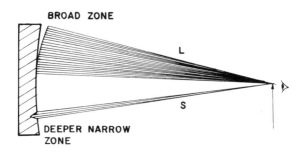

Fig. 26.11.   The effect of zonal size on the knife-edge test.

(S) of Fig. 26.11.] At (L) there is a large amount of light and at (S), a small amount. Experience can overcome this problem to a great extent.

Again, when it is necessary to use the knife-edge the surface quality is such that these extreme zones are not permitted and the surface quality improves. The misleading effects of these extreme zonal conditions disappear when a near perfect spherical surface of revolution is produced, say, better than $\frac{1}{20}$th wavelength. The general curvature can be perfect, however, yet badly zoned.

9.   TURNED-DOWN EDGE

The proper setting for the turned-down edge shown in Fig. 26.12 is at the center of curvature of the central zone.

In any test of a front-surface mirror there is one rule that must be remembered and that is to visualize the origin of the light as opposite to the

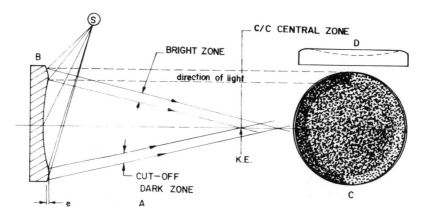

Fig. 26.12.   The effect of a turned-down edge in knife-edge testing.

direction of the knife-edge. Recall that Brown makes his cutoff left-to-right; therefore, reflective mirror slopes are illuminated from the right-hand side.

Actually, of course, the light source is near the image (on the left) at the center of curvature. It is less confusing, however, to think of it as *light* striking the surface at grazing incidence and from a direction *opposite* the cutoff. (See Fig. 26.12.) Remember, in reflective optics this is a fixed rule for the correct interpretation of lights and shadows. In Fig. 26.12 the knife-edge is at the center of curvature (c/c) of the mirror's central area. Some of the light from this area is intercepted. All the light from the left edge is intercepted and it is dark, whereas the light from the right edge enters the eye past the knife-edge and it is bright. (See Fig. 26.12.) Figure 26.12(D) is presented as flat because at the c/c of this area it looks flat. Note also (C) in Fig. 26.12. The vertical dividing line rotates as the knife-edge rotates at zones of revolution; they are reversed for each half of the surface at (C); in other words, if a bright zone appears on the right extreme edge, the zone will be dark or complementary on the opposite extreme edge. The turned edge is a zone of revolution; therefore the left side is dark and the right side, bright. The cutoff is again left to right.

9.1. Rules to Remember

1. If the knife-edge cuts off from left to right (the observer is facing the mirror), light will strike the surface from the right. If, however, it cuts from right to left, the reverse will be true. All lights and shadows are reversed.

2. The diameter of the mirror parallel to the knife-edge blade is the dividing line of the lights and shadows and a bright area on one-half will be a dark area on the other.

3. All slopes facing the light will be bright; all slopes facing away from the light will be dark.

After the observer becomes familar with these rules, it is not necessary to analyze them because interpretation becomes automatic. One situation that sometimes develops, however, must be guarded against; that is, the reversal of the interpretation of lights and shadows. The cause is unknown, but it is more prevalent among people with poor eyesight. This reversal will cause a bright area to appear dark and a dark area to appear bright. Generally, if the eyes are relaxed for a few minutes the condition will clear. Unfortunately, the observer is sometimes unaware of it. Therefore any abrupt change in the apparent surface quality should be viewed with suspicion and carefully analyzed before attempting correction.

Again assume that the central area is a good spheroid with a turned-up edge. The knife-edge is cutting from left to right. (See Fig. 26.13.) Note that the rays from the right side are intercepted by the knife-edge when it is at

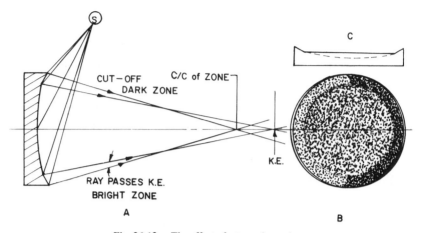

Fig. 26.13.   The effect of a turned-up edge.

c/c of the central area and the rays from the left edge pass it. Consequently the right edge is dark and the left edge is bright. An up-edge slope or a turned edge in relation to the central area is indicated opposite to Fig. 26.12. Here, again, part of the light from the central area is intercepted and part passes the knife-edge; hence the central area has a lighter shading than the edge but is uniformly shaded.

In these first examples it was assumed that the change in the slope was abrupt and that no zones accompanied the transition. This is rare, however. Invariably a groove will appear just inside the change of slope. (See Fig. 26.14.)

In (A) of Fig. 26.14 the ray tracing from a spherical surface with a turned down edge and a groove inside that edge is shown; (B) is a shadow pattern

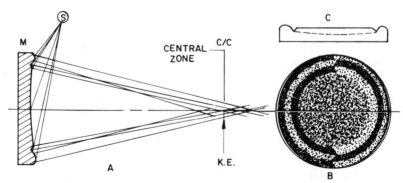

Fig. 26.14.   The groove typically accompanying slope change.

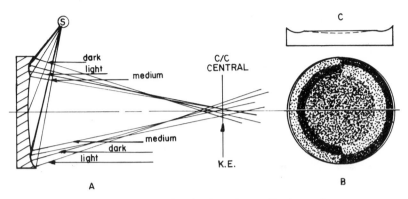

Fig. 26.15. A typical situation in testing a reflecting surface.

and (*C*), a side elevation. The sketch at (*C*) also indicates a flat surface instead of a curve. This, of course, is not true if the surface is spherical. On the other hand, if the knife-edge is at the *c/c* of a spherical surface, this surface will look flat to the eye. The dotted line indicates a curved surface. Figure 26.14 is not an exception. In fact, it is one of the most frequently encountered conditions in optical work. Actually, Figs. 26.12 and 26.13 are exceptions.

Again the rule rather than the exception is emphasized in Fig. 26.15. The ray tracing at (*A*) and the shadowgram at (*B*) will be self-explanatory if the cutoff is made left to right.

The grooves that occur are hard to remove and test with the knife-edge when it is necessary to measure elevation of a slope. In most cases they are quite narrow, making it difficult to determine their *c/c* relation for surface balance. Every precaution should be taken to prevent their occurrence.

Attention should be called to the fact that all the zonal conditions discussed have been based on the placement of knife-edge at the c/c of the central area which is assumed to be a spherical surface of revolution. This position however, is not always the most desirable when a decision must be made in regard the extent of local treatment that would be most advantageous. Also, much depends on the method of local treatment. In fact, because of the many variables involved, this question is best answered by experience.

10. SPHERICAL ABERRATION

All zonal conditions produce spherical aberration by preventing the light rays from converging to a point determined by object and images sizes,

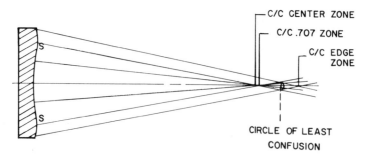

Fig. 26.16. The circle of least confusion resulting from spherical aberration.

focal length, and aperture. Rays that do not fall within this area are called aberrating and produce spherical aberration. In every optical system there is some spherical aberration. Therefore the area near the focus or center of curvature in which the diameter of the aberrating rays are the smallest is called the circle of least confusion. The knife-edge would be positioned at this point as a mean c/c in most instances. (See Fig. 26.16.)

The area indicated by dotted lines is the circle of least confusion.

To establish the knife-edge in this area three steps must be followed:

1. Focus the eyepiece on the edge of the blade.
2. Adjust the carriage until the sharpest image is seen in the eyepiece.
3. Swing the eyepiece to the side and observe the cutoff.

It must be remembered that these are not hard and fast rules to follow under all circumstances, but they do provide an average surface condition in the area of least confusion. When a spherical surface is within the tolerance of radius measurements (e.g., on test plates), an average or mean cutoff is established.

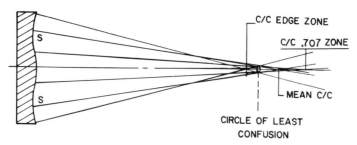

Fig. 26.17. The circle of least confusion and zonal foci for a turned-up edge.

It is assumed that the area S–S in Fig. 26.16 is spherical and has the desired radius. Also, note that the circle of least confusion is to the right of the c/c.

If the configuration on the mirror has a turned up edge, the circle of least confusion occurs inside, or to the left of, the desired radius of curvature. (See Fig. 26.17.)

After studying Figs. 26.16 and 26.17 it will be obvious that the position of least confusion can be elusive in relation to the desired radius of curvature, and the advisability of using this setting depends on the circumstances.

11. CENTRAL ZONES

In the discussion of zones only the edge has been considered. Now think about surfaces on which the outer edge is perfectly spherical and the central area is zoned. First consider the high center. (See Fig. 26.18.)

Figure 26.18 shows a spherical surface of revolution and its C/C at c/c. The central area is convex. With a median cutoff of the knife-edge from left to right the surface will look like Fig. 26.16. Here the knife-edge is at the C/C of the outer edge but inside the C/C of the central area. Note the position of the circle of confusion.

In Fig. 26.19 the same condition prevails but the center is concave. Note that the focogram at (B) in Fig. 26.19 has the same shading in the outer area, but in the central area it is reversed. This would apply to a medium cutoff, left to right. here the knife-edge is at the C/C of the outer zone but

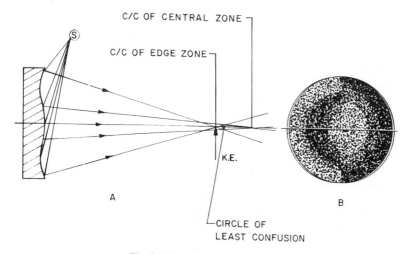

Fig. 26.18. The high center zone.

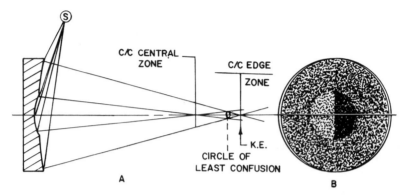

Fig. 26.19.   The concave or low center zone.

outside the C/C of the central zone. Also note the position of the circle of confusion.

With reference to a *medium cutoff*, this is a somewhat vague description because no standard or precise distance of the cuts made by the knife-edge in the cone of light has been established.

It would be difficult to set up a standard because of the variation in spherical aberration (circle of confusion) and the observer's opinion of the proper position for the knife-edge. The relation between the actual and required radii of curvature must also be considered. Consequently much depends on the observer's experience.

If the novice has survived the treatment so far, it may be encouraging to say that a perfect surface of revolution (spheroid), free of zones and astigmatism, and to the exact radius of curvature (test plate) has been achieved. It is not difficult to recognize a perfect surface. The imperfect ones cause the trouble.

The zones discussed so far have been simple in form; the multitude of variations and magnitude of zones and astigmatism have been disregarded.

It should be obvious from the discussion that prevention of zones is extremely important and every precaution should be taken to eliminate them.

One of the more important precautions is support of the work, the quality of which should be equal to the surface quality required. The polisher should be symmetrical in form and density. Avoid constant repetition of a *given stroke*. Note especially that a well-charged polisher and little water will help to produce smooth surfaces and sharp edges.

Although the problems of zones have not been covered in their entirety, the basic discussion should be helpful.

12. ASTIGMATISM

Astigmatism is undoubtedly the most serious problem in precision optics confronting the optician. Its causes are manifold and in most cases unsuspected until observed in the test. These causes include improper support, lack of symmetry in working the polisher, temperature effects, and variation in the density (homogeneity) induced by improper mixing of the glass which produces striae, seeds, and bubbles.

Support is important. It must show the same quality as that required for surfaces and must be stable to prevent rocking. The entire area must be supported; it must not be allowed to to rest only on a series of points such as wax and pitch buttons. It must be remembered that optically most glasses are extremely flexible.

Brown, it must be remembered, is describing the working of glass mirrors, ranging from 24 in. or larger, and not smaller pitched blocked elements for mass production.

Symmetry in working requires a constant change of the tool in relation to the work. Avoid unequal pressures unless for a specific purpose.

Temperature effects caused by handling, washing, sudden changes in temperature, and high-speed polishing can cause deformation.

Variations in the density of the glass also cause hard and soft areas to polish away at different rates.

All of these factors must be taken into consideration in processing high-quality surfaces.

Astigmatism is difficult to explain verbally or with sketches because of its many complex forms especially when associated with zonal inequalities and irregularities. Consequently, if the following discussion makes it possible for the observer to recognize it, it will be helpful. Prevention is the best cure, however.

A simple method of detecting astigmatism, no matter how bad the surface of the mirror, involves the eyepiece test, which is performed by defocusing the eyepiece inside and outside the best focus at a distance of 1/32 in. or so. If astigmatism is present, the diffraction pattern of rings will be small ovals on each side of the best focus. Information can be taken from these elliptical images and the planes of the ovals. Generally, the major axis of the ovals will be 90° but they can be rotated at any degree. The astigmatism, therefore, is determined by the two rotated planes in the ring pattern.

Among the many forms in which astigmatism appears is a definite cylindrical surface that extends edge to edge in one diameter except when there is deformation resulting from improper mounting or the lack of sufficient thickness in all diameters. The last two instances are not due to the way in which the work is done but how the work is supported, particularly when there is no uniform rigidity.

Modern glass blanks supplied to amateur telescope makers have a one-sixth (1/6) ratio of thickness to the diameter of the mirror and can be obtained "fine-annealed." The 200-in. mirror on which Brown worked was a prototype model with core-shaped recesses to make it lighter (see Appendix 15). The 60-in. space telescope for the Space-Shuttle has a lightweight egg-crate mirror and is in Baker's tertiary system.

13. QUALITATIVE AND QUANTITATIVE ANALYSIS OF ASTIGMATISM

For the purpose of illustration in the present discussion assume that a cylinder appears on one diameter. Consequently one diameter will have a longer radius than the diameter at 90°.

This cylinder deformation in the surface can be more readily detected by using the eyepiece, as described above.

In (A) of Fig. 26.20 the c/c is used for the short diameter and at (B) we have the tracing for two pairs of rays. If the knife-edge is positioned at the short radius (KE_1), the focogram will be seen at E with the horizontal cutoff. With the long radius diameter in the horizontal plane the shadow will start from the left edge of the mirror and show a dark shadow as the knife-edge intercepts the first rays. Then as the knife-edge enters the rays on the axis a shadow will appear as a band at the center of the c/c, long on the vertical diameter, then spread both ways from the vertical diameter, then spread both ways from the vertical dividing line toward the edges of the horizontal diameter.

With the proper amount of cutoff the focogram will be dark on the left edge, somewhat lighter in the central area, and bright at the right edge.

At (F) in Fig. 26.20, with the knife-edge positioned at KE_2 and the cutoff right to left the focogram is reversed. The right edge shades first, and so on. If the cutoff is made right to left, each focogram will be reversed.

At (D) in Fig. 26.20 the knife-edge is positioned at the mean c/c, or the "circle of least confusion." Here the knife-edge would travel farther for the total cutoff, and if the cutoff is left to right the focogram will start at the left edge and advance progressively across the face of the mirror toward the right, as indicated by the arrows at (G) in Fig. 26.20. If the cutoff is made from right to left, then, of course, the travel of the focogram is reversed.

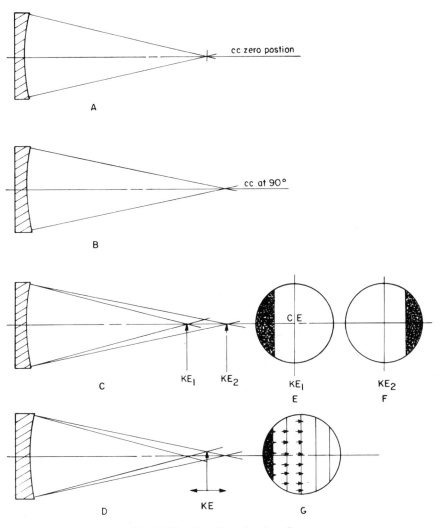

Fig. 26.20.   The effect of astigmatism.

Again a reminder: the greatest sensitivity is paralleled to the blade of the knife-edge, regardless of the direction of the cutoff. Therefore the astigmatic condition discussed has influenced the shadows on the horizontal diameter which is the long radius diameter.

The effect of the short radius diameter is minimized. The effect of astigmatism is qualitative, not quantitative; it reveals the presence of astigmatism. For a quantitative value of astigmatism. The approach to the problem must be changed. (See Fig. 26.21.)

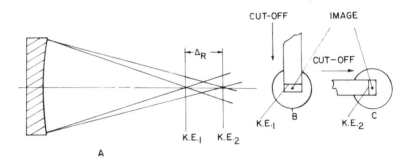

Fig. 26.21. A more quantitative measure next of astigmatism.

Again it is assumed that the long radius diameter is horizontal and the short plane, vertical. First rotate the knife-edge from the vertical cutoff [see KE_1 (A) and (B) in Fig. 26.20], and move it longitudinally to the smoothest cutoff, which will be inside the mean c/c; this will be the zero position for the vertical diameter; then make a note of the reading on the instrument. Next rotate the knife-edge to the horizontal position of the smoothest cutoff and again take a reading. The difference between the two readings will be the delta R (ΔR), or the approximately the difference in the radius of curvature of the two diameters.

It would be well to remember that because of the astigmatic condition of the surface the position of the cutoffs will not be exact and the Δ value will be an approximation. It will be close enough for practical purposes, however.

As and example assume that the value of ΔR is equally divided plus (convex) and minus (concave) about the mean c/c. This may not be absolutely true in all cases, but it will suffice.

$$A_{st} = R_1 - (R_1^2 - y^2)^{1/2} - R_2(R_1^2 - y^2)^{1/2} \tag{1}$$

where A_{st} is the difference in the curvature of the two diameters, R = the radius of curvature of 100 in., y = the semidiameter of a 6-in. mirror and ΔR = 0.020 in.
Then

$$R_1 = 100 - 0.01 = 99.99$$
$$R_2 = 100 + 0.01 = 100.01$$

By substitution obtain the following for h_1 and h_2: h_1 = 0.18018 and h_2 = 0.18015. Then $h_1 - h_2$ for astigmatism is 0.00003 in. or 1.36 wave of astigmatism.

Because the largest h_1 value may result from the shortest radius, it is important to solve for this first.

This problem may seem difficult, but actually the sagittal formula $h = R - (R^2 - y^2)^{1/2}$, where h is the depth at the center of the spherical curve of the semidiameter y and R is the radius of curvature of the mirror, is being used. This problem is taken up later.

Because this form of astigmatism involves two radii (R_1 minus for concave and R_2 plus for convex), divide the formula into parts and solve for h for each radius independently.

This form of astigmatism is not often encountered and usually only when testing an optical flat with a spheroid, but it does occur. Generally, it is a localized condition associated with zones of revolution. Here it is next to impossible to isolate the astigmatism or measure it quantitatively.

It can be measured by using the defocusing eyepiece. Defocus the eyepiece a like amount on each side of the focus and measure the major and minor axes of each set of ellipical patterns. The formulas are complex and are not discussed.

All this discussion of astigmatism could be avoided by stating that in precision optics astigmatism (wrapped surfaces) is not permitted, which is true. It is important, however, to be able to recognize it.

Other pecularities of the shadows that indicate astigmatism when the knife-edge is close to the c/c are shown in Fig. 26.22(A)–(F). One thing that all cases of astigmatism have in common is a lack of symmetry. Consequently always be suspicious of lack of symmetry in the shadows.

Another question to be solved in testing for astigmatism with the knife-edge is what contrast or amount of cutoff to use. About the only rule to be followed is that the best cutoff occurs when the lights and shadows can be seen distinctly.

The subject of astigmatism, like zonal inequalities, eyepiece-image interpretation, quantitative curvature analysis, correction procedures, and glass internal inequalities (strain), if discussed in its entirety would be too

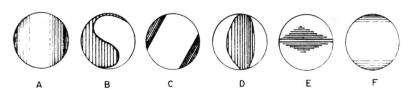

Fig. 26.22.   Shadowing of astigmatic surfaces.

voluminous for shop use, but it is hoped that this elementary introduction to astigmatism will be a helpful guide.

Do not mix up the forms of astigmatism that are a part of optics: the astigmatism discussed above is a fabrication fault; a different form is an aberration fault found mostly in off-axis image patterns for image forming systems.

Astigmatism is caused by the use of small nonflexible polishers during the localization of zonal irregularities or inequalities.

In Appendix 15 note that the small localizing polishers used by Hendrix and Cowan have a thin section of plywood between the metal holder and the pitch giving a slightly flexible condition. For small surface mirrors astigmatism can be removed easily by using a pitch lap of the same size without weights on the polisher. A slow one-third translating stroke is used. For larger optics which have some degree of paraboloidizing a solid soft-pitch lap held stationary by spring wire at three sectors can be used successfully. Here the solid lap which has been pressed overnight (by three hydraulic pads on the polishing lap) will actually self-center and iron out the irregular small zones without destroying the degree of correction already established. The bottom spindle is rotated a little faster than normal (5 rpm for large mirrors) and the anchored pitch lap has a small eccentric stroke of several inches. Only short runs should be made and watery polishing compound should be added. Of course, the pitch lap has small rippled facets in the pitch (see Fig. 23.8) to hold the watery polishing compound. Renew small rippled facets if they are compressed. The pitch must be on a lightweight aluminum holder.

14.   TO FIND *h* OF A CIRCULAR SEGMENT

Following the discussion of front surface reflections from a spherical surface of revolution it would seem desirable at this time to be able to find the depth or height of the curve of a spherical surface of a given radius and diameter. This means the distance from a straight edge place on one diameter to the deepest part of the curve. (See Fig. 26.23.) This depth of curve is the measurement given by the spherometer and is the *h* value to be solved, be it concave or convex.

The value of *h* can be found in a number of ways, but the one that follows meets most requirements. In Fig. 26.24 it will be noted that one-half the diameter D of y and line c–c bisects the angle formed by R_1 and R_2 which forms two right angles at X. One right angle is indicated by a-b-c. If the hypotenuse a were rotated around origin o from S and T, it would be

STRAIGHT EDGE

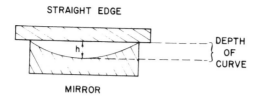

DEPTH
OF
CURVE

MIRROR

Fig. 26.23.   The sagittal depth, *h*.

exactly the same as $c + h$. If the length of a and c is known, then $a - c$ will equal h. To find $h = a - c$, but $c = \sqrt{a^2 - b^2}$, where $a = 10$ in., $b = 3$ in., and substituting and squaring, $c = 0.460678$ in. An electronic hand-held calculator with a memory would be handy.

A more acceptable form for this *h* value is

$$h = R - (R^2 - y^2)^{1/2} \qquad (2)$$

where *R* is the radius of curvature or *a* and *y* is *b* in Fig. 26.24. Then substitute the value of 10 in. for *R*, the radius of curvature, and 6 in. for *y*, the semidiameter. (See Fig. 26.24.)

This formula (2) applies to concave and convex surfaces as shown in Fig. 26.25 which illustrates the *h* value of a concave mirror.

When the depth of a curve *h* is known, it is possible to find *R* with the following formula:

$$R = D^2 = (2h)^2/8h \qquad (3)$$

where *R*, *D*, and *h* are the values shown in Fig. 26.24.

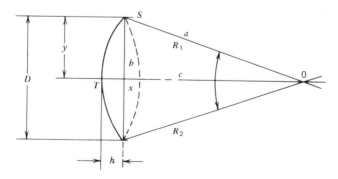

Fig. 26.24.   The sagittal theorem.

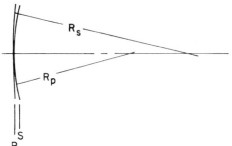

Fig. 26.25. The comparison of a sphere and paraboloid.

15. TO FIND *h* OF A PARABOLA

The parabola is an aspheric surfaces used in front surface work. The usual procedure is to make a spheroid with the required radius and then deform the spherical surface to its parabolic form.

There are three practical ways of deforming a spheroid to a paraboloid. In Fig. 26.25 R_S is the radius of the spherical surface and R_P, the radius of the parabola.

1. In this case the pole of the spherical surfaces is not altered because all the material is removed from the peripheral area. The resulting parabola will have a slightly longer radius than if the central area were removed. This lengthening of the focus is small and in most cases unimportant.

It is important now, however, in some of the new multiple-mirror telescopes (MMT). The first of this type is now in operation on Mt. Kitt. It has six paraboloids which must approach focal length for best imagery. To do this use the second method.

2. In the second method the arc of the parabola intersects the spherical curve at the 0.707 zone shown in Fig. 26.26, where R_S equals the radius of the parabola. The focal length will fall in an intermediate point between the first and third methods. This method of parabolizing is difficult because no

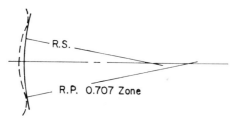

Fig. 26.26. A second comparison of a spherical curve and paraboloid.

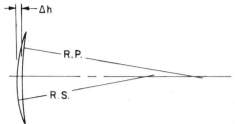

Fig. 26.27. A third sphere paraboloid comparison.

work is done (the polisher) at the 0.707 zone radius and a narrow edge zone is left' to be corrected. This is also difficult to do, although it calls for a minimum of glass to be removed.

The most successful polisher is the rose-leaf described in Appendix 14.

3. In the third method the parabolic curve intercepts the spherical curve at the extreme edge in Fig. 26.27, where R_S is the spherical curve and R_P, the radius of the parabolic curve.

This method requires the removal of material inside the 0.707 zone and is preferred by most opticians because it eliminates the edge zone problem and consequently is easier to perform. It also produces the shortest focal length for the paraboloid.

If the focal length is critical, this fact must be taken in consideration (using the second method) when determining the starting radius of curvature.

In all three methods the focal length will be approximately one-half the radius of curvature. (See Fig. 26.28.) This will be familiar because, as stated in Section 1, the angle of incidence equals the angle of reflection.

The function of the parabola is to bring the parallel rays of light to a focus point.

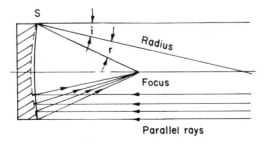

Parallel rays

Fig. 26.28. The focus of a paraboloidal mirror.

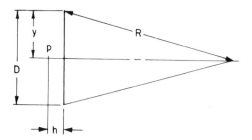

Fig. 26.29.   The parabolic *h* value.

The formula for calculating the *h* value for a parabolic curve is simple:

$$h = y^2/2R \tag{4}$$

where *h* is the depth of the parabola and *y*, the semidiameter of the mirror.

This formula can be used as a quick approximation of *h* because the spherical departure from the parabola to a sphere is small, usually only a few thousandths.

In making a paraboloid it is often desirable to know what amount of glass will have to be removed to deform the spherical curve to a parabolic curve, or the *h* values of formulas (3) and (4). Combine (3) and (4) into formula (5):

$$dh = R - (R^2 - y^2)^{1/2} - y^2/2R \quad \text{or} \quad dh = h_1 - h_2 \tag{5}$$

where *dh* is the difference in the depth of the parabolic and spherical curve.

In a spherical curve deformed at the center to create the paraboloid the *h* is less than the *h* value for the parabolic curve. This is true because the parabolic formula is based on the parabolic deformation shown in Fig. 26.29 and the parabolic *h* would not be subtracted from the spherical *h* value.

16.   PARALLAX SHIFT OF ZONES

Before further discussion of zones there is an important aberration to consider; namely parallax shift.

Parallax shift is the difference between the apparent and actual place of the object. This effect applies to the location of the crests or troughs of zones and is important in the computation of departures from a true surface of revolution.

The effect of parallax shift is to cause an apparent shift of the high and low points from the actual high and low points. The amount of parallactic error is governed by the separation of the source and image, the angle of

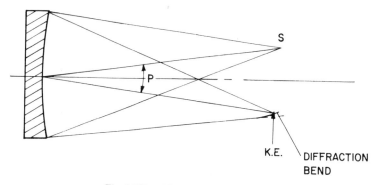

Fig. 26.30. The parallactic shift P.

convergence of the light rays and, a small extent, the bending of the light rays around the knife-edge. (See Fig. 26.30.) Here p is the parallactic shift caused by the separation of the source and image where there is some bending of the light rays around the knife-edge. This is diffraction. Because the effect of diffraction is relatively unimportant, disregard it.

The effects of parallax are shown in Fig. 26.31 as indicated by W and W_1 in figure (B) and X in (A). Locate the crest of the highest point of the zone in question. In order to work on the diameter on which the parallactic aberration is the greatest, this is considered the horizontal diameter.

First make the cutoff from left to right and decide on a certain degree of contrast. This is important because this degree of contrast must be duplicated as nearly as possible when the opposite cutoff is made. With the chosen degree of contrast mark the surface with a grease pencil on the leading edge of the dark shadow as indicated at S in (B). Next reverse the cutoff

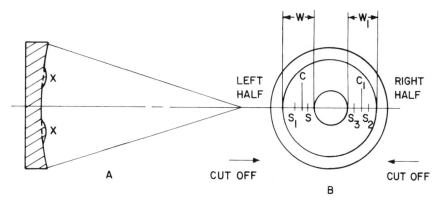

Fig. 26.31. The effect of parallax on cutoff.

and set the knife-edge for the same degree of contrast and again place a mark on the leading edge of the dark shadow, indicated by S_1 in (B). Now a half-way mark between S and S_1 is the highest point in this zone, indicated at C in (B). Note in particular that when the cutoff was left to right the shift of the shadow was toward the right for the left side of the mirror (or in the direction of the cutoff). When the cutoff was right to left the shift was toward the left (or in the direction of the cutoff) for the left of the mirror.

As a rule these zones are zones of revolution and it is not necessary to locate them on both sides of the diameter, but it is important to realize that the symmetrical patterns are opposed. However, if it is desirable, locate the opposite crest as well.

Again referring to (B) in Fig. 26.31, zone W_1 has been outlined and making the cutoff left to right the leading edge of the shadow is at S_2. Next, with the cutoff from right to left, the leading edge is at S_3 and the crest is at C_1. The shift of the shadow is in the direction of the cutoff.

This, in general, covers the problem for a "single" front-surface reflection. For multiple reflections or transmissions the displacement of the shadows or illuminated areas will vary; however, the procedure outlined will locate the high or low points and with experience all problems can be overcome.

Because the parallax will vary under different conditions, the important factor is the awareness of its existence. Shifts of 0.5 in. are not uncommon in 12-in. mirrors for one direction of cutoff or an overall displacement of 1.0 in. for both cutoffs.

Because this is a qualitative analysis of focograms, the use of Ronchigrams outlined in Appendix 14 shows that the correlation of focograms and Ronchigrams will remove this parallax shifting by near on-axis testing. To change the knife-edge from the focograms to Ronchigrams all that is needed is a small Ronchi grating over the larger slot. With a small extension of the grating the eye can view the Ronchigrams. A rather coarse grating of 80 lines/in. is sufficient to make this correlation. The crests can be marked with a grease pencil where the change of slope takes place.

Another method uses the single or double wire-testing setup by placing a single wire in one of the slots of the knife-edge (use a hair). Then observe a curved Ronchiline for any zonal condition and the parallax shift can be located. (See Chapter 28.)

17.   MEASUREMENT OF ZONES

The quantitative measurement of zonal errors can be made provided that certain surface conditions are satisfied. The mean center must be spherical and the zonal transitions must be smooth changes from the spherical surface

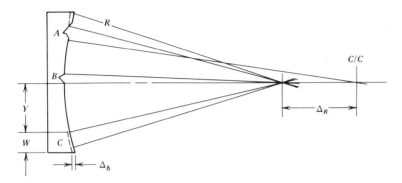

Fig. 26.32. Zones in a surface.

to the zone without a deep groove or sharp ridge in the area of transition. The crest and the lowest point of the zone must be carefully located by the procedure outlined in the discussion of parallax because the width of the zone is an important factor in computing the zonal error.

In Fig. 26.32 three different zones are shown. This is representative of the transitive zones and not the zones of revolution. Zones (A) and (B) would be difficult or impossible to evaluate, whereas the zone shown at (C) would be ideal. This can be a turned-down or a turned-up edge, an intermediate or low zone, or a raised or depressed zone.

The formula for the zone is

$$\Delta h = (y\Delta R/R^2)(W) \tag{6}$$

for a single reflection, where Δh is the departure of the zone from the spherical curve, ΔR is the intersection differences of the zonal rays on the axis from the mean c/c, y is the distance from the axis to the edge, and W is the width of the zone.

18. TURNED-DOWN EDGE

As an example consider the turned-down edge in Fig. 26.33. The spherical surface is 16 in. in diameter and has a 100-in. radius. These constants will be used throughout the discussions that follow.

It is assumed that the turned-down edge is a zone of revolution, that the highest point indicated by y has been located by the knife-edge, and that ΔR has been carefully measured with an arithmatical mean of five measurements for the zone width of W or 2 in.

Now consider Δh the magnitude of the error. $\Delta h = y\Delta R/2R^2(W)$. Divide 0.000024-in. by 0.000022-in. (wavelength value) to arrive at a 1.14 wavelength slope error for this zone.

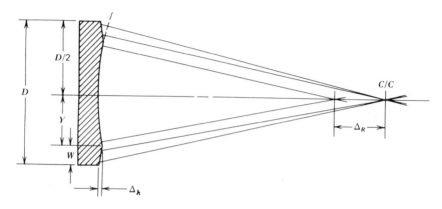

Fig. 26.33. A turned-down edge zone.

19. AN INTERMEDIATE ZONE

Figure 26.34 shows an intermediate zone and the points of measure for (y_1), (W_1) and (ΔR_1). This zone has two parts, indicated by (y_2), (W_2), and ΔR_2; therefore it is necessary to compute both halves of the zone of revolution.

A study of this zone will reveal that if the area indicated by (W_1) is deeper than the spherical surface the opposite side indicated by (W_2) will have a steeper slope to reach the spherical surface. Because of this, it is necessary to calculate both parts and take the mean of the two computations:

$$\Delta h = (y_1\Delta R^2)/(W) + (y_2\Delta R_2/R^2(W) \qquad (6)$$

where y_1 is 4 in., R is 100 in., W_1 is 1.5 in., ΔR_1, the mean of five measurements, y_2 is 5.5, and R_2 is -0.02 in.

$$\Delta h = (y_1\Delta R_1 W_1/R^2 + y_2\Delta R_2 W_2/R^2/R^2)/2$$

Then the total change is $0.000,012 - 0.000,0165 = 0.000,00142$ in., which is 0.78 wave, the mean error for this zone.

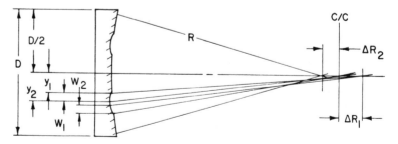

Fig. 26.34. An intermediate zone.

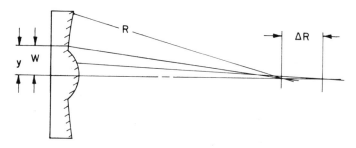

Fig. 26.35. A high center zone.

20. A HIGH OR LOW CENTER

When the zone occurs in the central area the measurements of (y) and (W) are quite different. (See Fig. 26.35.) Here (y) is the distance measured from the axis to the edge of the zone and (W) is measured from the edge of the zone to the axis. This applies to a high or low area:

$$\Delta h = (y\Delta R/R^2)(W/2) \tag{7}$$

Where y is 4 in., W is 4 in., ΔR is 0.01 in., and R is 100 in. By substituting and calculating 0.000,008 in., which is about 0.04 wave is obtained. This same procedure can be followed for calculating a low center.

21. OUT-OF-FOCUS KNIFE-EDGE TESTS

Out-of-focus or inside the center of curvature tests (Ronchi) are an important phase of knife-edge testing.

The application of the knife-edge in this manner is similar to the Ronchi test. The single edge produced by the knife-edge blade is similar to a Ronchi pattern and is sensitive to zonal inequalities. It is not quantitative but informative when used in conjunction with knife-edge and eyepiece tests at the c/c.

The first case is a turned-down edge and the cutoff is inside the c/c. (See Fig. 26.36.) Consider only inside-of-focus cutoffs for the following problems: in Fig. 26.36 at (B) the leading edge of the shadow of the blade is indicated by single lines for different amounts of cutoff of intermediate zones. The knife-edge is positioned approximately 0.125 in. inside the c/c of the central area.

As the shadow starts across the surface of the mirror, and because of the influence of the turned-down edge, the focogram at (f) will show a slightly concave front in the vertical plane. As the cutoff is increased to (g) the central area of the shadowgraph is bent toward the center line. Actually the

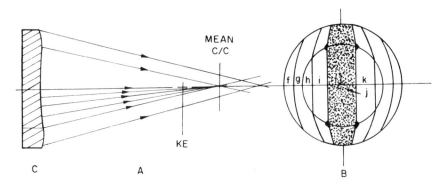

Fig. 26.36.   Turned-down edge viewed with a knife-edge inside the center of curvature.

focogram of the turned-down edge is curved instead of straight because a turned-down edge is curved toward the center line.

The leading edge of the shadow at (*f*) appears to be concave because the light rays in this area are reflected toward the left when facing the mirror and cut off from left to right. Then as the shadow enters this area of the spherical surface the leading edge becomes straight because the light rays converge at a single point and will be uniformly distributed to the area of the cutoff. [See (*C*) in Fig. 26.36.]

If the cutoff is continued until the shadow of the knife-edge blade's leading edge passes beyond the vertical line, the ends of the shadow will bend back toward the vertical center line as indicated at *j* and *k* in Fig. 26.36. Now if *j* and *i* were shaded in solid, they would form the characteristic shadow of the Ronchi test for this type of a zonal condition.

21.1.   A Turned-Up Edge

This condition produces the opposite effect at the edge. When the shadow starts at *f* in Fig. 26.37, the first appearance of the shadow is slightly convex or the focogram on the horizontal diameter is bent toward the vertical line until the shadow passes into the area of the spherical surface. It then becomes a straight line. In the area of the zone the shadow will bend away from the center line, as at *g*, *h*, and *i* in Fig. 26.37. After the shadow passes the center line the shadow will appear at *j* and *k*. Here, again, if the space between *i* and *j* is shaded the characteristic shadow of the Ronchi pattern for a turned-up edge will appear.

It must be kept in mind that the magnitude of the shadow deformation is a function of the magnitude of the zone; that is, if the zonal error amounts to several fringes, the deformation of the shadow will be pronounced, and as the zonal error diminishes so will the contrast of the shadow deformation.

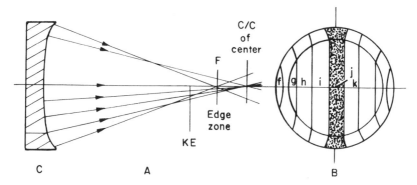

Fig. 26.37. Turned-up edge viewed with a knife-edge inside the c/c.

21.2. A Low Center

In a low center, where the outer edge area is a spherical surface of revolution, the leading edge of the shadow will be straight until it reaches the area of the zone. (See Fig. 26.38.) Then, of course, after the shadow has passed the center line it will take on the opposite curvature. If i and j are shaded the characteristic pattern of the Ronchi pattern for a low center will form.

21.3. A High Center

A high center will naturally be opposite that of a low center in the area of the zone, g to j in the spherical area forming straight lines. (See Fig. 26.39.) If the area i and j is shaded, there will be a Ronchigram for a high center.

The preceding examples of zonal aberration illustrate the influence of zonal errors on the leading edge of the shadow formed by the straight edge

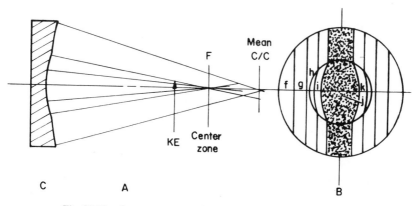

Fig. 26.38. Low center zone with the knife-edge inside the c/c.

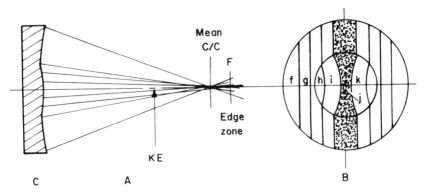

Fig. 26.39. High center zone with the knife-edge inside the c/c.

that intercepts the light rays reflected from the spherical surface of revolution and their use as a guide to the location of zones. They also verify many of the conditions that develop when the knife-edge is a mean c/c.

Although only a single zonal condition has been discussed in each instance, it is obvious that it would be useful when two or more zones are present. An analysis of the changes in slope indicated by the configuration of the shadow is all that is necessary.

In the case of several zones the shadow can become complicated. The important feature, however, is that in precision work they must be removed.

The value of the shadow pattern is in the isolation of zones for local treatment or in the location of a suspected zonal condition.

It can be mentioned that this test is not sensitive to astigmatic surface conditions when the mean center of curvature is used.

This application of the knife-edge blade as a Ronchi test is not considered a true test because it uses only one line.

As mentioned before, the knife-edge can be used as a Ronchi tester by simply placing a grating over the extented slot source. The Ronchigrams of the single and double zonal faults are described in Appendixes 12 and 14.

22. ASPHERIC OR CONIC SURFACES

The paraboloid is often required on aspheric surfaces of revolution. The following discussion considers some of the problems of testing such surfaces.

In the earlier sections the paraboloid of revolution and the procedure of deforming the central area from the spheroid to the parabola, preferred by most opticians, were dealt with. On this basis the testing of the paraboloid is reviewed.

There are several methods of determining the surface quality of the parabola; for example, the null flat test, and a null test method (Dall) that uses a planoconvex lens at the mean center of curvature, the center of curvature, and the star image test with the eyepiece.

Among the four methods mentioned the eyepiece test or the image quality test was used by Sharp, Newton, and others until the invention of the Foucault test.

Starlight was the source of parallel light and the image quality was examined with an eyepiece, not a highly satisfactory procedure because it was mainly "cut and try." In the hands of experts however, it offers a precise method of testing the inside-of-focus against the outside-of-focus imagery if the ring system is identical to the mirror and the lens system is well corrected.

See Welford's chapter in *Optical Shop Testing*, D. Malacara, Ed., Wiley, New York, 1978, which covers interferometric, knife-edge, and single-wire testing.

With the advent of the knife-edge the situation was greatly improved. The date on which the flat null test was first used by Clark and perhaps Draper and Fitz is obscure. It require a high quality optical flat. Certainly, the null flat method in which a plane mirror produced parallel light is one of the simplest for the determination of surface quality and has the same interpretation a spherical surface at the mean c/c. The difference is that the test is made at the focus of the parabola.

23. NEWTONIAN METHOD NULL TEST

In Fig. 26.40 the knife-edge is at S, the image at I, the diagonal flat at D, the paraboloid at A, and the plane flat at B. The light from the source S travels to D, to A, and then to B. From B the light rays are reflected over the same path and are focused at I. At P the dotted line indicates how the spherical surface would appear before parabolizing. After parabolizing the surface appears plane or flat at the c/c as a sphere would. The dead spot caused by the diagonal is of no importance because it is not often used in optical systems. The diagonal reflector at D makes it possible to place the source and knife-edge outside the light path between the parabola and plane. (See Fig. 26.40.) This is necessary because the light source causes turbulence in the light path.

By removing the plane mirror and substituting starlight a typical Newtonian-type telescope is obtained.

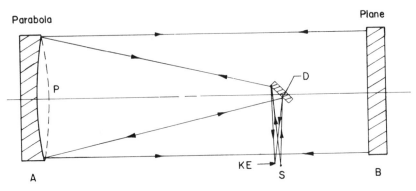

Fig. 26.40. The Newtonian Method Null Test.

It should be emplasized here that the knife-edge test will show an oblate spheroid focogram in this autocollimation optical test. (See Appendix 14.2.) The Ronchigrams are illustrated in Fig. A14.2.

There are two things to keep in mind in setting up Fig. 26.40. First, the optical flat should be set three or more diameters beyond the diagonal; otherwise the radius of curvature image may be observed instead of the auto-collimated image. (See Fig. 25.8*b*.) Second, it is important to center the reflected light source precisely on the mirror's center; if it is not exact, an astigmatic focogram will be observed.

24. CASSEGRAINIAN METHOD NULL TEST

When the parabola is to be used as a Cassegrainian-type installation, it can be setup for testing through the hole in the primary mirror. (See Fig. 26.41.)

In this case the plane mirror functions as a reflector for divergent light from the source and a reflector for the parallel light. As indicated, the rays

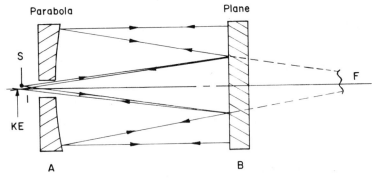

Fig. 26.41. Cassegrainian Method Null Test.

originate at *S*, then reflect from *B*, strike *A*, return to *B*, and are reversed over the same path to *I*. This is actually a double-folded path reflection.

Again a smooth cutoff indicates a good parabola. A useful feature of this method is its compactness for testing. Here the knife-edge focogram for an undercorrected mirror will show an oblate spheroid. (See Appendix 14.)

25. THIRD NULL FLAT METHOD

In Fig. 26.42 another setup for testing a parabola is shown and, as indicated, the plane mirror has a hole in its center. This is another Cassegrainian system.

The hold in the middle of the plane mirror must be large enough to provide clearance for the light rays. This method is also compact. The hole in the middle of the plane produces a dead spot in the center of the parabola which cannot be tested. However, in the area in which the parabolid has been deformed by removing excess glass at the center the surface is generally found to have the quality required.

With one of the tests outlined above the highest accuracy can be obtained, provided that the plane has the necessary quality.

The Dall null test can be used if the ratio of the parabola is *f*/3 or higher. Here the test consists of a planoconvex lens with a red filter set at an established distance. This test is described in *Amateur Telescope Making*, Book Three, p. 149, *Scientific American*, New York, (1953). This book contains reprints and publications of professional and amateur telescope designers and opticians like James G. Baker and Don Hendrix.

The precautions described in Chapter 24 must be used to keep the light source and knife-edge close together to prevent astigmatic off-axis imagery.

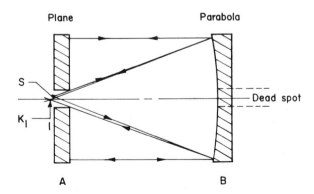

Fig. 26.42. A third null flat method.

26. OFF-AXIS PARABOLOID NULL TEST

The demand for off-axis paraboloids is increasing and the knife-edge test is the most practical method of determined surface quality. A brief description follows.

Figure 26.43 illustrates the principle. Here the knife-edge is positioned at (*A*), the focus of the off-axis of the sphere, indicated by the dotted lines; (*D*) is half-way between the angle of eccentricity. The angle of eccentricity is (*E*) and the plane mirror is positioned on the axis of eccentricity to permit the light rays from (*B*) to strike it at normal incidence.

The light path is marked by arrows. Although only one-half the light path is shown, it is completed by a return system of rays.

Again a smooth cutoff indicates a good off-axis parabola.

Some critical requirements in this test that should be considered, but first determine the nature of the deformation required.

When deforming a spherical surface to a paraboloid of revolution, the deformation is one with symmetry of revolution, but when making an off-axis paraboloid the surface of symmetry is destroyed and the pole of the surface shifts by an amount dependent on the eccentricity. The pole shifts from the center of the spherical surface to the side opposite the direction of eccentricity in the plane of eccentricity.

In Fig. 26.44 the spherical surface at (*Z*) is shown before correction was started. The dotted line at (*X*) indicates how the spherical surface would appear with the knife-edge before correction and *y* is the approximate shape of the finished surface. Of course, this is greatly exaggerated.

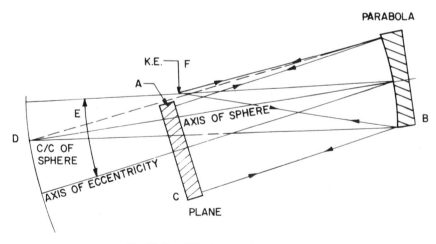

Fig. 26.43.   Off-axis paraboloid null test.

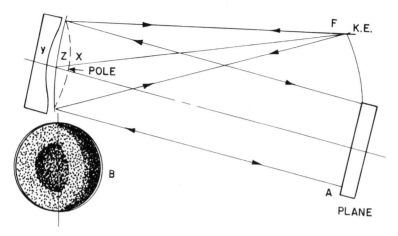

Fig. 26.44. The result of using a sphere in the off-axis paraboloid null test.

At (*B*) in Fig. 26.44 the focogram is similar to a special surface, setup off-axis. The shadow area indicates a high center if the cutoff is left to right as the observer faces the mirror. Note also that the shadow extends to the right of the vertical center line because the pole of the surface has been shifted by the loss of symmetry in the system.

It is obvious in this form of deformation that the finished surface cannot be a concentric surface of revolution. Consequently it is important when arranging the mirror for test that it always be placed in exactly the same position and orientation. This is accomplished by a permanent identification mark on the periphery of the mirror and a pointer on the cell holder that locates the mirror precisely in the cell. This is most important and cannot be neglected.

The appearance of the shadow, as indicated in the focogram at (*B*) in Fig. 26.44, is only representative of the deformation, for the magnitude of the contrast will depend on the *f*/ratio value of the system and the eccentricity introduced. The most conspicuous features are the pronounced high center and the pole slightly to one side of the vertical dividing line. Note also the Ronchigrams on both the sides of the focus.

The question how much material has to be removed is of secondary importance because this off-axis deformation does not result in a surface of revolution and must be produced with fractional-sized tools. The knife-edge will show where the corrections should be made as the work progresses.

It may be possible to design a machine that would generate this type of surface. The demand to date, however, has not justified the expense.

The Optical Center of Arizona has made some progress but little information has been given.

Here 10-in. or larger diameter mirrors are described. Smaller diameter off-axis mirrors are generally core-drilled in a larger diameter paraboloidal mirror.

27. PARABOLIZING BY TESTING AT THE CENTER OF CURVATURE

There are some advantages as well as disadvantages in using this method of testing a paraboloidal surface. A good plane or flat is not necessary. The problems of aligning the mirrors each time is eliminated and there is no limitation of size.

The disadvantages are that the surface is tested by diaphragm zones of small aperture and the use of masks which make it impossible to view the entire surface at one time. Some skill is required in determining the center of curvature of the various zones by equal contrast, and a mathematical problem is involved in the results. Testing by this method is far less costly, however, than making an optical flat of one-twentieth wave, and the surface quality of the paraboloid will be sufficient to satisfy most requirements.

In this method of testing the spherical surface is deformed according to the procedure outlined in Section 14, where the parabolic curve intersects the spherical surface at the periphery. (See Fig. 26.25.) This edge zone is important because the narrow area or band at the edge becomes the *reference* zone to which the parabolic surface is generated.

The star type of pitch polisher is generally used when the central area is equal to the 0.6 zone and the rays of the star extend out to the edge. (Review Appendix 14.)

Although it is not necessary to bring the spherical surface to a high degree of accuracy before starting the parabolization, it is important to give the edge zone a proper radius, free of astigmatism and with no turned edge. This method of parabolizing does not shorten the focal length.

To construct a suitable mask take a piece of cardboard or hardboard, $\frac{1}{16}$ in. thick and large enough to accommodate a disk the exact size of the mirror's diameter. If the parabola is a steep curve, the cardboard must be slumped to the curvature of the mirror by soaking it in water, until saturated. Then position it on the mirror. Next place the full-sized grinding tool on the cardboard and allow it to dry. Paint both sides with shellac and two coats of lamp black. It is advisable to paint the concave surface first so that the cardboard will hold its shape.

Next, cut the disk to the *exact* size of the mirror.

For the layout of the aperture openings, the size of which depends on the radius of curvature, it is never advisable to attempt to use openings smaller than 0.5 in. for radii greater than 50 to 60 in. because of the diffraction effect that restricts the effective aperture. For 60 to 120 in., 0.75 in. openings, and for greater than 120 in. the aperture openings should be approximately the radius divided by 225; for example, if the radius were 300 in., 300/225 = 1.3 in. This is not an absolute requirement because a 1-in. aperture could be used by an experienced observer. The important consideration is that the apertures be large enough for viewing.

Figure 26.45 is a layout for the typical mask. This layout must be as accurate as possible in relative size of the apertures, spacing, and concentricity. The apertures can be made with a sharp wood bit and the cardboard, backed with a smooth pine board.

After the openings are cut out the edges should be painted black.

In Fig. 26.45 (D) is the diameter of the mirror. When mounting the mask, it must be concentric to the axis and in contact with the mirror surface. This is important. Do not separate it from the mirror surface.

The spacing of the aperture openings should be such that the edges of the openings are separated at 0.125 in.

If the surface quality requirement is critical, two masks are often necessary and the apertures of the second mask will have their central openings half way between the centers of the first mask.

In Fig. 26.45 the aperture openings are shown on the horizontal diameter. This is important, for it produces the least amount of tension for the

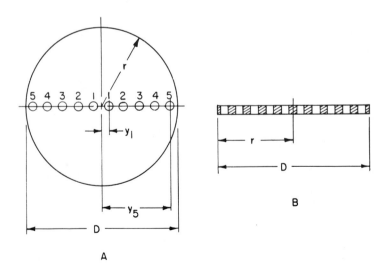

Fig. 26.45.   A mask for c/c testing of a paraboloid.

observer because the measurements are made on the diameter of the least deformation. When deformation is due to instability of the glass or its improper support, the maximum effect is usually in the vertical diameter and at a minimum on the horizontal diameter.

In general the support of large mirrors in vertical testing on a tilting machine has a steel band on one-half its periphery. (See Fig. 2.29.) Actually, when large mirrors are fabricated, testing towers are used and the mirror is married to the cell support in the mounting of the telescope. See Appendix 15 for a description of cellular supports.

28. CALCULATIONS OF ZONAL DEPARTURES

In Fig. 26.46 the mask is shown on the horizontal diameter and in the plane of the paper. The mask is in front of the mirror with the separation exaggerated.

The delta R_s (ΔR_s) is indicated for a parabolic surface only. For a spherical surface mirror (spheroid) there would be no delta change because all zones are the same. In Fig. 26.46 the cones of light designed (1 through 5) are for each half of the mirror. Note that five zonal holes for testing the paraboloid are shown.

The ΔR_s are the points on the axis at which the light intersects. The procedure is explained later in measuring these cones of light by y_1, y_2, y_3, y_4, and y_5 on each side of the zonal mask. (Only one side is shown; see Fig. 26.46.) For example, y_5 and y_5 would form one pair.

Measurement of ΔR_s is a measure of the changes in the *slope* of the mirror's surface at these points and their influence on the reflected rays at or near the c/c shown in Fig. 26.47.

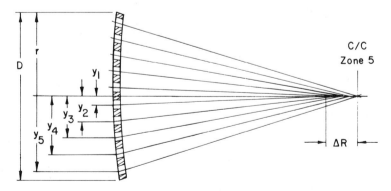

Fig. 26.46. A mask on a mirror for the determination of zonal departures.

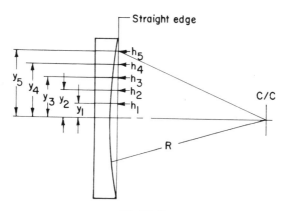

Fig. 26.47

Refer to Section 14, Fig. 26.28, which shows that a perfect spherical surface would reflect all the light rays to a single point at the center of curvature. We learned, also, in generating a parabolic surface, starting with the
extreme edge (the reference zone), that it is possible to test a parabola in
what is called the "center of curvature test."

29. USING THE MASK AND CALCULATING THE *h* VALUES (See Fig. 26.46.)

1. Placing the mask in front of the mirror surface divides it into a
number of circular zones. Then, when the surface is deformed into a parabolic surface, the intersections of pairs of cones of light on the axis are
separated by the calculated values of the *y* points if the paraboloid is
perfect. If the paraboloid surface is not perfect, there will be differences
between the calculated and theoretical ΔR_s. This difference gives the errors
of the surface which are plotted as wavelengths after their *summation*.

Most amateur telescope makers are frightened by this mathematical term.
Actually it is simply the crosswise algebric addition of the last two columns of
the final data sheet. More on this later.

2. Set up a hypothetical case: mirror diameter, 12 in., radius of curvature, 120 in., and focus of paraboloid, 60 in. The mirror is an excellent
spheroid. The mask has 10 one-in. circular aperture openings with a separation of 0.125 in. between the edges of each aperture. The centers of the
openings and the calculated parabolic values are listed in Table 26.1.

TABLE 26.1.   RUN NO. 1 (INCHES)

1	2	3	4	5	6
Zones					
5	8.675	−8.549	+.126	+.126	.000
4	8.620	−8.549	+.071	+.079	+.008
3	8.584	−8.549	+.035	+.044	+.009
2	8.556	−8.549	+.007	+.019	+.012
1	8.538	−8.549	−.011	+.004	−.015

Column 4 = 2 − 3 and 6 = 5 − 4.
Column 5 contains the calculated parabolic ordinates.
Column 6 lists the ΔR measured differences.

3.   Formula (4) has been used previously to find the sagittal depth or the h value of a parabolic curve at the center of the surface on the axis of the mirror. The parabolic formula (5) is $h = y^2 = y^2/2R$.

When the knife-edge and its source are moved together to measure the zonal departures or when the light source is stationary and the knife-edge is moved,

$$h = y^2/R \qquad (6)$$

where h is the parabolic value and R is the radius of curvature. Note that the parabolic h value is twice that in formula (5).

Using formula (5), y is the distance from the axis to the center of the aperture openings in the mask. These measurements are listed in Table 26.1 for R which is 120 in.

4.   Now arrange the parabolic ordinates in Table 26.1 for use in the calculations that will follow to find the slope of the aperture openings (1.0 in.) in the mask. It is good practice to write each calculated parabolic ordinate value in white ink below the mask opening. (See Fig. 26.48.)

5.   In Fig. 26.48 the center of curvature of the edge zone or y_5 (0.126 in.) and the movement of the knife-edge from this *reference zone* in ever *decreasing* increments is h (actual differences between each y zone). Hence as zone y_5 is used as the reference and all are measurements related to it.

Suppose, for example, that the knife-edge is set at the center of curvature for zone y_5. How much would it travel toward the mirror to reach the intersection of zone y_4? This incremental distance would equal 0.12604 − 0.7975 = 0.04629 in.; from zone h_4 to zone h_3 would be 0.03574, and so on. The total travel from zone y_5 to zone y_1 would be 0.12194 in. or approximately ⅛ in.

It must be pointed out here that the observed measurements for the outer zones will be quite accurate, whereas for the center zones they will be less so because there will be an increase in the measurement error as the angle of convergence diminishes. However, the probable error due to the paraxial zones (center) will not produce errors that exceed those of slope for the edge zone because zone y_5 decreases from the edge to the center. Actually, a 0.005 in. error in the central area would be permissible.

6. In the final computation of the *slopes* use Eq. (7). The individual errors can be cumulative if all the errors are positive or negative. If the errors consist of plus and minus terms, they can cancel one another in their summation.

The delta *R* differences must be changed into slope errors for various diameter aperture openings, but because the openings are 1 in. the size of other aperture openings must be considered.

The problem has three factors:
(a) The radius of the spherical surface before parabolizing.
(b) The calculated *y* values of each zone listed in Table 26.1.
(c) The parabolic ordinate calculated values.

The formula is

$$E = (yd\Delta R)/R^2 \tag{7}$$

where *y* is the central distance of the aperture opening in the zonal mask, *d* is the diameter of its aperture openings (which is 1 in. for the mask), ΔR is the difference between the parabolic ordinate value and the measured value using the knife-edge, and *R* is the radius of curvature or 120 in.

7. Because the knife-edge has a counter geared to the longitudinal screw which reads in thousandths of an inch, the value of the readings diminishes as the knife-edge moves from y_5, which is the reference zone.

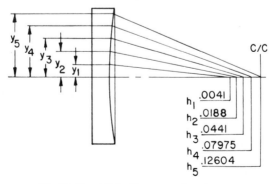

Fig. 26.48. ΔR for five zones of a mirror.

TABLE 26.2. SUMMATION OF RUN NO. 1

Zone	y	d	ΔR	R^2	E	Wave	$\sum$
y_5	5.500 × 1.0 ×	0.000	= +0.00000/14400	= +.0000000/21 × 10^{-6}	= +0.000	+0.000
y_4	4.375 × 1.0 ×	+0.008	= +0.03500/14400	= +.0000024/21 × 10^{-6}	= +0.115	+0.115
y_3	3.250 × 1.0 ×	+0.009	= +0.02925/14400	= +.0000020/21 × 10^{-6}	= +0.096	+0.211
y_2	2.125 × 1.0 ×	+0.011	= +0.02337/14400	= +.0000016/21 × 10^{-6}	= +0.077	+0.288
y_1	1.000 × 1.0 ×	−0.015	= −0.01500/14400	= −.0000010/21 × 10^{06}	= −0.049	+0.239

$\sum$, or summation; for example, 0.115 plus 0.096 equals 211 and 0.288 plus −0.049 is +0.239.

The numbers on the counter can be any number like 7543 or 2345, just so that the exposed numbers are not smaller than the largest calculated parabolic ordinate for the zero setting. A dial indicator with sufficient travel above 0.126 in. can be used if the knife-edge lacks a numerical dial.

Take an average of five readings at each of the five zones. Next prepare a chart similar to Table 26.2. Starting at a reading on the knife-edge in zone y_5 at 8675, subtract the parabolic value of y_5 which is rounded off to 0.126 from 8675 to give −8.549, a constant (note the decimal change for easier subtraction). If a change to a negative number is not made, it cannot be determined whether the measured zonal values is positive or negative. See zone y_1.

After the preparation of Table 26.1, Table 26.2 is used to summarize the degree of correction in a fraction of a wave for the paraboloid of a 12-in. $f/5$ mirror.

8. The fractional wave values in Graph A_1 are taken from Table 26.2 for the first run with the star polisher. Note that the solid line shows the smooth curve of the plotted values. The dashed lines will rotate as the correction proceeds. Note how the (X) in Graph A_1 marked by dashed lines in a rectangle outline is anchored at zone y_5 without any polishing in this area. The slope of the curve indicates that the star polisher was performing properly to produce a parabolic surface. Note the highest zone at y_2.

30. IMPROVEMENT OF THE FIRST PARABOLA RUN

To limit the paraboloid diffraction to $\frac{1}{20}$ wave, for example, reduce the ΔR values by more optical figuring with the same star polisher, but now perhaps by little more than one-half. The zonal summations are for y_5, 0.000, y_4, 0.057, y_3, 0.100, y_2, 0.134, and y_1, 0.157 wave. Plot these summations in Graph A_2 and note how "X" marked by a dashed line has rotated without any work being done in the zone y_5 area. A_2, the shape of the curve, indicates that the polishing tool was performing properly to produce a parabolic surface.

Some of the previous ΔR_s were reduced by a little more than one-half the gain in the second step and the correction was more than one-half the error from the first summation of the shape of the curve. (See Graph A_2.)

This procedure conforms to actual practice. It is desirable to make comparatively short runs when changing a sphere to a parabola to get the feel of the job.

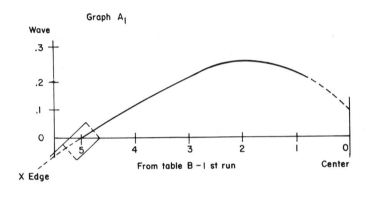

Graph A_1

From table B – 1 st run

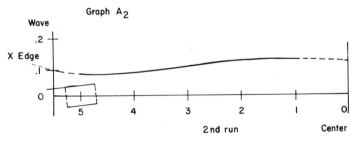

Graph A_2

2 nd run

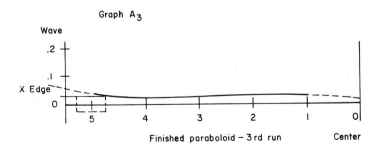

Graph A_3

Finished paraboloid – 3 rd run

If the requirement is to produce a parabola of $\frac{1}{32}$ wave, the ΔR_s would again have to be reduced. By reducing all the ΔR_s with the same star polisher (although this seldom happens) plot a hypothetical run in Graph A_3. The delta R_s, so reduced in this case, are approaching the limits of accuracy for measuring the delta R_s values.

Even to measure these values (ΔR_s) by Brown's method would be a problem, but it could be overcome by a stationary light source and a moving knife-edge. This gives twice the delta R_s values:

$$h = y^2/R \qquad (8)$$

where h, y, and R are identical to those in formula (6). Formula (7) remains the same.

31. OPTICALLY TESTING A PARABOLOID OF LARGE APERTURE

It is interesting to plot the summation of a rather large paraboloid that has gross slope faults of one wave and a number of intermediate scattered slopes in its aperture. The mirror has a relatively large aperture of 30 in. in an $f/2.97$.

Graph (B) in Fig. 26.49 shows the slope of 1.3 wave. (See Table 26.3.) It is apparent that it would be useful to repolish the mirror to a spheroid. This is not correct. If this is plotted against the parabolic curve (see Graph B), it would be well within reach of optical figuring with a specially made pitch polisher.

How to correct this surface and what caused it is of interest to the beginning optician and amateur telescope maker. First, what type of polisher caused this surface deterioriation? The pitch polisher in all probability was a star polisher with a diameter of approximately 12 in., which was poorly fitting and allowed to dry out. To correct this slope examine Graph B carefully. Note the two bulges. Determine where the glass is removed by the ring polisher shown in Fig. 26.49. Here the polisher is required to work harder on the crest of zones 4 and 10. Note the embossed facets of the polishers for best contact. The polisher is worked slowly on the mirror with a 2-in. stroke; the bottom spindle is rotated at 20 rpm and the translating stroke, perhaps at 20 rpm.

Only short runs of 10 to 15 min should be made and a new graph prepared from the delta R_s measurements.

32. CASSEGRAIN AUTOCOLLIMATION TESTS OF SECONDARY MIRRORS

The hyperbolic is only one conic section of the curves in the secondary mirror of the Cassegrain telescope. The primary mirror is a parabolic surface.

TABLE 26.3

30 in. $f/2.97$ Paraboloid R_0 178.281 in. 2 R_0 356.57
focus 89.14 R_0^2 3184.9

Zone 1 in. diameter	1	2	3	4	5	6	7	8	9	10	11	12	13
y	2.69	3.69	4.69	5.69	6.69	7.69	8.69	9.69	10.69	11.69	12.69	13.69	14.69
$y^2/2R_0^2$ or PO	0.020	0.038	0.062	0.091	0.125	0.167	0.212	0.263	0.320	0.383	0.451	0.525	0.605
Reading ΔL	−0.128	−0.036	0.036	0.112	0.155	0.178	0.207	0.250	0.311	0.395	0.472	0.544	0.605
differ.	+0.158	+0.074	+0.026	−0.027	−0.030	−0.011	+0.005	+0.013	+0.009	−0.012	−0.021	−0.019	0.000
$e \times 10^{-6}$	+13.3	+8.5	+3.8	−4.8	−6.3	−2.6	+1.3	+3.9	+3.0	−4.4	−8.3	−8.1	
Summation of $\sum e$.	−14.0	−22.5	−26.3	−21.5	−15.2	−12.6	−13.9	−17.8	−20.8	−16.4	−8.1	0.0	
	−0.7	−14.0	−22.5	−26.3	−21.5	−15.2	−12.6	−13.9	−17.8	−20.8	−16.4	−8.1	
Wave error	−0.3	−0.6	−1.0	−1.2	−1.0	−0.7	−0.6	−0.6	−0.8	−0.9	−0.8	0.0	

where y is the height or center of each marked zone laid out on the mirror.

R_0^2 is the radius of curvature of the central area. The parabolic ordinates are PO.

$y^2/2R$ is the theoretical displacement of the crossing point of the normal and axis from the center of the mirror (see Fig. 26.49). Note that the light source and knife-edge are a unit.

$\Delta(L)$ is the difference between the theoretical and actual dial indicator readings.

(e) is the height of inner edges of each zone with the respect to normal height, assuming that the outer edge is correct. Formula: $e = ydL/R^2$, where y = the zone distances, d = the 1 in. diameter zonal openings, L = the zonal differences, and R_0^2 = 31784.9.

(d) is the width of zone openings in the mask with 13 holes.

$\sum e$ is the integration of the zonal heights. Note the alternate addition of each sum to the next zone.

λ is the angular error of the mirror at zone y (See Fig. 26.49B.)

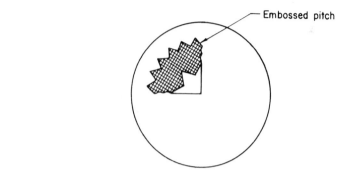

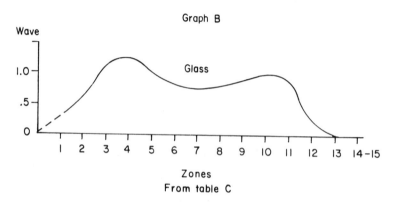

Fig. 26.49. Slope faults in a mirror (see Table 26.3).

The hyperbolic secondary is usually ground and polished to a precalculated spherical mirror and then figured in conjunction with the parabola in a parallel light setup. The knife-edge gives a flat focogram and the Ronchi test, a Ronchigram of equidistant and parallel lines if the secondary figure is a truly figured hyperbolic. (See Appendix 14, Figs. A14.1 and A14.2.)

When the spherical secondary is placed in the optical setup shown in Fig. 26.50, the focogram with the knife-edge or Ronchi test shows an oblate spheroid. To figure the oblate spheroid null in this optical setup use a rose-leaf polisher of perhaps 0.707 diameter.

Figure 26.50 illustrates the conventional setup for testing a hyperbolic secondary mirror and a parabolic primary mirror. The primary mirror should be aluminized because the imagery is faint.

The primary mirror is at (A), the secondary mirror, at (B), and the aluminized flat, at (C). The focus of the combined mirrors is at f' and the knife-edge and light source, at I and S; f is the focus of the primary mirror, P', the distance from the surface to the focal plane of the system, and P, the distance from the surface of the hyperbola to the primary focus at f.

First decide on the diameter and focal length of the primary mirror. Setup a hypothetical design and assume that the primary will have a 12-in. diameter and a 30-in. focal length.

First decide on the amplification ratio: 4, represented by A in (10), is generally a good choice:

$$P = b + f/A + 1 \qquad (10)$$

where P is the distance inside the focus of the primary mirror, b, the distance of the final focus from the vertex of the primary mirror (generally $\frac{8}{10}$ of the diameter of the primary mirror), f, the focal length of primary mirror, and A, the amplification ratio of the system. Then $P = 10 + 30/4 + 1$ or 8 in.

The diameter of the secondary mirror is $12/30 \times 8 = 3.2$ and should be about 3.5 in. The radius of curvature from (11) is

$$R_c = (e + 1) P \qquad (11)$$

where $e = A + 1/A - 1$. If A is 4, $e = 5/3 = 1.666$ and $R_c = 2.666 \times 8$ or 21.328 in. The total focal length is 30×4 or 120 in.; this makes the system $f/10$.

The hole in the primary must be large enough to accommodate the incident and reflected cones of light in testing. For this size primary a 3.5-in. hole would be satisfactory. Make the secondary mirror of drilled glass from the primary.

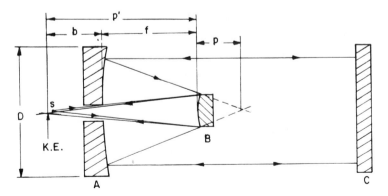

Fig. 26.50. Testing a secondary mirror.

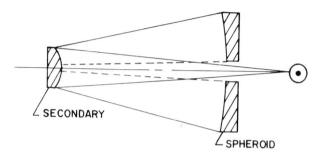

Fig. 26.51. Hindle's test of a Cassegrain.

33. TESTING THE SECONDARY BY HINDLE'S METHOD

Hindle's method makes use of a spherical mirror a little larger than the primary in place of the parabola, with the radius of curvature approximately equal to the focal length of the primary. The number of reflections are cut down from six for the Cassegrain setup in Fig. 26.51 to three for the spherical setup in Fig. 26.52. More information is available in Hindle's article in *Amateur Telescope Making*, A. I. Ingalls, Ed., *Sci. Am.*, New York, 226–227 (1953).

This test has been used for secondary mirrors in the Ritchey-Chretien design telescopes which have hyperboloid primary mirrors. This system has two hyperbolic mirrors that provide a wider field than the conventional Cassegrain system.

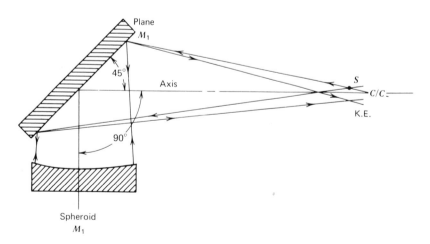

Fig. 26.52. Hindle's test of a spheroid.

34. KNIFE-EDGE TEST OF PLANE MIRROR

To test a flat or plane with the knife-edge a master spherical surface is required.

The master spheroid must have the highest quality and a long radius of curvature. Generally, an R/D of 20 will suffice when R is the radius of curvature and D, the diameter of the test spheroid.

The quality of the spheroid is important because there are three reflections involved, and although there is only one reflection from the spheroid all errors in the spheroid are doubled. Consequently for quality work the spheroid can have no visible zones and should present a smooth uniform cutoff with the knife-edge. It should be free of astigmatism and coated with a high-reflecting surface of aluminum.

The spheroid should be mounted in its cell and carefully checked to determine that the mirror has a uniform coat of aluminum of excellent quality and has not been deformed by the cell. This, of course, must be done by direct observation at the center of curvature. If high-quality planes are required, do not try to use a spheroid that has visible zones in the aperture. The cells for both elements should be sturdily constructed and have suitable adjustments for alignment. Makeshift mountings are not profitable in the long run.

The diagram in Fig. 26.53 illustrates the conventional setup for testing a

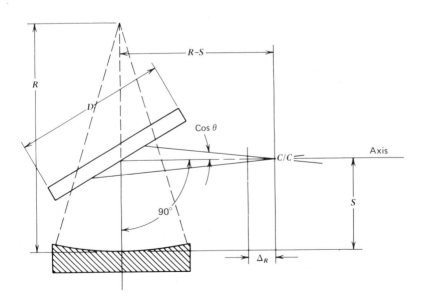

Fig. 26.53. Plane mirror testing with a spherical mirror.

plane mirror with a spheroid. The knife-edge is positioned at S and the radius of curvature of the spheroid measures from c/c to M_1 and M_2.

The interpretation of lights and shadows is identical to that resulting from viewing a spherical surface directly at its center of curvature; the exception is that the cutoffs must be made in both directions, vertical and horizontal, by rotation of the knife-edge. If a perfect spheroid and the plane are set up together and the cutoffs can be made in all positions of the rotation, the shading will be uniform; one seldom finds two mirrors this perfect, however. The slightest curvature, if plus or minus (convex or concave), introduces spherical aberration into the system and this is readily detected because the three reflections produce high sensitivity.

Because the interpretation of the shadows has been covered in previous problems (see also Appendix 14), there is only one noteworthy condition; that is, the plane mirror will appear as an oval or ellipse because it is viewed at a 45° angle in the horizontal plane.

35. DETERMINATION OF THE CURVATURE OF A PLANE

To measure the general curvature of a plane when tested in conjunction with a spheroid is somewhat more complicated than testing a spherical surface because of the three reflections, the 45° of the plane mirror, and the extreme sensitivity.

As a rule it is possible to set up the system so that the plane mirror is almost exactly at a 45° angle to the axis of the system, in which case a somewhat simple formula can be used. If this is not possible, however, the angle should be known. Also the distance from the pole of the spheroid to the pole of the plane mirror should be known within 0.25 in.

First, the basic formula and then the simplified procedures:

$$h = 0.03125 \, (D/R)^2 \, (\cos \theta / \sin \theta^2) \, (\Delta R R / (R - S)) \tag{12}$$

where h is the departure, positive or negative. If it is positive, it is convex; if it is negative, it is concave; 0.03125 is a constant, $\cos \theta$ at 45° equals 0.70711, $\cos^2 \theta$ or $(0.70711)^2$ equals 0.5, ΔR is the difference between the vertical and horizontal cutoffs, and S is the separation of the two mirrors in Fig. 26.53.

Note ΔR this difference in the cutoff of the vertical and horizontal planes. To arrive at this value the procedure is as follows: set the knife-edge for the vertical cutoff; make the cutoff from the top downward; and move the knife-edge in or out longitudinally until the most uniform shading is obtained. Take the reading of the counter again; if the knife-edge was moved in from the setting, the results will be positive.

It is desirable to be systematic in this procedure. Always make the vertical cutoff first, then the horizontal cutoff; for example, suppose the reading for the vertical cutoff were 7895 (on the counter) and for the hori-. zontal, 7875; 7895 − 7875 = 0.020 in. and is negative. If the vertical reading were 7895 and the horizontal cutoff, 7915, then 7915 − 7895 = 0.020 in. and R is positive.

If the various factors in formula (12) are taken in sequence, the value ΔR will equal the curvature of the plane. A constant 0.03125 applies to all problems encounted with formula. $(D/R)^2$ is a variable and the ratio of the diameter of the plane to the radius of the test spheroid.

Cos $\theta/\sin \theta^2$ equal $0.70711/0.70711^2$ or $0.70711/0.5 = 1.4142$; this is also a constant when the plane mirror is positioned at an angle of 45° to the optical axis.

It is important to know the 45° angle accurately

This or any other tilt angle, as Figs. 26.53 illustrate, must be known within ¼°. It is obvious that the greater the tilt angle, the larger the plane that can be checked and the more sensitive the test.

Δr is a measured value depending on the change measured at the center of curvature.

Simplify Eq. (11) by using two numerical constants already known for a 45° angular tilt:

$$h = 0.03125 \times (D/R)^2 \times 1.41412 \times \Delta R \times R/(R - S) \qquad (13)$$

For example, say that

$$D = 10 \text{ in.} \qquad R = 240 \text{ in.}$$
$$\Delta R = 0.2 \text{ in.} \qquad S = 20 \text{ in.}$$

Then by substituting in the last term $R/(R - S) = 1.091$ and $(D/R)^2 = 0.01736$. Now it is known that ΔR is 0.2; finally there is multiple magnification:

$$h = 0.03125 \times 0.001736 \times 1.4142 \times 0.02 \times 1.091 = 0.000,0016 \text{ in.}$$

divided by 0.000021 = 0.076 wave or approximately ¹⁄₁₀ wave. This curvature is in the 10-in. diameter plane, and if ΔR is negative the plane will be a concave surface; if ΔR is positive, the plane will have a convex curve.

It is important that the diameter of the spheroid be larger than the plane to give it full coverage. The diameter of the test spheroid does not enter into the calculation.

The *approach so far* would give the approximate curvature for one diameter only—the horizontal diameter—which is designated D_1.

To follow through to the completion of the problem it is necessary to rotate the knife-edge to 90°. *Never disturb the spheroid setup.* Again, take a series of five readings at each position of the knife-edge in the vertical position which is then averaged. These ΔR readings are used in the calculations of the new 90° rotation D_2, and this is identical to the horizontal D_1 calculations.

36. CALCULATING THE ΔR_s CHANGE DUE TO ASTIGMATIC FOCOGRAMS

Only an approximation of the curvature for diameter D_1 is given because there is a curvature in plane and a different one in another. It is possible to correct the ΔR_s for the two diameters. This correction is the difference between the vertical and horizontal cutoffs for the first setting; V equals the difference between the second cutoff when the knife-edge is rotated 90° to the horizontal plane. (See Fig. 26.54.)

Figure 26.54(A) will appear elliptical because the round plane is being tested at an angle of 45° in relation to the spheroid. The plane mirror should be marked to place it in the same position in the test setup: (B of course, is the foreshortened ellipse measured in the direction of the longer plane D_2; (A) is the initial position in the horizontal knife-edge cutoff and the astigmatic focogram is due to the difference in the curvature in the second elliptical plane. This approximate change is given in formula (13) and is evaluated in the next section. When the knife-edge is rotated 90° for a vertical cutoff, it again gives only an approximate value:

$$\text{Diameter } D_1 = 2H/3 \, (+) \, V/3$$
$$\text{Diameter } D_2 = H/3 \, (+) \, 2V/3 \tag{13}$$

where H is the horizontal and V, the vertical cutoff. If the plane mirror were perfect however, the cutoff on both diameters would be zero within the limits of accuracy of the test. In other words, if the plane mirror being tested has a large curvature error, say 12 to 15 wavelengths, the test calculation will give an answer within a wavelength or so. As the surface quality of

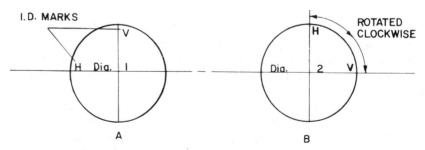

Fig. 26.54.   Positioning a mirror to examine astigmatism.

the plane mirror improves, the test and calculated results become more accurate. When the results indicate a 0.25 wavelength error or less, the results are accurate. This peculiarity is true in most optical testing.

37. THE NEW ΔR_s

As indicated in Fig. 26.54(A) and (B), assume that the plane mirror has been checked for the two diameters and that the ΔR_s was obtained by the counter of the knife-edge. Next set up an example: D_1 the ΔR value is 0.02 in. for the horizontal cutoff and D_2 the ΔR is $-.015$ in. for the vertical cutoff:

Then ΔR for $D_1 = 2H/3$ (+) $V/3$ or $D_1 = 2 \times 0.02/3 = 0.013$ $V/3 = -0.005/3 = -0.0016$ in.; then $D_1 = 0.01$ (+) $-0.0016 = 0.0084$ in. or ΔR for D_1. Then for $D_2 = H/3$ (+) $2V/3 = 0.015/3 = 0.005$ in. and $2 \times -0.005/3 = -0.003$; $D_2 = 0.005$ (+) $-0.003 = -0.002$ in.

Now ΔR_s can be used in formula (12), where $h = 0.03125 \times 0.001736 \times 1.4142$ $\Delta R \times 1.091$.

Of course, there will be an h value for two diameters and the difference would indicate astigmatism; for example, say, for the above example, $D_1 = 0.03$ wave; $D_2 = -0.005$ wave. The astigmatism would then equal 0.025 wave.

38. IRREGULARLY SHAPED ZONES

In testing irregular shapes or surfaces it must be kept in mind that the D value in the formula is measured on the horizontal. Therefore, if a surface 6 by 8 in. were being tested, the D_1 value would be the smaller diameter and D_2, the larger. This must be kept in mind because of the foreshortened effect of the 45° (or any other angular setup); each surface will be an ellipse to the observer and each diameter is proportional; therefore it is important that each diameter be identified for correct placement in the setup.

39. MEASUREMENT OF ZONES INVOLVING THREE REFLECTIONS

The measurements of zones when testing a plane in conjunction with a spheroid are identical to the procedure described for a spherical surface except for the calculation. The crest must be located by left-to-right and right-to-left knife-edge cutoffs on the horizontal diameter. This must be done on the horizontal diameter only. The formula is

$$h = y\,\Delta R W/(2R^2) \tag{15}$$

where h equals the depth or elevation of the zone from the general curvature

y, w is the width of the zone, and ΔR is the difference between the mean center of curvature on the horizontal diameter and the mean center of the slope.

In locating the crest or trough of a zone with two or more reflections because of parallax in the system, a single mark on the plane seen through the knife-edge will appear double.

The parallax that causes this secondary mark is the result of the separation of the source and the image. When the source is on the right side of the axis, as the observer faces the setup, and the image is on the left, the apparent secondary grease pencil mark will be displaced to the left. There the outline of the zone for both cutoffs should be the apparent right-hand mark for each part of the zone. Also this double-line effect indicates the importance of keeping parallax to a minimum in the knife-edge test.

For an example consider a turned-down edge on a 10-in. diameter piece:

R = 100 in., the radius of the master spheroid.

W = 1.5 in., the width of the turned down edge.

y = 3.5 in., measured from the axis 3.5 + 1.5 or $\frac{1}{2}D$ of the diameter and ΔR is 0.01 in.

Substituting in formula (15),

$$3.5 \times 0.01 \times 1.5/(2 \times 10,000) = 0.00000262 \text{ in.}$$

which is slightly more than 0.1 wave when divided by 0.000021 in.

This example could have been a turned-up edge or a zone between the center and periphery. The only difference in dealing with three reflections is that the divisor in the calculations is always $2R^2$ instead of R^2 for single reflection. Otherwise, the section on the measurement of zones can be used as a guide.

40. THE NULL TEST FOR PLANES AND PARABOLAS

The working of planes or parabolas by one of a number of null tests, when elements are mounted as a coaxial system, has many advantages. The only important disadvantage is the dead-center area caused by the shadow of the diagonal; in a Newtonian type setup, a perforating hole in the plane or parabola, or a collimator for a test setup the perforation is not objectionable. In the first instance the diagonal casts a shadow over the central area of the parabola and in the second, with a perforating hole, only a portion of the collimating light is needed for the testing purposes. If the full aperture of the parabola is required, it will be necessary to resort to an off-axis paraboloid.

For general shop purposes the off-axis paraboloid is undoubtedly the most desirable collimating unit and also the most difficult to make. The coaxial system in which the plane mirror is perforated will have many valuable applications. (See Fig. 26.55.)

In either large or small optical shops off-axis parabolic mirrors are seldom used as collimators. A 6-in. off-axis parabolic collimator is about the largest that would be available, if at all, because of its cost. Since the advent of lasers laser illumination with small ball bearings or silver balls has become practical. (See Appendix 16.) It is obvious that a 2-mm ball source with its thin rod holder would cast a minute central shadow.

Here the positions of the parabola and the plane, the knife-edge source, and image are illustrated. The areas that can be used for testing are indicated at (TA), called "test areas" and w, "work to be tested."

This is a compact system and in most cases the operator can adjust the work without assistance. For many purposes the plane M_2 is not necessary; for example, front surface work or planoparallels and right-angle prisms or elements that can reflect parallel light internally. For examination of glass for image quality, homogeneity, or focal length the plane mirror is not necessary and the source can be positioned on the axis of the parabola, thereby eliminating parallax.

A method of testing achromatic objectives in which a plane mirror functions as a collimator is discussed later.

It is apparent from the study of Fig. 26.55 that this setup can be used to test plane or parabola; in any case one of the mirrors must be of excellent quality for the desired results.

An important consideration in this method is the alignment of the mir-

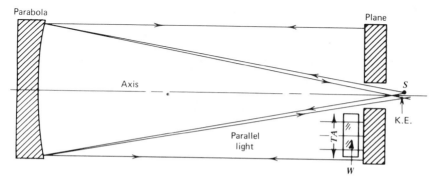

Fig. 26.55. Testing with a perforated plane mirror.

rors. The parallelism must be exact or astigmatism and coma will result to a degree not acceptable in precision work.

One of the simplest methods of completing this alignment is to attach cross hairs of heavy black thread to the reflecting periphery of each mirror; one is on the vertical diameter and the other, on the horizontal. [See (A) and (B) of Fig. 26.56.] These cross hairs can be attached to the cells with masking tape or lab-wax. They should be as close to the surface as possible.

For preliminary alignment the knife-edge is moved to one side and the surface of the mirror, illuminated with a table or desk lamp. If the alignment is out, three or more cross hairs will appear. Now align the mirrors until the cross hairs are superimposed or close together. This will suffice for the first step. In making these adjustments it might be necessary to shift one of the mirrors at right angles to the optical axis. If the test bench lacks guide rails, a center line can be drawn on the bench with a grease pencil or felt pen and the vertical cross hairs, replaced with a plumb bob line. The bob is then centered on the center line of the bench and exactly on the vertical diameters.

It is important that the axes of the mirrors be in the same plane of elevation or at an equal height.

With this preliminary adjustment completed, the knife-edge can be moved into position. Now the cross hairs and their silhouettes will be seen. Here the effects of parallax are evident and it is impossible to superimpose all the cross hairs and their silhouettes. Close adjustment will reduce their separation to a degree that only three cross-hair pairs will be present. Because the adjustments are the "cut and try" form, no specific procedure can be given. When proper adjustment is achieved, however, the silhouettes of the cross hairs on the plane will be positioned between the silhouettes of the cross hairs on the paraboloid.

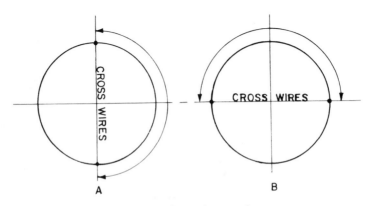

A   B

Fig. 26.56.   Cross-wire mountings.

In certain stages of the alignment procedure second- and third-order diffraction shadows of the cross hairs will appear, but they will disappear when the mirrors are in close alignment.

Because this is a null test with proper alignment of the system and a high-quality component, a flat or uniform cutoff will indicate a high-quality mirror.

The location of zones as a guide to figuring can be done as described. A $10\times$ eyepiece is useful to test image quality.

Because a test procedure of this kind would be justified for surfaces of only the highest quality, zones or other inequalities are not permitted; consequently visible determination with the knife-edge and eyepiece, can carry the correction of a surface to the most critical requirements.

BIBLIOGRAPHY

Focault, L. M. "Description des Procédés Employé pour Reconnaitre la Configuration des Optiques," *C. R. Acad. Sci. Paris*, **47,** 958 (1858); reprinted in *Classiques de la Science*, Vol. II, Armand Colin.

Porter, R. W. "Mirror Making for Reflecting Telescopes," in *Amateur Telescope Making*, A. G. Ingalls, Ed. Munn, New York, 1928, pp. 1–19.

Strong, J. *Procedures in Experimental Physics*, Prentice-Hall, New York, 1943, pp. 72–77.

Texereau, J. *How to Make A Telescope*, Doubleday, Garden City, NY, 1963, pp. 95–150.

27

Testing Achromatic Lenses

Marcus Brown and A. S. De Vany

The testing and figuring of achromatic lenses of large diameter and high quality is without doubt one of the most important phases of prcision optics. As a rule, two or more uncemented elements, 6 in or more in diameter are involved.

> The designing of lenses is not within the scope of this book. Modern designers have computers with programs that correct coma, spherochromatism, astigmatism, and so on. Also, more than 300 different glass types with excellent homogeneity and refractive indices are now available from which to choose. The Clark Brothers had only six.

Special attention must be given to the method of mounting, both the optical flat (for optical testing) and the elements of the objective lens. When the aluminized optical flat and assembled lenses are mounted close together, it becomes an autocollimator in conjunction with the knife-edge. The plane mirror must be an optical flat of high quality mounted in a cell holder designed especially for large apertures. For smaller apertures the flat is often raised on two wooden supports separated at a 60° included angle. (See Fig. 27.1.)

It is important that precise parallelism be maintained between the front surface of the first element of the objective and the reflecting surface of the plane mirror. This is easily accomplished by making a plastic ring spacer whose parallelsim is checked with an autocollimator and planoparallel plate. Lack of parallelism will introduce astigmatism.

1. THE HOLDER AND THE AUTOCOLLIMATION MODE

Support of the plane mirror is important and must prevent distortion.

Another method places the flat on a one-sixth thickness diameter ratio plate with points of support 60° apart at the lower edge, indicated on Fig.

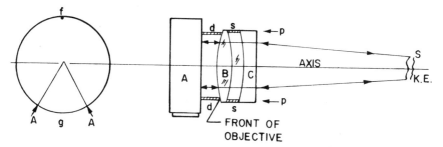

Fig. 27.1. The use of an optical flat in testing large aperture lenses. The test flat may be supported in a 60° wooden block at *A* and *A*.

27.1 as (*A–A*), and a single point at *f*. At these three points the flat mirror should be held in place with leveling screws and spring tension (with pressure directly behind or in front of the leveling screws).

It has been shown the order of supports for high quality optical elements function but in this order: (1) a well-designed cell with compensating lever supports at the periphery inside the cell; (2) a steel-band support that covers one-half the peripheral area of the mirror; and (3) the two-point 60° support.

In Fig. 27.1 note the lines (*A–A*) or the 60° supports and the plastic cylinder spacer that defines the parallelism of the front surface of the lens in relation to the front surface of the mirror. This spacer can be made parallel with an autocollimator used in conjunction with a planoparellel. Note also the spacer between the two elements shown by *s–s* and the pressure points (ball-bearing spring plungers) for holding the assembly in the cell. A better solution for mounting the lenses involves the cell in which it will be mounted, provided that it has three nylon contacts between the cell and lens (front and back) and a suitable mounting for holding the cell in place in front of the mirror. In either case the *d–d* parallelism must be maintained between the assembled lens and its spacer and the testing flat mirror.

Most two-element lenses (crown and flint) are separated by only a few thousandths of an inch and Scotch Tape spacers may be used to advantage. However, a lens may have as many as four or more elements and this calls for greater care in mounting and positioning spacers to prevent distortion.

The 6-in. Clark-type objectives have spacers approximately 0.25 in. thick. The 40-in. objective at Yerkes Observatory has a spacer about 10 in. thick.

Another important rule to remember is that the front of the lens that faces the incident light must also face the flat mirror. (See Fig. 27.1.) The reason for this is that when a lens is set up for testing it is necessary to view it as it would be in actual use, and the light seen at the focus is the light that is reflected from the plane mirror, not the light that is incident from the source. The plane mirror acts as a source of parallel light at infinity. Consequently the front surface of any lens, regardless of design, should be positioned in front of the plane mirror as it is in actual use.

It is advisable to keep the front surface of the lens as close as possible, say about 0.25 in. to the plane because light passing through an uncorrected lens will not produce parallel rays and will cause them to strike the surface of the flat at an angle instead of at normal incidence. The reflected angle will be doubled and greatly exaggerated errors will be made. The greater the distance between the lens and the plane, the greater the exaggeration. Therefore the separation of mirror and lens should be as small as possible. If the lens were perfect, the separation would be of no importance because it would function correctly at any distance if the incident light were parallel rays.

It is customary in production work to acquire as large a number of glass blanks as possible from the same melt to be used in a parallel design of lenses. The designer gives the necessary radius curves spacings and lengths to the maker. In that way the manufacturer is relieved of much of the old cut and try method. In a new design, however, it is sometimes possible to provide correct curves, spacings, and focal lengths. Research then becomes necessary.

As pointed out earlier, in the 1950s electronic computers were not so accessible as they are now. Today a lens designer with a programmed computer can, in a few days, arrive at a lens design that could not be done in several months with logarithms and desk calculators. It is often necessary, when the current supply of glass of a certain melt is gone and a new supply is required, that it vary, say, by one or two points in the fourth decimal place. The lens radius curve and thickness and its spacing will have to be altered and new test plates may have to be made. This is the reason why camera lens are so costly to make.

In the intial test of the lens diamond-scribe deep reference marks on the outer edge of each lens element and always assemble with these reference marks. This is particularly important when correcting for variations in the refractive index.

When making a high-quality lens, test plates should be proved for all surfaces. The test plates should never be more than 10 in. in diameter or they will be too heavy to handle and can cause digs and scratches. Do not try to get by with poor quality surfaces on the assumption that they can be corrected in the final stage of figuring. This would be difficult and a great deal of unnecessary work and expense. As stated before, when a lens system contains two or more elements, there are certain areas in which there will be some inequalities in the refractive index that must be corrected to arrive at one-quarter wave or better.

Concave surfaces can be tested at their centers of curvature with the knife-edge test if test plates are lacking.

There are a number of ways to test convex surfaces on the larger elements: in one the convex surface is set up to form a concave reflective surface and is tested through the second surface. A foreshortened effect here is due to the thickness of the glass, but this apparent concave surface can be tested best with a Ronchi grating placed over the slot aperture of the knife-edge with a red filter and using the foreshortened radius of curvature. If an apparent high or low area is observed, it can be due to the area being tested as a concave surface or of both surfaces contributing to the surface fault observed. To identify it as a hill or hole rub lightly with the middle finger for a minute or two on the area at the location of the zone; if it is a hill, the Ronchigram will bulge; if it is a hole, it will appear to be partly filled. A turned-down or turned-up edge is also tested. Any surfaces fault observed means that both convex (if the element is biconvex) surfaces must be tested until no surface fault is observed.

2. OPTICAL FIGURING

When starting the correction of the entire assembly select one surface only for the necessary correction. This is important because in most cases the final correction will involve small tool polishers and finger spotting—small tools for zones of revolution and finger spotting for local index variations.

It is said that "for the final rub, Alvan Clark could find no sufficiently soft cloth, and so he used his bare thumbs! This also insured detection of any particles of grit which might have gotten into the fine polishing powder. George Davidson noted that, when he last saw Alvan Clark, in 1885, the optician's thumbs had actually burst open from this punishing technique." This is taken from Deborah J. Warner, *Alvan Clark & Sons-Artists in Optics*, Smithsonian Institution Press, Washington, D.C., 1968 p. 27.

Again remember the scribed reference marks on the edges and always assemble the lenses in the same position in relation to one another.

Much care is required in assembling the elements in its cell because a crooked element is very difficult to remove, especially the flint element with its leading sharp concave edge. A soluble vegetable oil (olive oil) can be rubbed on the sides of the cell and lens for easier sliding. A rubber "plumbers friend" is the best way to pick up the elements and place them in the cell.

For a newly designed, high-quality lens it is important to check the glasses carefully for strain (cross polaroids), striae, and variations in the refractive index (see Chapter 5). Do not assume that the glass is good enough because the manufacturer says so. Take the time to find out yourself and be sure. The possible correction in a lens assembly is limited; consequently the best glass is required.

If in the final correction the surface on which the work is being done becomes so badly chopped up that it makes the final correction impossible, it is advisable to reestablish a surface of revolution and start over.

It was assumed that the degree of correction for the whole lens was perfect. This does not happen, no matter how carefully the radii and glass type are chosen: the lens will invariably be over- or undercorrected when tested in a parallel light setup. (See Fig. 27.1.) First make the overall correction by working the concave surface to final correction; next, the front convex surface can be localized or retouched according to Brown's description. When local or spotting correction is required, the objective can again become under- or overcorrect; then the concave surface must be reworked to return the correction to null. (See also Appendix 15.)

3. THE LIGHT SOURCE

In dealing with refraction, a monochromatic source is necessary. It is impossible, however, to produce a perfect monochromatic or one-color source, although a nearly monochromatic source can be used with narrow band filters in connection with a mercury or halogen sodium source. The mercury green line provides a fair contrast but the sodium D line creates the greatest.

The knife-edge lens barrel can be replaced with one that has a receptacle between the two condenser lenses in which the filters can be inserted. See Fig. 27.2, where p is the focus, L_1 and L_2 are lenses, FH is the filter holder, and S is the source. The filter holder and lens barrel should be a rigid unit, interchangeable with the conventional lens barrel. Make the filter holder of

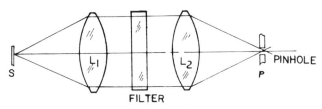

Fig. 27.2.   A source for knife-edge testing of lenses with a position for a filter.

a size that will permit the filter to cover the entire beam of light and also make the filters removable.

4. TESTING THE LENS

In testing an objective lens a rule to remember is that all lights and shadows that indicate imperfections in the system are the opposite of those for reflection; for example, what would appear as a turned-up edge on a reflecting surface would be a turned-down edge for refraction. More explicitly, assuming that the cutoff is made left-to-right and on the right side of the lens where a broad bright ring appeared with a complementary dark ring on the left side, this would indicate a high edge. (See Fig. 27.3.) The rays are deflected away from the focal point (F) because the edge is thicker and the rays, due to refraction, are bent away from the normal. [See (B) in Fig. 27.3.]

Actually, at (A) this would be a divergent lens and the rays would separate on passing from it. This figure was used to illustrate the effect of a change in slope in the area of the edge; (B) is a true effect of a wedge and a deflection of the rays from the normal. This can also occur at (A), however. It is important to have all surfaces in a doublet lens true in curvature and revolution out to the extreme edge before attempting correction of the entire assembly.

In Fig. 27.4 a double convex lens has a source of light at (S); one focus would theoretically form an image at (I); the second conjugate focus is indicated by ($r_1 - r_2$). At t the thickness increases by a high edge. Consequently the emergent ray r_4 would not be deflected so much as ray r_2 and therefore would focus at (L). The opposite effect would be true at (S), as indicated by r_6.

Figure 27.5 is a diagram of a lens with a high edge. A source at (S), one conjugate focus at (I), the second for the central part of the lens, is

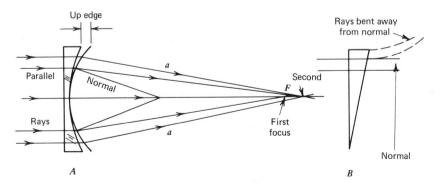

Figure 27.3. The effect of a turned-up edge in lens testing.

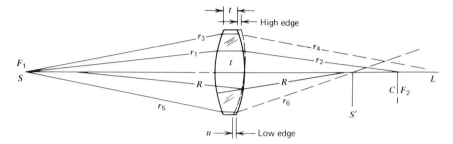

Figure 27.4. Surface-error effects in a biconvex lens test.

illustrated by (r_1, r_2, r_3, r_4). The focus at (X) for the rays (r_5, r_6, r_7, r_8) caused by the increased thickness at the edges indicated by the dotted lines at (y_1 and y_2). Now suppose that the light rays intercept at (I), which would be the normal setting in the central area. Then ray (r_8), cut off first, would cause a shadow to start moving in from the edge at y_2, and as the cutoff continued the shadow would travel toward the center until the knife-edge reached the axis at the intersection of the rays ($r_2 - r_4$), when the central area would shade over simultaneously.

By stopping the knife-edge at this position the crest at y_1 would remain a bright crescent. To form a comparison with the mirror, the indication would be a badly turned-down edge.

It is important that the lensmaker keep this in mind when testing auto-collimated lenses with the knife-edge. The lights and shadows will indicate the opposite in curvature effects, as they do when mirrors are tested.

Figures 27.3 and 27.4 are not correct because the setup is not used in actual practice. It is used merely as an explanation. Actually, there would be parallel rays coming from the direction of the $C_1 - F_2$ conjugate focus.

In selecting a surface for final correction, when possible, choose a crown surface. This surface is easier to work than flints.

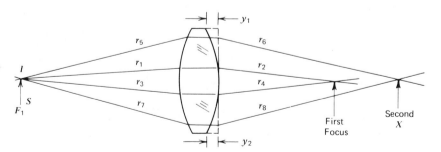

Fig. 27.5. A biconvex lens with high edge.

This is a matter of taste, although some exotic flints used in astrographic cameras may be subjected to staining, and flint surfaces should not be retouched. In some types of objective lens the inner surface of the crown and flint may have identical radii; therefore it would not be a good practice to choose the inner crown, for if it and the inner flint are the same, the crown radius could be tested interferometrically against the knife-edge-tested spherical flint surface.

In this case, for the Clark type objective, the front surface is the better surface for zonal correction with small polishers or thumb localizing and the back surface becomes the correctiong surface to change the overall correction.

Many amateur astronomers have converted war-surplus lenses and large old-type Petzval projectors into photographic and richest-field telescopes by refiguring them optically and testing them in a parallel light setup. Where the correction is to be done will vary in each lens system. The Cooke triplet requires a zone to be turned down on its back surface. In the Gregory-Maksutov the 0.707 zone needs to be lowered. Question: what glass element should be refigured to arrive at a null figure in this type of cadioptric system? Answer: figure the corrector lens with a rose-leaf polisher or the concave mirror. The corrector lens has a steep radius that would be hard to change but it could be done; the concave mirror would require only one-quarter of the work.

For testing the image quality it is important to use a good quality eyepiece to prevent the reintroduction of chromatic problems that are not actually present in the lens itself. The Hastings triplet, Possel symmetrical, or similar eyepiece are good choices.

For the optician the manufacture of precision lenses is the ultimate in optical work. The problems include all aberrations that are known to optics. If the optician had a complete understanding of the cause and effects of all the aberrations, he would be a highly trained theoretical specialist far removed from the grinding and polishing of lenses.

Many of the causes of these aberrations can be eliminated by preventive measures in the manufacture of the components before the assembly for corrections. This would eliminate the need for a complete study of the complications.

The following is a list of some of the preventative measures that can be taken:

1. Before starting to grind and polish the lens components examine them carefully for strain, striae, seed, bubbles, and variation in the index. To do this it may be necessary to parallel the glass plates and then grind and semipolish the surfaces for inspection.

2. Select only the best blanks and allow enough stock for radius curves, thickness, centering, and edging to size.

3. All elements should have exactly the same diameter for final assembly.

4. All elements should be made with precision with a three-stud tool fixture and dial indicator. Large elements should be edged and centered precisely before grinding and polishing.

5. Grind and polish all surfaces as near to perfection as possible, free of zones and astigmatism in particular. Take care to support the glass element in its fabrication.

6. Work the elements to the exact nominal thickness given in the formula or blueprints.

7. Be careful in optical centering and edging to center each element is optically in relation to the periphery and exactly the same. This is very important.

8. Scribe heavy reference marks on the edge (V-type marks are excellent) of each element and always mount the elements in exactly the same orientation in the cell or mounting when testing the final figuring.

9. Select one convex surface of the crown element for localizing with smaller polishers or thumb figuring. The last surface of the flint is the better surface for the final correction. Some opticians prefer the concave surface of the flint because it can easily be changed to the desired conic surface and this is easily identified by knife-edge or Ronchi testing.

10. Do not try to make zonal corrections on two surfaces. You will be asking for trouble. The Clarks actually did use two surfaces for zonal correction.

11. When testing, allow sufficient time for heat effects to dissipate before testing and before resuming work after testing.

12. In figuring do not overdo the required zonal figuring (after all one-quarter wave is generally the given tolerance). This is important because at this stage a great deal of work and expense has been put into the job and overdoing it takes off glass that is hard to put back on; only computers can put glass on optical surfaces.

Following these procedures will make a difficult job surprisingly less difficult.

When the focal length is critical and the initial test indicates that it is too long or short, it is advisable to ask the designer what component to work on to make the correction. The correction for focal length should be made as a surface of revolution by changing the radius of curvature. Do not attempt zonal corrections until after the focal length is corrected. In this instance smooth surfaces on all components will be appreciated.

In the knife-edge test of the assembled lens interpretation of the lights and shadows is identical to that of front surface testing except that the lights and shadows are reversed for reflective elements.

The eyepiece is important for the determination of image quality, and a retical scale in the eyepiece is useful for measuring image diameters. The use of the eyepiece for determining the errors in a lens system, such as astigmatism, coma, spherical aberration and improper collimation, is important. It is suggested, however, that for this information the worker consult a published source. One of the best is Taylor's *The Adjustment and Testing of Telescope Objectives*, published by Cooke, Troughton and Simms, Ltd., York, England. This book is out of print, but it is reproduced in *Amateur Telescope Making*, A. G. Ingalls, Ed., Munn and Co., 1935, pp. 428–436.

When the function of the lens is critical, to be sure that it fulfills the requirements the designs must give the focal lengths of the dominant colors—red, yellow, green, and blue. The separation of the focal lengths is often great enough that they can be measured; for example, if the focal length of the mercury green line is the most important, the lengths of the other colors would help to determine the quality of the lens, for they should conform to calculated or theoretical focal lengths.

BIBLIOGRAPHY

Twyman, F. *Prism and Lens Making*, Hilger and Watts, London, 1968, pp. 546–554.

Warner, D. J., Alvan Clark & Sons, Artists in Optics, *U.S. National Museum, Bulletin* **274**, 25–30 (1968).

28

The Double- and Single-Wire Tests

Double- and single-wire testing of aspheric conic reflectors and autocolli-mated lens systems merits full treatment of its history and application. In essence, the double-wire test shows some similarity (De Vany) to interfero-metric testing because the translating wire patterns form quasi-interfero-grams of the slope departures of a lens system being tested (Ward et al.) in an autocollimating setup with an optical flat. The double- and single-wire tests are similar to the knife-edge and Ronchi tests (Brown, Section 26.21). The first wire in the double-wire test (the second wire serves as a referencing wire) and the single wire in the single-wire test form a bifurcating or Ronchiline encircling each zone when the single wire and its pinhole source are moved together. This is an off-axis setup; another on-axis testing apparatus is described. In both cases no aperture masks are required for measuring the slope departures of aspheric conics or for optical image-forming systems.

1. BRIEF HISTORY

The history of these tests involves many individuals. The single-wire test has been used in conjunction with a slit and pinhole source. Many amateur tele-scope makers build their telescopes about $f/8$, and the Ronchiline or bifurcating encirclement does not occur with a pinhole source unless the aspheric mirror is around $f/5.5$. The odds are 6 to 1 that the right combina-tions will be obtainable for single- or double-wire testing. King used the single-wire test with a slit source in 1934 in which the gauge slots were expanded with Ronchi patterns by moving the single wire with a stationary slit source. This was a quantitative test, but it was handicapped by difficulty in observing the line of demarcation of the Ronchi patterns and gauge slots. Yeagly in 1938 used the single wire and a pinhole source but his mirror was $f/8$ and therefore he could not observe the Ronchiline formation. Gaviola and Platzeth (1939) used the single wire with the slit source which has since become known as the Platzeth test. This test is fully explained in D. Malacara, *Optical Shop Testing*, Wiley, New York, 1978.

It was not until around 1965 that Jim Kent, Jr., a teacher of optical fabrication at Citrus College, Glendora, CA, fully developed the single-wire test while working at a space firm in the same area. His findings were published in 1968. This test was used to measure the slope departures of the 90 in. mirror (229 cm) by Loomis. Kent further revealed to the author (1968) that in testing any conic aspheric surface a simple relationship exists between its eccentricity and the sagittal curve. This will be described at length in another section.

Summarizing the history of the single wire test it can be seen that many experimenters made use of the pinhole source with the single wire testing but failed to observe the Ronchiline formations. Cox's hot wire test failed because the f/ratio was f/7 and then there are moving currents around the single wire, if what were used as a single wire.

2. SINGLE WIRE TESTING

The single-wire test designed by Kent consists of a mounted moving 0.008-in. (0.203 mm) single wire and a 0.005-in. (0.127-mm) stationary pinhole. The light source and the test wire are slightly off-axis and the pinhole must be strongly illuminated by a condenser lens system. (See Fig. 28.1.) The bifurcating Ronchiline will encircle each zone which is premarked with a felt pen as it is moved along the axial plane of the aspheric concave mirror. Set the Ronchiline precisely on the premarked zones. (See Fig. 28.2.)

The quantitative measurements of the zonal departures of the test mirror depend on the test setup. If the pinhole is stationary and the test wire is moved, this gives the y^2/R values or twice the sagittal values of the parabolic measurements. If the pinhole and the single wire are made into an assembled unit, then moved, the parabolic slope departures equal $h^2/2R$, where R is the radius of curvature and y is the zonal height or the radial premarked zone; for example, Table 28.1, which is similar to Table 26.3, illustrates the measurement of the zonal departures and the final quanti-

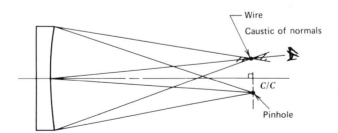

Fig. 28.1.   The optical setup for the single-wire testing of conic aspheric surfaces.

TABLE 28.1

30 in. $f/2.97$ Paraboloid R_0 178.281 $2R_0$ 356.57
focus 89.14 in. R_0^2 31784.9

Zone 1 in. diameter	1	2	3	4	5	6	7	8	9	10	11	12	13	14	15
y	1	2	3	4	5	6	7	8	9	10	11	12	13	14	15
$y^2/2R_0$ or PO	0.003	0.011	0.025	0.045	0.070	0.101	0.137	0.179	0.227	0.280	0.339	0.404	0.474	0.550	0.631
Reading ΔL	0.010	0.025	0.030	0.040	0.066	0.092	0.132	0.182	0.220	0.274	0.338	0.410	0.470	0.542	0.631
differ.	+0.007	+0.014	+0.007	−0.004	+0.004	−0.009	−0.005	+0.005	−0.007	−0.006	−0.001	+0.006	+0.004	0.008	0.000
$e \times 10^{-6}$	−0.02	+0.8	+0.6	−0.5	+0.6	−1.6	−1.1	−0.7	−1.9	−1.8	−0.3	+2.2	+1.6	−3.5	0.0
Summation of	−3.80	−4.6	−5.2	−4.7	−5.3	−3.7	−2.6	−3.3	−1.2	−0.6	−0.3	−1.9	−3.5	−0.0	0.0
$\sum e$.	−3.82	−3.8	−4.6	−5.2	−4.7	−5.3	−3.7	−2.6	−3.3	−1.2	−0.6	−0.3	−1.9	−0.35	0.0
Wave error	−0.2	−0.2	−0.2	−0.2	−0.2	−0.2	−0.2	−0.1	−0.1	−0.05	−0.03	−0.04	−0.09	−0.01	0.00

where y is the height or center of each marked zone laid out on the mirror.

R_0 is the radius of the central area or zone 0. The parabolic ordinates are PO.

$y^2/2R$ is the theoretical displacement of the crossing point of the normal and axis from the center of the mirror (see Fig. 28.3). Note that the light source and double wire move together.

$\Delta(L)$ is the difference between the theoretical and actual dial indicator readings.

(d) is zero because the laid out marks are zonal values or y.

(e) is the calculated values $yd\Delta L/31784.9$, Example zone 7: $e = 7 \times .005/31784.9$ equals 0.0000011 in. or 1.1×10^{-6}.

$\sum e$ is the integration of the zonal values or heights. Note the alternate additions of each sum to the next zone, assuming that zone 15 is zero.

λ is the angular error of the mirror at zone y. See Fig. 28.3.

Wave error   is $\sum e$ divided by 21.6.

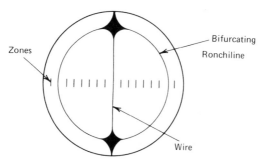

Fig. 28.2. The bifurcating or Ronchiline formation formed by the single- or double-wire test units which can be placed to encircle any premarked zone by movement of the test unit along the axial plane.

tative analysis of how well the optical figuring of the large $f/3$ paraboloidal mirror has progressed. Note that the paraboloid still needs further figuring to 0.05 wave. Because this is the limit of the off-axis test unit for faster $f/$ratios, the on-axis test unit should be used. [See Fig. 28.3 *(upper)*.]

3. THE DOUBLE WIRE TEST

My double-wire test is similar to the single-wire test. It was published before the single wire (1967) and further described in 1970. Actually, the

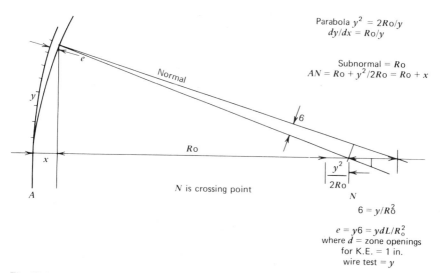

Fig. 28.3. The geometry of the measurement of the zonal slope departures with the double- or single-wire or knife-edge testing units.

test was used as early as 1963 by Tom Griffith of the Northrop Corporation and me for making large wind tunnel optics.

The double-wire test is a novel quantitative and qualitative method of testing a reflector or an image-forming system. This method makes several correlations between focograms, Ronchigrams, and interferograms. This is discussed in Chapter 29. Figure 28.4 illustrates a converging beam from a well-corrected autocollimated lens system that passes through a focus and diverges. Its two fine wires are placed at equal distances inside and outside the focus and translated at right angles across the light cones; it is immediately apparent that the magnified, projected images of each zone of the entrance pupil or aperture is being compared. A well-corrected system will project two equally intense, parallel zones at any zone in which the wires are placed.

In (b) of Fig. 28.4 see the results in a system that is undercorrected for spherical aberration. The pattern is unlike that described for a well-corrected system. The curved lines or Ronchilines have unequal radius of curvature but are reversed in their relationship to the optical center of the exit

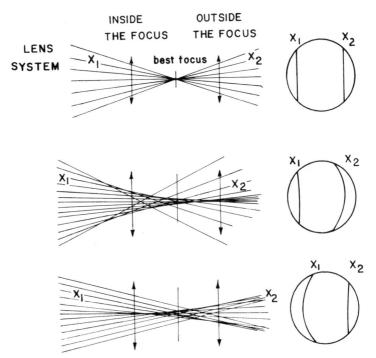

Fig. 28.4. ronchiline formation for the double-wire test showing the degree of correction of the spherical aberration.

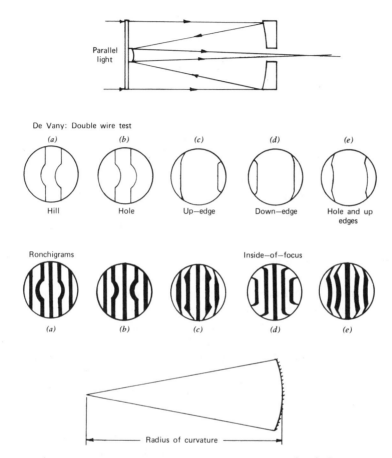

Fig. 28.5. Ronchi test patterns for zonal aberration in two types of optical system: autocolli-mated system is illustrated. Testing concave mirrors inside the focus becomes outside the focus and vice versa.

pupil. The results of a system that is overcorrected for spherical aberration are shown in Fig. 28.4(*c*). Again the lines are reversed and have unequal curvature.

Most opticians figure an optical system by dividing the aperture into two equal area sections, usually comparing the central area from the center to the 0.707 zone with the remaining space out to the periphery. Furthermore, the optician is confronted with scrambled irregularities due to poor figuring or to variations in the refractive index in the optical glass. In Fig. 28.5 some of the common errors that occur during the optical figuring of a representa-tive optical lens system with parallel light formed in conjunction with an optical flat are shown. Note the correlation between the wire-test patterns with the left-side Ronchigrams; they are also similar to the quasi-interferograms discussed in Chapter 29.

Figure (a) illustrates a hill in the central area with a well-figured area at the peripheral zone.

Figure (b) is a pattern from a central hole with a well-corrected peripheral zone.

Figure (c) shows a well-corrected central area with a turned-up edge the pheripheral zone, and figure (d) contains a well-corrected central area with a turned-down edge.

Figure (e) illustrates a central hole (overcorrected) with a turned-up edge.

Although other examples could be drawn for other configurations, the illustrations give the principle involved.

The double-wire test can also be used as a knife-edge by using only the first wire as the blade. The correlation of the knife-edge shadows (focograms) with the double-wire patterns is instructive. The second wire is too near the eye to cause diffraction effects during testing.

The interpretation of the focograms is well known. (See Chapter 26.) In a refractive element system figured null in a parallel light setup consider that the lights of the slope departures come from the same direction as the knife-edge. In testing any reflective optical element think of the light as coming from the opposite side of the knife-edge blade no matter where the actually light source is located. With the knife-edge alone a great deal of experience is required to interpret the observed focograms correctly; but with the simultaneous use of the double wire or Ronchi test the interpretation is greatly simplified.

The double-wire test setup is similar to the single-wire. In Fig. 28.6 two setups are shown: first, is an on-axis tester which consists of a 45/45 coated membrane pellicle between the two mounted wires and the pinhole source is in turn reflected by the partially aluminized surface toward the test surface. Here a bright halogen source is required with a fast condenser system. This is a test unit with a light source and double wire; second, if the stationary light source is wanted, it becomes an off-axis tester, whereas parallax may prevent testing $f/2.5$ or faster paraboloids or any other conic optical mirror surfaces. Off-axis single- or double-wire testing has been used for $f/3$ and slower parabolic or conical surfaces. It must be remembered that when a moving light source is used with a double-wire unit the y^2/R values are measured, but with the stationary light source and moving double-wire test unit the y^2/R values are measured. Again y is the premarked radial zone of the mirror and R is the radius of curvature.

It is apparent that the first wire of the double-wire tester forms the bifurcating Ronchiline which actually maps the axial portion of the caustic

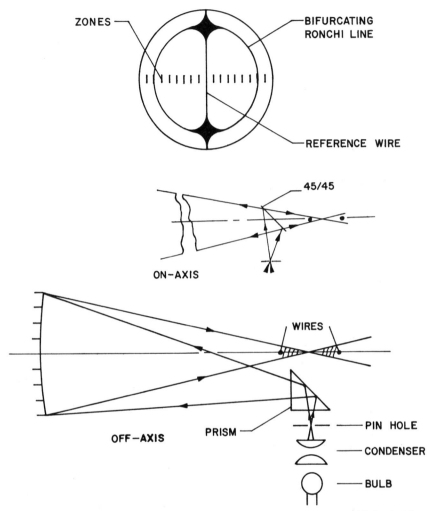

Fig. 28.6. The upper part shows the makeup of the on-axis double-wire tester with the double-wire, pinhole source, and pellicle assembled in one test unit. The lower part shows the off-axis double-wire tester which consists of two units.

of the normals to the principle axis. (See Fig. 28.1.) The second wire serves as a referencing wire guide to prevent the translating bifurcating circle from developing oval off-axis patterns, as it would if the travel test unit were translated off-axis.

The test unit is easily made; it consists of three brass tubes, two of which have an 0.008-in. (0.20-mm) wire soldered to one end. These tubes can be rotated into a third tube until they are vertical in the same plane. The distance between the two wires can be varied to suit the cone angle of the f/number of the system under test; for example, with an $f/16.2$ system the two wires were set approximately 0.75 in. (19 mm) apart. Some effort must be expended to set the double wires in azimuth for an autocollimated system like a Schmidt-Cassegrain when checking its optical correction for zonal spherical aberration. A 100× microscope will be helpful when the necessary adjustments of the double-wire test unit and the returned image from the telescope are made. The two images of the double wire are found near the central area of the exit pupil, which can be located by sliding the test apparatus along the axial plane on the optical bench until both lines have equal intensity. Then bring the lines to the center of the exit pupil with a slight rotation of the rod that holds the double wires. As shown in Fig. 28.7 a transverse screw on the holder with the double wires moves them across the image plane at any zonal area. (See Fig. 28.2.)

The illuminated pinhole is approximately 0.003 in. (0.07 mm) diameter

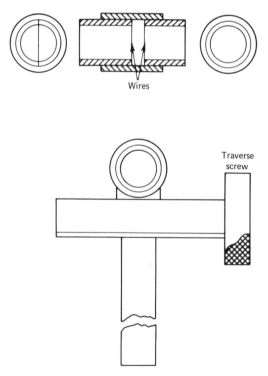

Fig. 28.7.   The assembly of the double-wire testing apparatus.

and must be strongly illuminated for the bifurcating encirclement to be observed on an uncoated mirror.

4. QUANTITATIVE TESTING

In Section 2 a method of testing a parabolic concave mirror of large aperture approximately $f/3$ was discussed briefly. It is necessary to use the optical setup in the upper diagram of Fig. 28.6 because a stationary off-axis pinhole and the test wires could produce errors due to parallax. This is near the limit for the use of the off-axis mode, and it must be realized that the single or double test wire must be ⅟32 in. (1 mm) from the light source. Loomis and other opticians have used this off-axis test with success for $f/3$ conic surfaces. Of course, the on-axis tester would be more better but it requires a high-intensity illuminator for the pinhole source and a wire for the bifurcating Ronchiline to be observed on a bare surface mirror.

Testing an aspheric system with on- or off-axis apparatus is similar to the knife-edge for the mirror at its radius of curvature or for the autocollimated system shown in Fig. 28.6 (*upper*). In a double-wire test of aspheric mirrors the first wire splits into a Ronchiline encirclement at any radial zone (premarked with a felt pen with some degree of accuracy). This is dependent on the position of the test apparatus along the axial plane of the aspheric mirror. Here it is necessary to anchor the test stand and the mirror and, of course, the test apparatus which may have to be moved at the most 0.25 in. (6.4 mm). The second wire serves only as a reference line for the sliding unit which must travel only a short distance to measure the slope departures of the test mirror. With the on-axis tester the pellicle and the second wire is important because it keeps the pinhole and the two wires on-axis and constantly checks the bifurcating central line of the centrally located and positioned Ronchiline encirclement as it measures the deep sagittal slope departures of $f/3$ or faster aspherics.

1. Mark 0.5-in. (12.5-mm) radial zones with some degree of accuracy across the mirror, starting at dead center of the mirror, with a felt pen.

2. Locate the Ronchiline or bifurcating pattern caused by the first wire on the mirror.

3. Move the test apparatus along the axial plane or radius of curvature and observe how precisely the Rochiline can be positioned on the premarked radial zones.

4. Locate the second reference wire near the eye, using it to observe whatever translation is taking place as a number of inner zone and outer zones are checked.

5. Record the measured zonal departures (the average of five readings), starting at the extreme edge which is generally assumed to be corrected or to have zero measurement error. Of course, measurements are not taken if a turned edge (up or down) is known to exist. A turned edge often appears when the overall summation of the nearly completed mirror ($\frac{1}{10}$ wave) is graphically displayed, in which case assume another area such as the 0.707 zone as the zero for measurement. The fact that the Ronchiline can be positioned on the extreme edge, or in any other zone, is an important aspect of this test because in the conventional use of masks the aperture openings had to be have some area (see Zones 14 and 15 in Table 26.3) and this is a cause of error. Again Wadsworth's effect is not apparent if the measurement of the zonal slope departure with either of the two wire tests described.

Compare Table 28.1 with Table 26.3. Note that no masks with holes are required when the single- or double-wire test is used. In Table 28.1 the premarked zonal areas were 1 in. (25.4 mm) apart, starting in the middle. Observe that the worst zonal slopes are in the middle area of the mirror. Now because the mirror has a 6-in. hole in the middle reestablish the reference point at perhaps zone 4, calling it zero, and then redo the summations. This problem is left to the reader to do as an example. Does the paraboloid mirror meet the tolerance of $\frac{1}{16}$ wave?

5. DIFFRACTION LIMITED SYSTEMS

McGuire has developed a simple formula for diffraction-limited systems:

$$s = N^2 Cd \tag{1}$$

where s is the calculated tolerance shift, C is a constant, depending on the makeup of the optical test mode and the geometry of the system. When the light source is fixed and the test wire unit is moved along the optical axis, C has a value of 0.0026 in. (0.065 mm). The value of C becomes 0.0013 in. (0.0325 mm) when the source moves. Both numbers are used for testing reflectors at their radius of curvature. The use of the larger C value $-.0026$ in. (0.065 mm) gives twice the value when the smaller value 0.0013 in. (0.0325 mm) is used. This is restricted to aspherics about $f/3$ or slower because it is difficult to place a fixed light source on-axis in the test apparatus for the faster aspherics. N is the f number of the cone, one-half the inverse of the numberical aperture.

For autocollimation with an aluminized flat moving with the knife-edge or single or double wires in a situation in which the light strikes the aspheric twice the value of C is also 0.0013 in. (0.0325 mm). If *only* the knife-edge or single- or double-wire test apparatus is moved in an autocollimation setup, C takes the value of 0.0026 in. (0.065 mm).

Kent has shown that the distribution of intersections with the optical axis of normals to any conic curve is equal to the maximum sagitta of the conic curve times the square of the eccentricity or

$$D_n = c^2 a_s \qquad (2)$$

where D_n is the distribution of the normals, e is the eccentricity of the conic curve, and a is the aspheric sagitta.

Use this equation to find the wire or knife-edge shift between the center and the edge zones for any conic surface by first calculating a_s from the series expression for sagitta:

$$a_s = r^2/2R + r^4 (1 - e^2)/8R^3 \qquad (3)$$

where r is the semidiameter and R is the radius of curvature. For a parabola with $e = 1$ the shift becomes the familar $r^2/2R$ or r^2/R, depending on the optical setup. From Eq. (3) it can be seen that the shift for any conic is nearly proportional to the square of its eccentricity; for example, to figure a Dall-Kirham ellipsoidal primary mirror of eccentricity 0.90, 200-mm radius, and 240-mm diameter by zonal measurements to an accuracy of $\frac{1}{32}$ wave calculate the sagitta as 3.60 mm (taking only the first term of Eq. (3) or 2 × 3.60 mm × 0.9^2 = 5.84 mm. Allowable error from this shift is given by Eq. (1) as $(2000/240)^2$ × 0.065 × $\frac{1}{32}$ = 0.15 mm. This is less than $\frac{1}{40}$ wave of the overall zonal shift and shows that it is not a simple matter to figure aspherics by measurements alone. It shows that null optical setups, null lens, or testing by immersion, described by Holleran, produce smoother figured aspherics. Shift tolerance calculations provide the quality assurances of the aspheric in fractions of a wave. The double-wire, or Kent's single-wire test, is a combination of Ronchi and Foucault tests, both of which are superior to the Plasket and Gaviola. This is shown by the remarkable capability of the measurements compiled by different observers of Loomis's 90-in. (183-cm) paraboloid figures, using the single-wire test.

The double-wire test was designed in 1963 when two 32.5-in. (82.5-cm) paraboloids were made for wind-tunnel optics.

6. TRANSMITTING GLASS SYSTEMS

Suppose we decide to make a 12.5-in. Cassegrain with a primary focal length F_1 of 50 in. and an amplification factor of four for an effective focal length F' of 200 in. For the distance d from the primary mirror's front surface to the final focus, we choose 12.5 in. giving 37.5 in. for S, the separation between the mirrors, in this formula:

$$S = F_1 (F' - d)/F_1 + F' \qquad (4)$$

This places the secondary a distance p = 12.5 in. from the final focus and p' = 50 in. from the final focus (the amplification A equals p'/p). From this the secondary's radius of curvature can be calculated by

$$R_1 = 2pp'/p' - p \tag{5}$$

In our example R_1 comes out 33.33 in. If R_2 is chosen as $1.2R_1$, it is 40 in., making the focal ratio $f/10$ for a secondary a secondary as large as 4 in. in diameter (3.6 in. is about right in this example). This is an easy figure to achieve and test.

6.1. The Secondary Test Formula

We now need to calculate the test's longitudinal aberration, LA, or the amount of the knife-edge (and light source) movement required to test the center and edge zones of the secondary. When the figure is satisfactory, the marginal zone will test farther away from the mirror by the calculated amount. The following definitions are required for setting up the formula:

C_1 = curvature convex secondary = $1/R_1$
C_2 = curvature concave back = $1/R_2$
T = center thickness of secondary
C = curvature = $C_1 - C_2 + C_1C_2T$
N = index of refraction of glass used for the wavelength of light source used
P = $C(N - 1)$ A = amplification
Q = $(A + 1)^2/(A - 1)^2$
L = $(T - C_1T)/(C_1 + P)$
y = radius of the zone being tested

For a primary mirror that is $f/3.5$ or slower the LA or longitudinal aberration is given by

$$\tfrac{1}{2}NQC_1^2 + NCP(1 - C_1T)(NC + C_1)y^2/(C_1 + P)^2 \tag{6}$$

This expression is derived from the formula (10) for primary longitudinal spherical aberration that is given on page 85 of A. E. Conrady's *Applied Optics and Optical Design* (Dover, 1957), with the hyperbolic term installed at the reflection. The same formula is the genesis of the useful and perhaps familiar G-sum method of calculating spherical aberration.

In the present case the optical system is very simple, the light from the slit being autofocused on itself. This permits including explicitly the mirror thickness T without unduly complicating the formula. Providing that R_1 and R_2 are not too different, so there is very little refraction at the second surface, the new expression is unusually accurate for a primary aberration formula. The knife-edge and light source should be as close together as one can make them in order to keep off-axis aberrations from influencing the test results.

In our example $C_1 = 1/33.33 = 0.03$; $C_2 = 1/40 = 0.025$. If the secondary mirror is made of pyrex finished 0.6 in. thick and tested with a yellow filter, the value of N (at the dominant wavelength of 5893 Å) is about 1.474. This permits calculating $G = 0.00545$. Next, $O = 0.002583$; $Q = 25/9 = 2.778$; and $L = 0.982/0.032583 = 30.14$. This last number is the distance in inches from the mirror's back surface (in front in the test) to the knife-edge, but the image of the slit reflected from this surface is twice as far away. Hence there is no chance of reflection interfering with the shadows seen on the hyperboloidal surface.

When the values above are substituted into the formula, LA turns out to be $(0.0521 + 0.0007)y^2$. The result is written as two terms to demonstrate the relative magnitudes of the Q or hyperboloidal term and the spherical aberration due to the two refractions (ingoing and outgoing) at the spherical surface. The first term is almost 75 times the second!

In the 12.5 in. Cassegrain of 200 in. effective focal length the marginal ray from the primary strikes the secondary about 1.8 in. off-axis. When this value is used for y, the hyperboloidal LA shift is calculated 0.1688 in. and the refractive term adds 0.0023 in. The total is 0.1711 in., which is the distance the knife-edge and light source are being moved away from the secondary when the peripheral zone is being tested instead of the central (on-axis) one. The very large numbers suggest that the test shadows will be very strong and definite.

For the intermediate zones the distance will be less; for example, the 70 percent zone ($y = 1.26$ in.) gives 0.0838 in. However, since we are testing by light reflected internally, there is no direct way to apply zonal masks. Because the light is diverging from the test point, the zonal value of y is on the hyperboloidal surface. This effect can be eliminated by marking y on the convex surface with a fiber-tip marking pen. The shadow crests will then be made to match the respective markings and these will be easily wiped off the polished surface.

This section is used by permission from *Sky and Telescope* issue of January 1970 and by the author John L. Richter of Los Alamos, NM.

This article points out a method of using the knife-edge at the precalculated zonal slopes for making the hyperboloidal surface. This method suffers from the fact that it is difficult to determine the y's when the shadows are on the premarked zones. It should be apparent how much more easily the Ronchiline of the bifurcating pattern can be set with the double- or single-wire test. This method of making a secondary mirror for a Cassegrain telescope is probably the simplest and the most direct way an amateur telescope maker can make the secondary. Figure 28.8 shows an optical setup with a double wire tester that must be a single unit because the calculations are definitely easier for this type of moving tester.

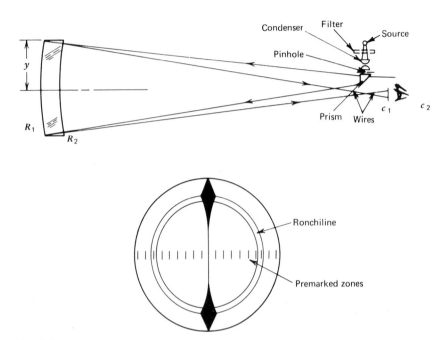

Fig. 28.8. The Ritcher double- or single-wire test for testing through the glass for making a hyperbolic secondary for a Cassegrain telescope. Calculated slopes are involved.

7. SUMMARY

It has been shown that wire tests have taken over the measurement of zonal slope departures because of the number of inherent errors in the Ritchey mask. The double-wire test has one advantage over the single in that a reference wire checks that the bifurcating pattern with its own center line matches a second reference wire. This is essential when testing aspherics of low f/ratios. It must be kept in mind that any conic surface has its own eccentricity value, and even if this is a negative number in a design it becomes a positive number when mathematically squared in Eq. (3). Therefore an ellipsoid is always less than 1 (100%) and a hyperboloid is greater than 1 or (100% plus) if the paraboloid is 1 (100%).

BIBLIOGRAPHY

Bell, L. *The Telescope*, McGraw-Hill, New York, 1922, p. 40.

De Vany, A. S. "Spherical Aberration Analysis by Double Wire Testing," *Appl. Opt.*, **6**, 1073–1075 (1967).

De Vany, A. S. Supplement to "Spherical Aberration Analysis by Double Wire Testing," *Appl. Opt.* **9,** 1720–1721 (1970).

Gaviola, E. "On the Quantitative Use of the Foucault Knife-Edge Test," *J. Opt. Soc. Am.,* **26,** 163 (1936).

Kent, J. J. Jr., private communication, 1967.

Loomis, D. A. *Newsletter,* Optical Science Center, University of Arizona, **3,** 75 (1968).

Loomis, D. A., private communication, 1971.

King, J. H. "A Quantitative Optical Test for Telescope Mirrors," *J. Opt. Soc. Am.,* April 1934.

King, J. H. "A Quantative Optical Test for Telescope Mirrors," in *Amateur Telescope Making Advance,* Book One, A. G. Ingalls, Ed., *Sci. Am.,* 104–108 (1948).

Selby, H. H. "Notes on the Optical Testing of Aspheric Surfaces," in *Amateur Telescope Making, Book Two,* A. G. Ingalls, Ed., *Sci. Am.,* 132–138 (1959).

Ward, J. E. et al., "Lens Aberration Correction by Holography," *Appl. Opt.,* **10,** 896–900 (1971).

Yeagly, H. L. "How to Make a Telescope," *The Sky* (fore-runner of *Sky and Telescope*) April 1934.

29

Some Aspects of Interferometric Testing and Optical Figuring

Examination and correlation of interferometric and interferential fringe patterns have shown that a general classification can be established for optical figuring.

The following interferometric and interferential (Ronchi) arrangements can be used for testing various telescopic cameras and single components and, in turn, can be combined with autocollimation setups. In Fig. 29.1 a test unit is shown in which a number of interferometric units can be placed for testing a mirror being figured to a paraboloidal mirror. Also shown is a Schmidt-Cassegrain corrector plate. Any polarizing interferometer, Koester prism, Murty and Saunders cube, or Ronchi grating setup can be positioned as illustrated. Because these setups in Fig. 29.1 have reflective or refractive elements (the Schmidt-Cassegrain has both), observe the interplay in making correct interpretations of the many interferometric modes. Many strange and wonderful things occur in reflective optics during optical testing.

The interferometers compared and tested were the following: Koester wavefront reversing, Saunders and Murty cube, Murty's parallel plate with a laser source, modified Franco-Veret, and a newly developed (De Vany) Babinet compensator interferometer. This device, shown in Fig. 29.2, uses the assembled Babinet compensator instrument with which many optical shops measure strain or incorporate in wind tunnels. The diagrams show a demountable unit that can be placed over the opening of the Babinet compensator. It consists of one polaroid (glass) cemented on a small plano-parallel with a small right-angle prism to direct the mercury pinhole toward the optical component being tested. A second polaroid is cemented to the right-angle prism and must be rotated before the cement dries while a concave mirror is tested at its radius of curvature. The optical elements must be kept small because the opening of the compensator limits the size of the components. Mine are mounted on a 10-mm by 10-mm square. It is possible to use the Nicol prism already in the eyepiece if the lenses are removed.

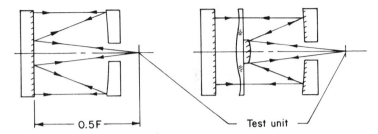

Fig. 29.1. Optical testing of two types of image-forming system: a reflective and a refractive with reflective elements.

A reed-type mercury source with a 0.125 mm diameter pinhole completes the assembly. This forms a complete unit for evaluating strain and homogeneity of glass and for interferometric testing. The testing of homogeneity of glass is described in Chapter 5.

The value of interferometers is generally rated in the following order: Ronchi, Murty cube (see Fig. 29.2), Koester, Franco-Veret, and Babinet compensators, and Bates shearing. Opticians prefer to use a course Ronchi grating (60 lines/in.) at the onset of figuring an optical system with a large amount of spherical aberration. [See Fig. 29.3 (*upper*).] Lower figures show

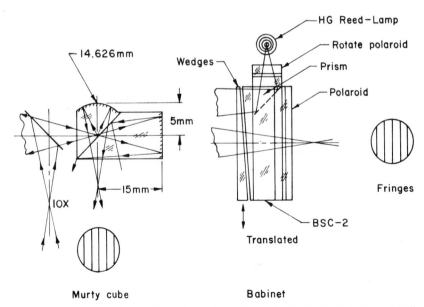

Fig. 29.2. Several interferometric testing units easily made for testing interferometrically image-forming systems. Murty cube uses a laser and Babinet, a mercury light source.

UNDER CORRECTED

HILL   HOLE   OVER CORRECTED

Fig. 29.3. A number of Ronchigrams for various amounts of spherical aberration. All are glass transmitting patterns and inside-of-focus. A corrector plate is the transmitting element.

representative inside-of-focus Ronchigrams during the optical figuring of the *light-transmitting* element, such as Schmidt corrector plates in a parallel light mode. (See Fig. 29.1.) In the final figuring a finer Ronchi grating is used, generally 100-line. Saunders and Murty cubes and Franco-Veret, Babinet, and Murty's compensators project radial-sheared interferograms similar in many aspects when testing autocollimated lens systems with an autocollimating flat or lens collimator. These interferometers help to correlate the interpretations of Ronchigrams and interferograms during optical testing.

Optical engineers prefer the Franco-Veret and Babinet compensator type because spherochromatism can be measured with narrow-band filters. It can also be used for checking the homogeneity of glass in Murty's interferometric test with a spherical mirror. (See Chapter 5.) Observe in color the pollutants that apparently float around in optical quartz and other glasses. It is a beautiful sight when one color is spread over the glass blank: the tints can actually be changed by a slight movement of the compensator's sliding wedges.

1. BUILDING A RONCHI AND KNIFE-EDGE TESTER

The Ronchi tester is easily made with an American Optical microscope illuminator. This bright illuminator is often required for bare-surface

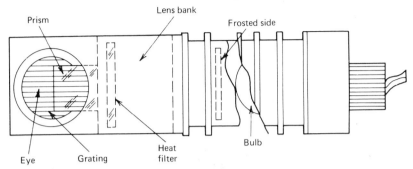

Fig. 29.4. A combination Ronchi and knife-edge tester: Note the slit source near the edge of small right-angle prism. The Ronchi grating is shown horizontal; it must *always* be parallel to the slit during testing. The knife edge is a second mountable unit with a small razor blade in the inner ring.

optical testing, and its brightness can be controlled. In Fig. 29.4 the Ronchi grating (interchangeable) is seen over a light slit ruled near the edge of the right-angle prism. For better illustrative purposes the Ronchi grating is placed at right angles to the parallel lines of the grating. Of course, the lines of the grating must *always* be parallel to the slit source during use. Various Ronchi gratings of high or low frequency can be placed over the opening.

A knife-edge with a close on-axis tester can be added if a small section of razor blade is cemented close to the edge of the right-angle prism with a 1-mm overhang. Both testers can be mounted on a three-axial adjustable sliding unit.

Consider some illustrations in Fig. 29.3. The upper figure is the Ronchi pattern with a 60-line grating of a Schmidt-Cassegrain system before the corrector plate (planoparallel) has been figured. Here the system is placed in the parallel beam of a collimator. This is actually the inside-of-focus Ronchigram from the two reflective mirrors in a system that has an amplification factor of 8. The pattern often has to be fine-ground with various types of glass-ring tool; the ground corrector plate can be made transparent with a few drops of kerosene and wiped dry with lens tissue.

The lower figures are of interest; note the far-left figure in which a hill appears in the corrector plate of the Schmidt-Cassegrain system (inside-of-focus Ronchigram for glass components being figured). Also note the sharp hole in the central area of the corrector plate, and another corrector plate that is slightly overcorrected.

2. CORRELATION BETWEEN RONCHIGRAMS AND FOCOGRAMS

In this section a representation of the Ronchigrams in Fig. 29.5 and the focograms in Fig. 29.6 are developed to reveal the symmetries and other

RONCHIGRAMS

Autocollimation (inside of focus). Parallel sections (box) refracting surfaces

Autocollimation (outside of focus). 45° sections (box) reflecting surfaces

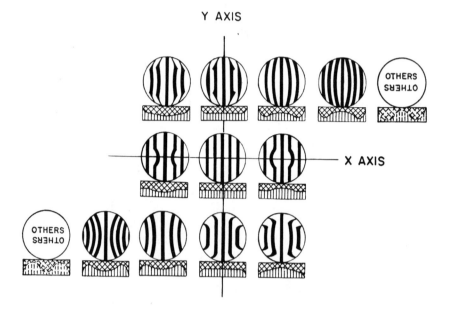

Y AXIS

X AXIS

Autocollimation (inside of focus). 45° sections (box) reflecting surfaces

Autocollimation (outside of focus). Parallel sections (box) refracting surfaces

Fig. 29.5. These Ronchigrams show high or low areas (in box) of residual spherical aberration for refractive and reflective glass elements.

relations in an array of patterns around one central null figure. Shown are the Ronchigrams and focograms of a lens system with spherical aberration within two equal areas. Then some aspects of the correlation between Ronchigrams and focograms and how this information can be of value in figuring lens systems and single components in the optical shop are discussed.

Consider a lens system or a concave reflector with residual spherical aberration within two equal areas—a central area (center to 0.707 zone) and a peripheral area (0.707 zone to edge). Denote the eccentricity of each area as

REFLECTING SURFACES

FOCOGRAMS O.7O7 ZONE

K.E.——►

AUTOCOLLIMATION OR

RADIUS OF CURVATURE

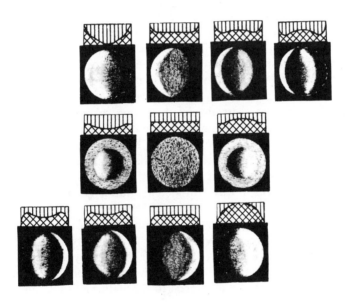

AUTOCOLLIMATION

K.E.——►

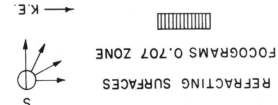

FOCOGRAMS O.7O7 ZONE

REFRACTING SURFACES

S

Fig. 29.6. Focograms showing high or low areas (in box) of the residual spherical aberration for reflective and transmitting lens systems.

455

k_c and k_p for the center and periphery, respectively. Any system can be described by the ordered pair $(k_c \; k_p)$. A parabola has eccentricity equal to -1; a lens system that is a parabola in both zones is denoted $(-1, -1)$. Because spherical aberration is the only concern here, it is defined in terms of eccentricity in relation to the parabola: aberration of the central zone A_c is its eccentricity less the eccentricity of a parabola; that is, $A_c = k_c - 1$ (-1); therefore the parabola $A_c = 0$. By adopting the same definition for spherical aberration of the periphery set the parabola as the origin $(A_c = 0, A_p = 0)$ of a two-dimensional vector space in which the aberration of any lens system is represented. The vector representation offers a means of establishing the pattern of correlation between Ronchigrams and focograms. Consider first the Ronchigram shown in Fig. 29.5. The Ronchigram in the center (null figure) is for a parabola with zero aberration in the central and peripheral zones. Now let the horizontal axis represent the aberration of the central zone in relation to the parabola. Then to the right of the origin is the positive aberration of an ellipsoid central zone with eccentricity greater than -1. To the left of the origin the central zone of the system becomes hyperbolic $(k_c < -1)$ with negative aberration. If the vertical axis represents aberration of the peripheral zone, the systems with positive aberration in this zone will lie above the origin, whereas those with negative aberration will lie below it. Each Ronchigram on the unit square represents the end point of a vector and obeys the laws of vector addition; for example, the Ronchigram at point $(-1, 1)$ is the positive inverse at point $(1, -1)$. In other words, a lens with a configuration $(1, -1)$—positive aberration in the central zone, negative in the peripheral zone—*corrects* a lens with configuration $(-1, 1)$. Any lenses with Ronchigrams lying on a line through the center are additive inverses in this sense. Likewise, other Ronchigrams, such as one at $(0, 1)$, may be obtained by the addition of the Ronchigrams in the square; for example, the lens at point $(1, 1)$ and $(0, 1)$. This vector addition may be of interest to the mathematically minded but it does show that optical figuring takes place by glass subtraction. Also, by the addition or subtraction of negative and positive powers a lens system obeys vector space addition.

In the focograms shown in Fig. 29.6 it can be seen that the same vector space representation holds here as well. In this case the horizontal axis indicates whether the central zone is "high or low" if the value is to the right or left of the origin. The vertical axis does the same for the peripheral zone. Thus, for example, the lens that gives the focograms at point $(-1, 1)$ is low in the central zone and high in the peripheral zone, as shown by the profile exhibited in the cross-sectional drawing below the focogram. By using Figs. 29.5 and 29.6 it is a simple matter to correlate Ronchigrams and focograms.

These focograms are formed by a knife-edge apparatus in tests of a reflecting or refracting surface of a lens system (glass surfaces) in autocollimation or parallel light. The focograms describe the systems with or without spherical aberration. The light source is on the right-hand side and the focograms are formed by making the cutoff of the knife-edge blade left to right for refractive lens systems in parallel light. In other words, the slope departures in the glass surface of the transmitting element are *always* illuminated in the *same* direction as the knife-edge blade is traveling. To view the focograms for autocollimation or radius of curvature (self-autocollimation mode) of a mirror or mirrors (reflective elements) invert Fig. 29.6 and make the cutoffs left to right but keep in mind that the light is *always* illuminated at the left of the moving knife-edge blade. It is important to remember several points when testing: all slope departures are illuminated by the light source position, either real (for glass transmission) or imaginary (reflective surface) position, which for the knife-edge is on the right-hand side and all slopes *away* from the light will be dark. Also, the diameter of the reflective mirror or transmitting lens is the dividing line of the lights and shadows, and a bright area on one half will be bright on the opposite half. This is further explained in Chapter 26.

Near each focogram and Ronchigram in Fig. 29.5 or 29.6 is a small box that shows the apparent slope departures for each optical setup (De Vany 1965, Malacara and Cornejo, 1974). Note the 90° vertical sectional lines and compare their contours with the left half-sector of the Ronchigrams in Fig. 29.5. For autocollimation or radius of curvature testing of reflective elements 45° cross-sectional lines show the apparent slopes of the optical or reflecting surface.

In summary, the Ronchigrams provide information on the focal plane errors. Thickness of the Ronchi band is the magnitude of the slope departure change of the glass errors.

Second, the knife-edge reveals the contour error in the glass of a lens system qualitatively.

Third the interferogram contours glass errors and therefore must be differentiated to arrive at focal plane errors.

3. INTERPRETATION OF INTERFEROGRAMS

First examine the customary approach used to interpret the interferogram. The point positions along the axis and the displacement therefrom of the reference and system under test are adjusted to give bands that are as straight as possible. (See Fig. 29.7.)

$$\text{Wave error (peak)} = p/4d$$
$$\text{Wave error (peak-to-peak)} = P/2d \qquad (1)$$

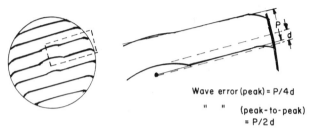

Wave error (peak) = P/4d

"   "   (peak-to-peak)
= P/2d

Fig. 29.7.   Calculation of wave-peak or wave-peak-to-peak errors.

In the typical interferogram for a system with spherical aberration a straight line is drawn through the average extensions of the adjacent bands. The formulas give the peak and peak-to-peak errors as they relate to the spherical wavefronts for the lens system under test.

To weight these errors to determine their relative effect on the image-forming capability of the lens a root-mean-square (rms) value must be taken. This is more laborious. The interference band is divided into a number of segments and the amplitude as well as the radius of each of these segments is used in formula (2):

$$\text{Wave error (rms)} = p/2n \left((r_1 a)^2 + (r_2 a_2)^2 ...\right)^{1/2} \qquad (2)$$

where r is the radius aperture, n is the number of rings overall, and a is the amplitude.

The number of rings in the pattern can be used to derive the actual displacement between the focus of the two mirrors. (See Fig. 29.8.)

$$\Delta f = 2n\lambda F^2 \qquad (3)$$

where Δf is the change in focus, n is the number of rings, and F equals R/r. (See Fig. 29.9.)

It is apparent that the spread between the reference mirror and the test mirror in the test system is described simply by counting the number of concentric bands in the interferometer. The next question is how to cor-

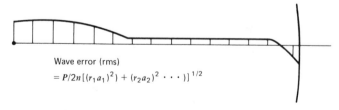

Wave error (rms)

$= P/2n[(r_1 a_1)^2) + (r_2 a_2)^2 \cdots)]^{1/2}$

Fig. 29.8.   Calculation of the wave error or rms.

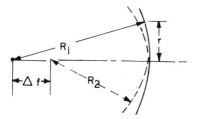

Fig. 29.9. Calculation of the delta (Δf) change of the two wavefronts.

relate the number of interference bands when they are straight, with the lateral displacement of the focus of the test mirror with respect to the reference mirror:

$$d = n\lambda F/2 \qquad (4)$$

where d is the displacement between the fringes laterally, n is the number of fringes, λ is the wavelength of the light source, and F or R/r is the focal ratio of the cone. (See Fig. 29.10.)

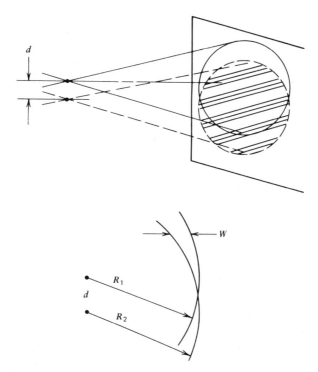

Fig. 29.10. Two wavefronts that form later-sheared interferograms projected on a frosted screen.

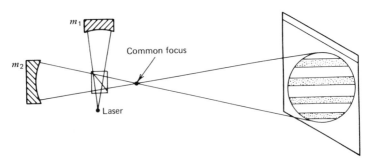

Fig. 29.11. Similar to Fig. 29.10 but showing the LUPI (laser unequal path interferometer). A lens system (not shown) would be placed in the divergent rays of this setup, which would form parallel rays to the screen and the interferograms on the frosted screen.

The preceding description of the interpretation of an interferogram for peak, peak-to-peak errors, and for the rms relates to two fronts as they arrive at a screen or are projected onto the objective. However, can this straightforward approach be used to define characteristics of the lens such as circle of confusion, distribution of focus, and the concentration of the light in primary maxima? The answer in no!

The following shows what the interference bands produced in the interferometer can actually tell us. A simple interferometer used as an illustration to develop the relationships is shown in Fig. 29.11.

Note that if the foci of the two mirrors are coincident, the two cones will appear to radiate from the same source. If two cones are perfectly spherical, then no bands will appear on the screen. The screen will be dark or light, depending on the phase relation of the two cones of light as they arrive on the screen.

Now if mirror m_2 is moved slightly closer to the screen, its focus will also be closer to the screen (see Fig. 29.11) and, correspondingly, the radius of the cone measure from the focus of the screen will be shorter for mirror m_2 than for mirror m_1. (See Fig. 29.12.)

Because the radius of the two cones will be different as they arrive at the screen, a bull's-eye pattern will be developed by the interference of the waves at the screen.

Needless to say, it is not necessary that the bull's-eye pattern be concentric with the center of the interferogram of the lens system to derive useful information. As a matter of fact, as shown in Fig. 29.13 (*upper*), the offset distance of the bull's eye is a significant factor in the evaluation of a typical interferogram.

Shown in Fig. 29.13 (*lower*) is an illustration of an interferogram of a lens

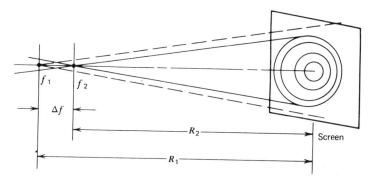

Fig. 29.12. The two different (R_1 and R_2 values) wavefronts from an uncorrected lens system, using the laser interferometer in Fig. 29.10, which forms concentric circular interferograms.

system with a center zone that focuses at a shorter distance than the outer zone. The reference mirror has been set so that its focus is coincident along the axis with the test mirror but is displaced laterally. The number of rings is then counted within this circle. Formula (3) for Δf is used to determine the focal spread along the axis for two zones of the lens system.

The same procedure for interpreting more complicated interferograms can be applied as long as the glass errors are in concentric zones.

Fig. 29.13. Same as Fig. 29.12 but with the concentric interferogram off-center.

4. CORRELATION OF RONCHIGRAMS AND INTERFEROGRAMS

Ronchigrams and interferograms can be correlated further by quasi-Ronchigrams. In this section an interpretation of the correlation is offered: quasi-Ronchigrams are mirror-transitive images of interferograms. Among the many methods of testing spherical aberration by analysis of the wavefront each has it limitations. The several types of interferometer used are the nonshearing represented by the LUPI (laser unequal path interferometer), Twyman-Green, Hendrix and the shearing type by Bates, Saunders and Murty's cube, and Koester. (See Fig. 29.14.)

Consider the residual spherical aberration. The image-forming system is tested on-axis with parallel light formed in conjunction with an autocollimating double reflective hand mirror and placed upright on this line.

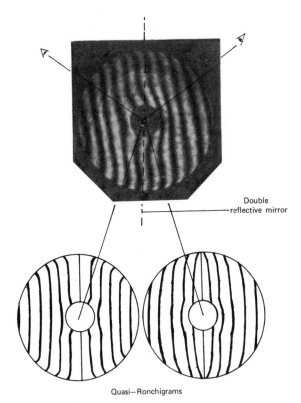

Double
reflective mirror

Quasi—Ronchigrams

Fig. 29.14. Quasi-Ronchigrams formed by a doubly reflective mirror placed upright at the central area and the resulting quasi-Ronchigrams (inside-of-focus on the left and outside-of-focus on the right).

Observe the symmetrical quasi-Ronchigrams. The eye will see the inside-of-focus Ronchigram on the left and the outside-of-focus on the right. Of further interest is the testing of a spherical mirror at its radius of curvature by the LUPI; because the radius of curvature is a self-autocollimating setup, all reflective optics of one or two components are exactly the same in any collimation setup with parallel light. Here we have a reversal interpretation of an interferogram for spherical aberration as in Fig. 29.14. The central zone out to the 0.707 zone is now overcorrect (low), the chosen reference 0.707 zone is still reasonably corrected, and the peripheral is undercorrect (high) with the extreme edge badly turned down. The Ronchigrams, in turn, are also of reversal interpretation: the inside-of-focus must be considered outside-of-focus and *vice versa*. It is apparent that the quasi-Ronchigrams are mirror transitive images with or without spherical aberration. It can be further shown that an interferogram can be drawn from the inside-of-focus or outside-of-focus Ronchigram by simply drawing slope departures across the aperture circle. To do this cover the right side of the inside-of-focus element. For the LUPI interferometer a slight tilt is introduced to form the desired number of fringes in the symmetrical interferogram. In this section the many problems of shearing with tilt or rotary shearing around a principle ray, centering, and defocusing to interpret high or low zonal departures cannot be discussed. Of primary interest are the symmetrical interferograms and mirror transitive images of the symmetrical quasi-Ronchigrams.

It is assumed that the projected laser interferogram being photographed or observed is properly oriented. It is essential that the wavefront information from the interferogram (high or low) for the correction of the residual spherical aberration be examined. This is determined by defocusing the interferometer from the established reference zone. Figure 29.14 is the symmetrical interferogram from the LUPI which shows that several zones of residual spherical aberration are present in a lens system. Among these zones of spherical aberration the central zone out to the 0.707 zone is undercorrect, the chosen reference 0.707 zone is reasonably corrected, and the peripheral zone is overcorrect, with the extreme edge badly turned up. A dashed line drawn across the aperture of the interferogram divides it into parts. Take a small Ronchigram and draw identical contour slopes or cover the left-hand side of the outside-of-focus Ronchigram. The interferogram developed is the same as Fig. 29.14.

Perhaps it would be well for beginners to know how to determine high or low areas. To interpretate high or low areas in an interferogram focus the LUPI (any interferometer) on the best chosen reference zone (here the 0.707) until the fringes are straight and parallel; then defocus the LUPI by touching the collimating flat and causing the fringes to be straight and parallel in the central zone or for any other zone. For high or overcorrect zones (transmitting element in an autocollimation mode) defocus to a shorter light-path distance by touching the collimating flat on its back toward the lens system; for low or undercorrected zones defocus to a longer path distance by touching the collimating stand in front. Hence all light paths are relative to where the autocollimating flat and the lens system receiving the parallel light are touched. It is interesting to note that when concave mirrors at their centers of curvature or one- or two-mirror systems in an autocollimation setup are tested, low areas cause the LUPI to defocus to a shorter distance (again touching the flat or one of the mirrors), whereas high areas cause a defocusing to a longer path distance. Where the optical element being tested is touched, is again relative to the reference zone. (See Appendix 21.)

5. CORRELATION OF INTERFEROGRAMS BY DOUBLE WIRE PATTERNS

Figure 29.15 contains a number of double-wire patterns similar to those in Fig. 28.4. For a lens system being tested by a parallel light setup with a double-wire tester note the translation of the double wires at each zone scanned. The diagram on the far left in Fig. 29.15 (*upper*) shows an undercorrect zone near the 0.707; the diagram on the far right shows an overcorrect zone. The center diagram (*upper*) illustrates the bifurcating Ronchiline formed by the first wire in the testing unit; the second wire keeps the testing unit from rotating off-axis. The lower diagrams illustrate the double-wire patterns for

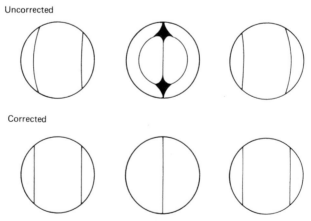

Uncorrected

Corrected

Fig. 29.15.   Double-wire test patterns from a lens system (glass transmitting elements) tested in parallel light, both showing the degree of correction.

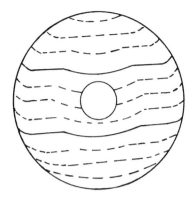

Fig. 29.16. A Quasi-interferogram formed by transitive mirror images from the double-wire test diagrams (see Fig. 29.15).

a well-corrected lens system being tested in a parallel light setup with an autocollimating flat.

J. E. Ward et al., in *Applied Optics*, April 1971, p. 896, describe the correlation of Ronchigrams by double-wire patterns.

> The results of the De Vany test are shown in the figure. The double wire apparatus was moved in the beam, and successive photographs were made of the corrected and uncorrected results. The uncorrected results appear in the top row of the photographs in the figure, with the corrected results below. We note that the expected cubic curves appear in the uncorrected version. The background fringes are due to the coherent illumination of the larger fringes of the hologram. These images of course cause some degeneration of the images formed. In the corrected results, the De Vany wires are projected as straight lines in all portions of the beam.

Ward's figures of the double-wire test are similar to the diagrams in Fig. 29.15.

Figure 29.16 can be developed from Fig. 29.15 if the entire exit pupil is scanned by the double-wire tester. Again place a small, double reflective mirror on a quasi-interferogram developed by the scanning double-wire test pattern in its central area. The reflective of the quasi-interferogram forms the inside-of-focus Ronchigram; the right half is the outside-of-focus Ronchigram. The dashed lines show the pairs of double-image patterns as the double-wire apparatus scans across the exit pupil of the lens system in a parallel light setup.

Can use be made of these quasi-Ronchigrams developed by mirror transitive images of interferograms? Answer: Yes. If the interferograms are correctly orientated in their photographed prints (perhaps with a strip of masking tape on the component being tested) and the proper optical setup is identified as a reflecting or transmitting element it can be determined whether the zonal departures are high or low. Take, for instance, Fig. 29.14 for a planoconvex light-transmitting lens being optically figured and tested with parallel light and an optical flat. Note the quasi-Ronchigram formed by the mirror's transitive image in the lower left-hand side. This is always considered as the inside-of-focus Ronchigram. Here the central area is low, the 0.707 zone reasonably well corrected, and the extreme edge turned up. The lower figure at the right-hand side is the outside-of-focus Ronchigram pattern and it will verify what is observed in the left-hand lower figure.

Again consider an example from reflective optics: a concave mirror is figured to a paraboloid and tested with parallel light formed by an autocollimating flat; it is retested at its radius of curvature which if closely examined is a *self*-autocollimating device. An interpretation of the quasi-Ronchigrams in Fig. 29.14 is the following: the lower left-hand diagram is the inside-of-focus Ronchigram pattern, the central area is high, the 0.707 zone is reasonably well corrected, and the extreme outer edges are high. Because the lower right-hand figure is always the outside-of-focus Ronchigram, it can verify the interpretations observed for the other side.

6. CONCLUSIONS

The general classification of fringe patterns observed with several interferometers is but a partial description of a large category of various types of interferometer that might be found in optical shops. *Optical Shop Testing*, edited by Daniel Malacara, Wiley, New York, 1978, should be in the hands of all opticians, optical engineers, supervisors, and advance amateur telescope makers. The book is mathematical for those who wish further explanation.

BIBLIOGRAPHY

Anderson, J. A. and Porter, M., *Astro. Phys.*, **LXX,** 170–78 (1929).

Baker, J. G. "Optical Systems for Astonomical Photography," in *Amateur Telescope Making, Book Three*, A. G. Ingalls, Ed. Sci. Am., 1–34 (1953).

De Vany, A. S. "Some Aspects of Interferometric Testing and Optical Testing," *Appl. Opt.*, **4,** 831–835 (1965).

De Vany, A. S. "Spherical Aberration Analysis by Double Wire Testing," *Appl. Opt.*, **6,** 1073–1075 (1967).

De Vany, A. S. Supplement to "Some Aspects of Interferometric Testing and Optical Figuring, *Appl. Opt.*, **9,** 1219–1220 (1970).

De Vany, A. S. Supplement to "Spherical Aberration Analysis by Double Wire Testing," *Appl. Opt.*, **9,** 1720 (1970).

De Vany, A. S. and Art De Vany, Jr., "Patterns of Correlation between Focograms and Ronchigrams," Boletin Del Instituto De Tonanttizintla, **1**, 295–297 (December 1975).

Francon, M. "Polarizing Interferometers Applied to the Study of Isotropic Objects," National Physic Laboratory, Symposium No. 11, Inteferonmetry, Her Majesty's Stationary Office, London, 129–132 (1959).

Murty, M. V. R. K. "The Use of a Single Plane Parallel Plate as a Lateral Shearing Interferometer with a Visible Gas Laser Source," *Appl. Opt.*, **3**, 531 (1964).

Ronchi, Vasco, "Forty Years of History of a Grating Interferometer, *Appl. Opt.*, **3**, 1041 (1964).

Twyman, F. *Prism and Lens Making*, 2nd ed., Hilger and Watts, London, 1952, pp. 431–438.

Veret, C. "Polarizing Intermeters," in *Metrology*, Ed. P. Mollet, Ed., Pergamon, New York, 1960, pp. 305–308.

Ward, J. E. et al., "Lens Aberration Correction by Holography," *Appl. Opt.*, **10**, 896–900 (1971).

Appendix 1

Types of Prism

Prisms serve one of two major functions in optical systems. In spectro-scopes, spectrographs, and spectrophotomers dispersing prisms are used to separate the different wavelengths (color). In visual instruments reflecting prisms reflect, deviate, or reorient an image or beam of light.

Prisms can be subdivided into two classes—dispersing and deviation. The discussion begins with dispersing prisms.

In addition to glass there are other light transmitting materials from which dispersing prisms are made: lithium, barium, calcium fluorides, rock salt (sodium chloride), potassium iodide or bromide, and natural quartz. Most of these crystals are grown by a special process and are available on the market. An ever-growing number of crystals is being developed each year for use in specific spectral regions. Each of these crystals requires special handling during grinding and polishing and often the fabrication methods are confidential.

1. DISPERSING PRISMS

A limited number of dispersing prisms is available for spectroscopes and spectrophotometers, which can be made of crystalline optical transmitting materials. All glass angles of the Wadsworth prism are at 60°. (See Fig. A1.1.)

The Littrow and the Abbe prisms have 60–30–90° glass angles. The Littrow differs in that the 30° face is coated with silver or aluminum for retro-reflecting the light rays through the prism. (See Figs. A1.2 and A1.3.)

The constant-deviation prism has four sides; its glass angles are shown in Fig. A1.4.

For transmitting light in the wavelength region beyond 15 μ silver chloride, KRS, or thallium-bromide-iodide, cesium bromide, and cesium iodide are available.

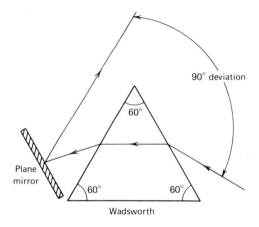

Fig. A1.1. Wadsworth prism.

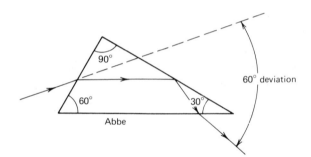

Fig. A1.2. Abbe prisms.

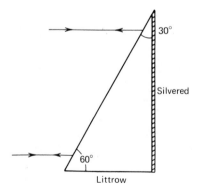

Fig. A1.3. Littrow prism.

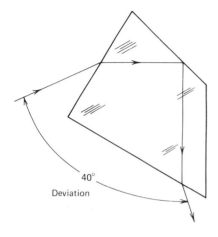

40°
Deviation

Fig. A1.4.   Deviation prism.

2.   DEVIATION PRISMS

This section treats the design data of many types of deviation prism made from crown glasses, generally of BSC-2 or BK-7. Other types of glass used include the flint in Dove prisms and a few achromatic wedges or prisms. These prisms reflect transmitted light rays internally, and the light rays must be confined in the prism by a number of glass reflecting angles that must be silvered or aluminized for testing. This is discussed more thoroughly here. Any attempt to classify the great number of prisms leads to many exceptions and contradictions because some have interchangeable uses in collimated, divergent, or converging light paths. Also some combinations of prisms are cemented together.

The deviation prism is shown in Figure A1.4.

The meaning of normal, reverted, inverted, and inverted-reverted images is shown in Fig. A1.5. These terms are used continuously in the descriptions of prisms. It is important to realize that an objective or lens inverts and reverts an image and to erect that image an assembly of proper prisms or

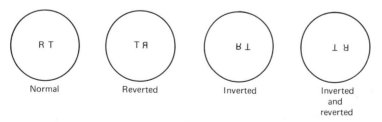

Normal   Reverted   Inverted   Inverted
                                                      and
                                                      reverted

Fig. A1.5.   Image nomenclature.

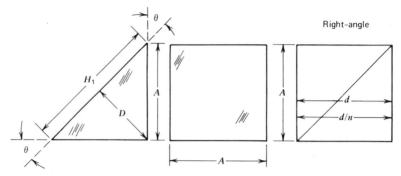

Fig. A1.6. Right-angle prism. Given: $A = 1.0$, $n = 1.517$, $\theta = 45°$, $C = 0.707A$, $H = 1.4142$, $A = 1.4142$, $d = A = 1.0$, $d/n = 0.6592$.

roof prism must be used. Again in some panoromic sights a right-angle or other prism may be placed in front of the lens to receive parallel light from the image. In the descriptions following the prisms the terms *invert* and *revert*, are used in reference to an object with the normal eye without any lens or mirror system in front of the prism or assembly of prisms.

Right-Angle Prism. This single prism deviates the line of sight through a 90° angle. The image will be inverted when the prism is held in a horizontal plane with one face viewing the object (Я); the object is reverted when the prism is held at a 90° to the horizontal plane.

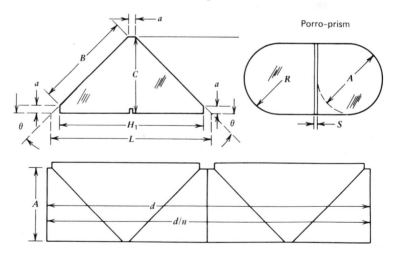

Fig. A1.7. Porro-prism system. Given: $A = 1$, $n = 1.517$, $\theta = 45°$, $a = 0.1$, $S = a/2$, $B = 1.1442A = 1.4142$, $C = 1.0 + a = 1.1$, $H_1 = 2A + a = 1.1$, $L = 2A + 3a = 2.3$, $d = 2(2A = 3a) = 4.50$, $d/n = 3.0324$.

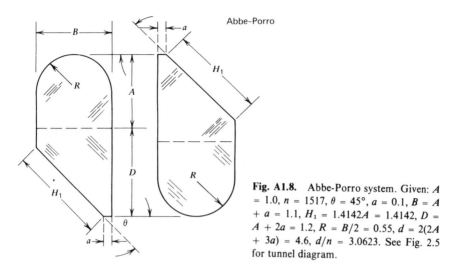

Abbe–Porro

Fig. A1.8. Abbe-Porro system. Given: A = 1.0, n = 1517, θ = 45°, a = 0.1, B = A + a = 1.1, H_1 = 1.4142A = 1.4142, D = A + 2a = 1.2, R = $B/2$ = 0.55, d = 2(2A + 3a) = 4.6, d/n = 3.0623. See Fig. 2.5 for tunnel diagram.

Porro-Prism System. M. Porro designed this prism (two are required for the system) in 1850. It is a right-angle prism and a central slot is cut across the middle to stop counterreflections (ghost images). Two prisms placed 90° to each other produce an erected image. Binoculars use this system, the precision of the reflecting-angles are not high.

Abbe-Porro System. Abbe modified Porro's system. This prism consists of two similar prisms cemented together but is not used today because four

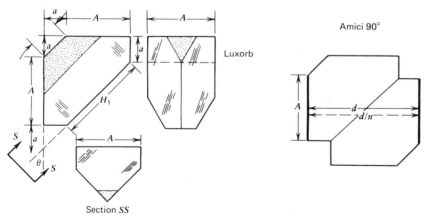

Amici 90°

Luxorb

Section SS

Fig. A1.9. Amici prism (90°). Given: A = 1, n = 1.517, θ = 45°, d = 1.7071A = 1.7071, H_1 = 1.4142A = 1.4142, d/n = 1.1253, a = 0.25 A.

polished surfaces are required. The two prisms cemented side-to-side result in an erected image that gives a vibration-proof system. The 45° glass reflecting angle must have high precision.

Amici Prism. This single prism, designed by an Italian astronomer (1784–1863), inverts and reverts an image and at the same deviates the line of sight through an angle of 90°. It is widely used in modern telescope systems for zenith and ground observations.

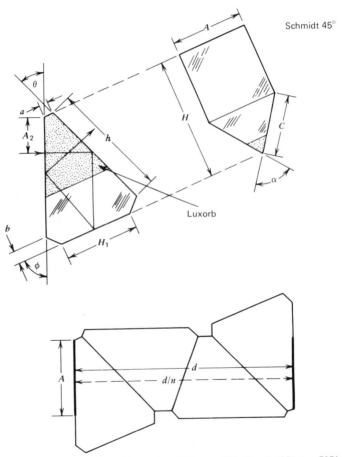

Fig. A1.10. Schmidt prism (45°). Given: $A = 1.00$, $\alpha = 90°$, $C = 0.7071A = 7071$, $a = 0.10$, $h = 1.4142A + 0.5412a = 1.4683$, $b = 1.8478a = 0.1818$, $n = 1.5170$, $H_1 = 1.0824A = 1.082$, $d/n = 2.2506$, $\theta = 45°$, $H = 1.4142A + 2.389a = 1.6531$, $d = 3.1442A = 3.4142$, $\phi = 67°30'$.

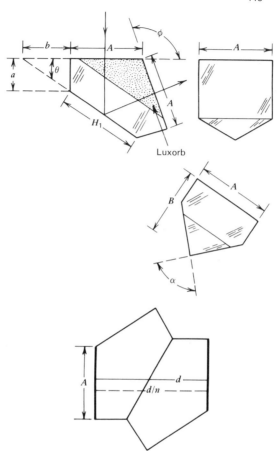

Fig. A1.11. Frankford Arsenal No. 1 (115°). Given: $A = 1.00$, $n = 1.517$, $\alpha = 90°$, $\theta = 32°30'$, $\phi = 115°$, $H_1 = 1.1857A = 1.1857$, $a = 0.7071A = 0.7071$, $B = 0.9306A = 0.9306$, $b = 0.7320A = 0.7320$, $d = 1.5697A = 1.5697$, $d/n = 1.0347$.

Schmidt Prism. This prism is similar to the Amici prism. It will revert and deviate the image and at the same time deviate the line of sight through an angle of 45°. Both prisms have a 90° roof angle that must be held within 2 sec or arc.

Frankford Arsenal No. 2 (60°) and No. 1 (115°). These two prisms are similar to the Amici or Schmidt prism. The difference being the deviated line of sight for the observer.

Frankford Arsenal No. 6. This prism is difficult to make because it consists of a single piece of glass and has a 90° roof angle of high precision. It will deviate the line of sight through an angle of 90° in the horizontal plane and through an angle of 60° in the verticle plane. The prism will invert the image.

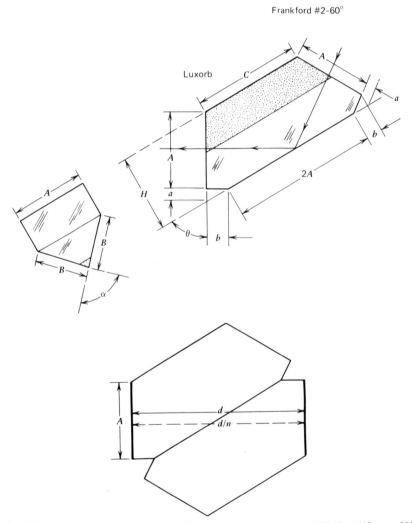

Fig. A1.12. Frankford Arsenal No. 2 (60°). Given: $A = 1.00$, $n = 1.5170$, $\theta = 60°$, $\alpha = 90°$, $a = 0.1547A = 0.1547$, $b = 0.2680A = 0.2680$, $C = 1.4641A = 1.4641$, $H = .866(A + a) = 1.0$, $B = 0.7071A = 0.7071$, $d = 2.2690A = 2.2690$, $d/n = 1.4951$.

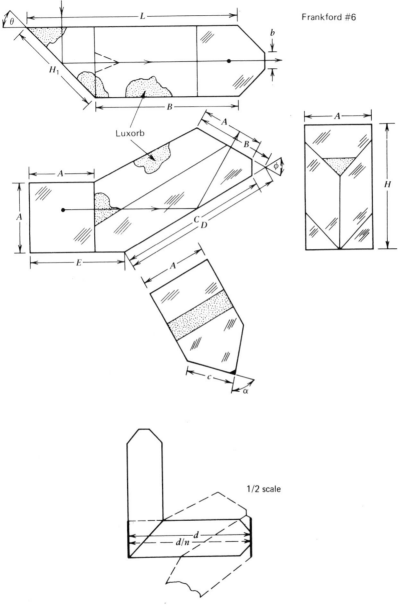

Frankford #6

Luxorb

1/2 scale

Fig. A1.13. Frankford Arsenal No. 6. Given: $A = 1.0$, $n = 1.517$, $\phi = 60°$, $\theta = 45°$, $\alpha = 90°$, $c = 0.7071A = 0.7071$, $B = 1.2071A = 1.2071$, $D = 2.4142A = 2.4142$, $C = 2.2071A = 2.2071$, $E = 1.5774A = 1.5774$, $H_1 = 1.4142A = 1.4142$, $L = 3.4888A = 3.4888$, $H = 1.8107A = 1.8107$, $d = 3.6681A = 3.6681$, $F = 2.1488A = 2.1488$, $d/n = 2.4180$.

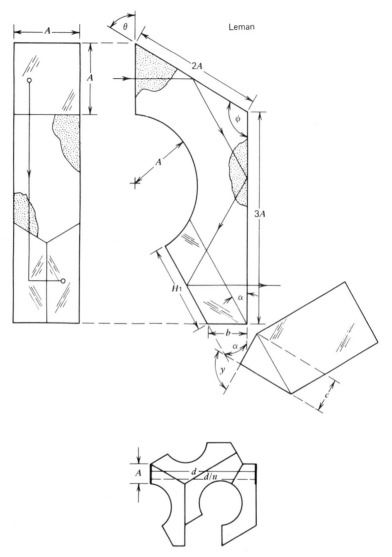

Fig. A1.14. Leman prism. Given: $A = 1.00$, $H_1 = 1.3099A = 1.3099$, $n = 1.517$, $c = 0.7071A = 0.7071$, $\alpha = 30°$, $b = 0.5774A = 0.5774$, $\theta = 60°$, $d = 5.1962A = 5.1962$, $y = 90°$, $d/n = 3.4253$, $\phi = 120°$.

Leman Prism. This prism is largely used in visual instruments. Its compactness is useful in small telescopes of long focal length. The glass path is long (3.4 times its aperture) and its homogeneity must be of excellent quality.

Penta-Roof Prism. This prism, which is similar to the penta prism, will invert and revert the image. The image remains in the same plane. The prism has limited use.

Penta Prism. This prism will neither revert nor invert the image but will deviate the line of sight through an angle of 90°. It has a large range of use in testing tooling fixtures and optical systems. Range finders contain one prism at each end. Two sides (H_1) are silvered.

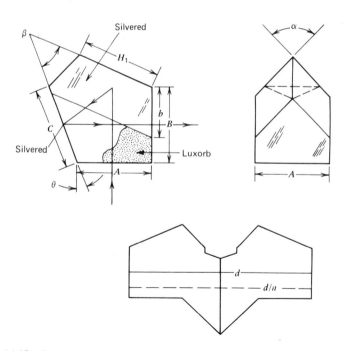

Penta, roof prism

Fig. A1.15. Penta-roof prism. Given: $A = 1$, $a = 90°$, $H_1 = 1.0824 = C$, $B = 45°$, $b = 0.875A$, $\theta = 22.5°$, $B = 1.125$, $d = 2.2508$, $d/n = 3.375$.

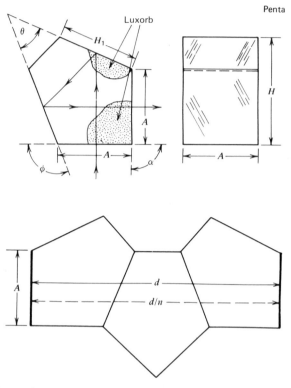

Fig. A1.16. Penta prism. Given: $A = 1.00$, $n = 1.517$, $\alpha = 90°$, $\theta = 45°$, $\phi = 67°30'$, $H_1 = 1.0824A = 1.0824$, $H = 1.4142A = 1.4142$.

Frankford No. 3. This prism is made of one piece of glass. The line of sight is deviated upward at a 45° angle in the horizontal plane and at the same time through an angle of 90° in the vertical plane. Inverted and reverted images are observed if the observer stands at right angles to the line of sight. The prism has limited use.

Frankford No. 4. This prism is made from one piece of glass and must have its bottom surface silvered or aluminized.

Frankford No. 5. All the internal reflecting angles of this prism are greater than 42° and require no metallized reflective coatings. It is made of one piece of glass. The line of sight is deviated through an angle of 90° in the horizontal plane and simultaneously through an angle of 60° in the vertical plane. The side-viewed object is reverted and inverted.

Frankford No. 7.   This prism is similar to the Frankford No. 5. It differs in that the line of sight is deviated through an angle of 90° in the horizontal plane and simultaneously through an angle of 45° in the verticle plane. The side-view image is neither inverted nor reverted.

Corner-Cube Prism.   This prism has several names and is made from an optical square element with superimposed entrance and exit faces. It returns

Frankford #3

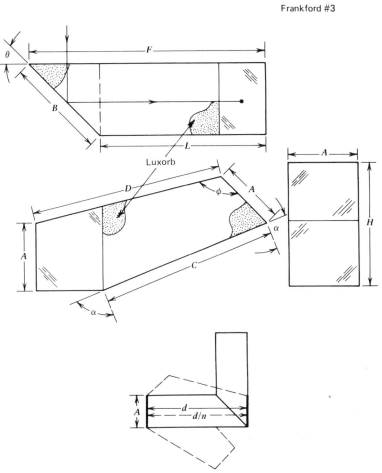

Fig. A1.17.   Frankford No. 3. Given: $A = 1$, $n = 1.5170$, $\theta = 45°$, $\phi = 120°21'40''$, $\alpha = 67°30'$, $B = 1.4142A = 1.4142$, $C = 2.6131A = 2.6131$, $D = 2.7979A = 2.7979$, $E = 2.4142A = 2.4142$, $F = 3.4142A = 3.4142$, $G = 1.7071A = 1.7-71$, $d = 3.4142A = 3.4142$, $d/n = 2.2506$.

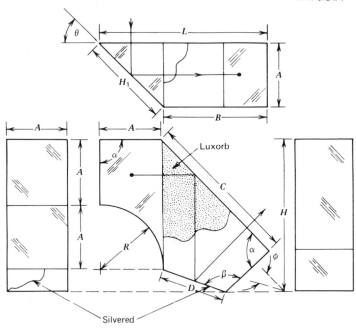

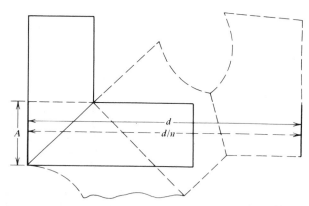

Fig. A1.18. Frankford No. 4. Given: $A = 1.00$, $\phi = 22°30'$, $\alpha = 90°$, $n = 1.5170$, $\theta = 45°$, $\beta = 112°30'$, $H_1 = 1.4142A = 1.4142$, $C = 2.4142A = 2.4142$, $D = 1.0824A = 1.0824$, $B = 1.7071A = 1.7071$, $H = 2.4142A = 2.4142$, $L = 2.7071A = 2.7071$, $R = A + 1.00$, $d = 2.4142A = 4.4142$, $d/n = \cdot 2.9098$.

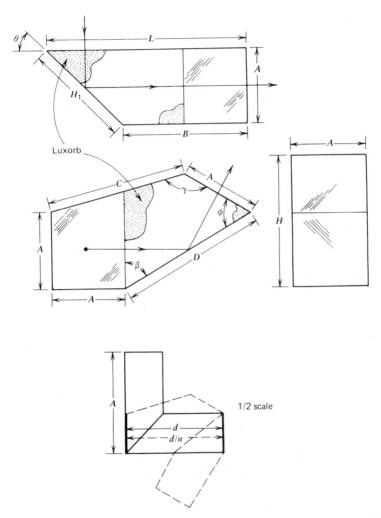

Fig. A1.19. Frankford No. 5. Given: $A = 1.0$, $n = 1.517$, $\beta = 60°$, $y = 135°$, $\theta = 45°$, $L = 2.7321A = 2.7321$, $C = 1.9318A = 1.9318$, $D = 2.000A = 2.000$, $B = 1.7321A = 1.7321$, $H_1 = 1.4142A = 1.4142$, $H = 1.5000A = 1.5000$, $d = 2.7437A = 2.7437$, $d/n = 1.8086$.

an image to its original source. Nests of 30 or more prisms were placed on
the moon at several positions during the Apollo program for laser ranging.

Harting-Dove Prism. Note that if the apex corner of a right-angle prism
is moved beyond its center a Harting-Dove prism is created. It is made of
one piece of glass. The image will be inverted if the prism is lying on its base
and reverted if it is resting on its side. It is a direct-vision prism and must be
used in parallel light. Note the shortening of its total length as the higher
index of refraction values are increased.

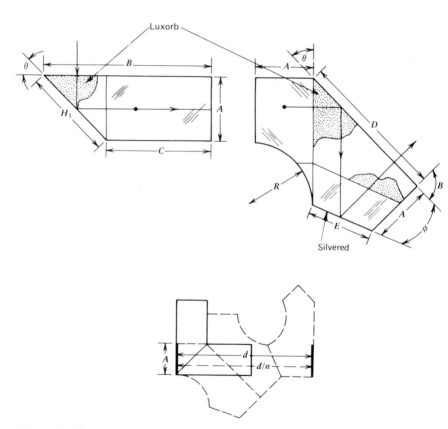

Fig. A1.20. Frankford No. 7. Given: $A = 100$, $n = 1.517$, $\theta = 45°$, $\phi = 22°30'$, $\beta = 90°$, $\alpha = 112°30'$, $H_1 = 1.4142A = 1.4142$, $B = 2.7071A = 2.7071$, $C = 1.7071A = 1.7071$, $D = 2.4142A = 2.4142$, $E = 1.0824A = 1.0824$, $H = 2.4142A = 2.4142$, $R = A = 1.00$, $d = 2.4142A = 2.4142$, $d/n = 2.9098$.

Corner-cube prism

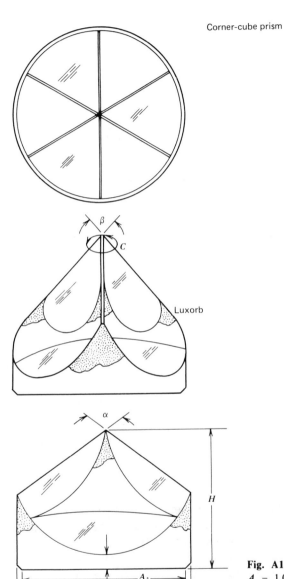

Fig. A1.21. Corner-cube prism. Given: $A = 1.0$, $A_1 = 0.97$, $C = 3$ (90°), $\beta =$ vector angle, $\alpha = 90°$, $a = 0.09A$.

Harting–Dove

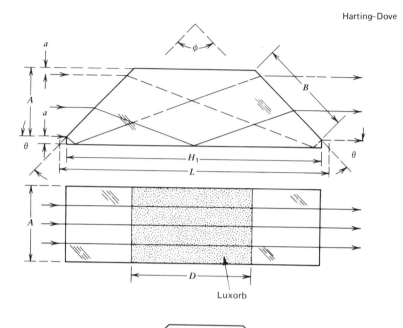

Fig. A1.22. Harting-Dove prism. Given: $A = 1.00$, $a = 0.05$, $n = 1.517$, $\theta = 45°$, $\phi = 90°$, $B = (A + 2a) [(n^2 - \sin^2\theta + \sin \theta)^{-1/2}/(n^2 - \sin^2\theta - \sin \theta)^{-1/2}] + 1 = 4.2271 (a + 2a) = 4.6498$, $C = B - 2a = 4.5498$, $D = B - 2 (A + 2a) = 2.4498$, $d/n = 2.4499$, $E = a + A/\cos a = 1.4142 (A + 2a) = 2.4498$, $d = n (A + 2a)/\sin \theta [(n^2 - \sin^2\theta - \sin \theta)^{-1/2}] = 3.3787 (A + 2a) = 3.7165$. Note: see Table A.

Double Dove Prism. This twin prism consists of two Harting-Dove prisms. The two inner faces are cemented together after they are silvered. This method cuts the length of the single Harting-Dove prism in half. It is used in parallel light and performs as a single Harting-Dove prism often in conjunction with celostats and plantearium projections to offset the rotation of the plane of the sun's image or the projection of the star field by moment of the reflecting flat.

Pechan Prism. This prism can be used in converging or diverging light paths and is similar to the Harting-Dove. Because of its long light path, a

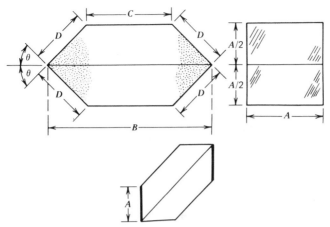

Fig. A1.23. Double-Dove prism. Given: $A = 1.00$, $n = 1.5170$, $\theta = 45°$, $B = A/2\,(n^2 - \sin^2\theta - \sin\theta)^{-1/2} + 1 = 2.1136A = 2.1136$, $C = B - A = 1.1136$, $D = A/2\cos\theta = 0.7071A = 0.7071$, $d = nA/2\sin\theta\,(n^2 - \sin^2\theta - \sin\theta)^{-1/2} = nAC = 1.6893$, $d/n = 1.1135$.

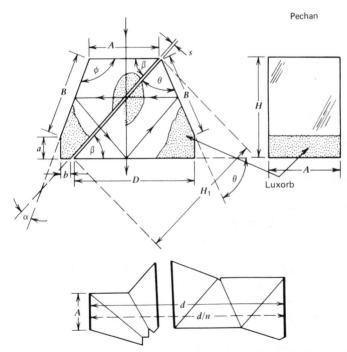

Fig. A1.24. Pechan prism. Given: $A = 1.00$, $\beta = 45°$, $\alpha = 22°30'$, $\theta = 67°30'$, $\phi = 112°30'$, $n = 1.5170$, $a = 0.2071A = 0.2071$, $b = 0.1414A = 0.1414$, $s = 0.002A = 0.002$, $B = 1.0824A = 1.0824$, $H_1 = 1.7071A = 1.7071$, $H = 1.2071A = 1.2071$, $D = 1.8284A = 1.8284$, $d = 4.6213A = 4.6213$, $d/n = 3.0464$.

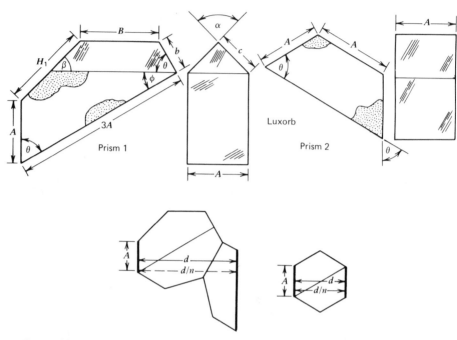

Abbe prism—Type A

Fig. A1.25. Abbe prism, Type A. Given: A = 1.00, ϕ = 30°, α = 90°, n = 1.517, θ = 60°, β = 45°, H_1 = 1.4142A = 1.4141, B = 1.3094A = 1.3094, c = 0.7071A = 0.7071, b = 0.5774A = 0.5774, assembled length (L) 3.4644A = 3.4644, d = 5.1962A = 5.1962, d/n = 3.4253.

great reduction in the overall length of the instrument is achieved. Surface B must be silvered. It will invert and revert the image, depending on the orientation of the prism in the instrument. The prisms in the assembly are slightly separated. The 22.5° glass-reflecting angle must be accurately made because of the double reflection of the internal images.

Abbe Prism, Type A.   This prism inverts and reverts the image but will not deviate the line of sight and is often called a "direct vision prism." It is used in expensive, straight, twin-barrel binoculars. Note that this is a cemented pair of prisms; prism 2 is cemented at 3A on prism 1.

Abbe Prism, Type B.   This prism is like type A except that three prisms are cemented together instead of two. Two of the prisms are twins with 60° entrance and exit faces.

Carl Zeiss System. This optical assembly consists of the three single prisms shown in Fig. A1.26. The objective lens can be placed in front or between P_1 and P_2. As rule it is placed between P_1 and P_2 for better protection of the flint glass in the objective. This system will invert and revert images but will not deviate the line of sight. The line of sight is displaced by an amount that depends on the separate distance between the prisms for the objective.

C. P. Goerz System. This prism system is similar to the Carl Zeiss which consists of three single prisms. It inverts and reverts the image but will not deviate the line of sight. Note the similarity of the two systems by which the 60° single prism in the Zeiss is incorporated in the roof prism of the Goerz system.

Wollaston Prism. W. R. Wollaston, an Englishman, designed this prism between 1766 and 1828. It is made of a single piece of glass and will neither revert nor invert the image but will deviate a light beam through an

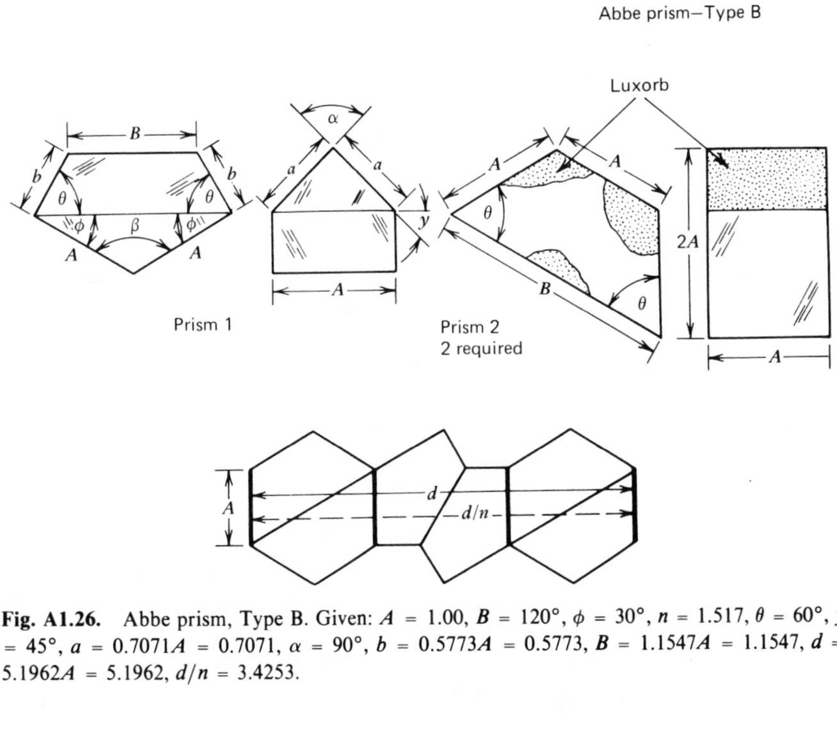

Fig. A1.26. Abbe prism, Type B. Given: $A = 1.00$, $B = 120°$, $\phi = 30°$, $n = 1.517$, $\theta = 60°$, $y = 45°$, $a = 0.7071A = 0.7071$, $\alpha = 90°$, $b = 0.5773A = 0.5773$, $B = 1.1547A = 1.1547$, $d = 5.1962A = 5.1962$, $d/n = 3.4253$.

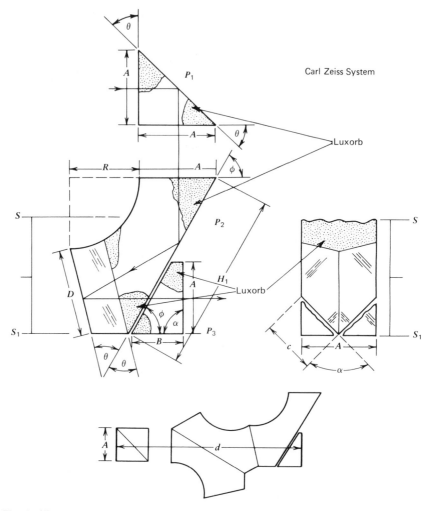

Fig. A1.27. Carl Zeiss system. Given: $A = 1.00$, $n = 1.5170$, $\theta = 45°$, $\phi = 60°$, $\alpha = 90°$, $H_1 = 2.3800A = 2.3800$, $B = 0.7320A = 0.7320$, $D = 1.1800A = 1.1800$, $c = 0.7071A - 0.7071$, $R = A$.

angle of 90°. It is not installed in military sights because of its unfavorable shape. Draftsman and artists make use of this instrument for copying. Note that if the prism is properly placed in relation to the observer's eye at about 10 in. from the copy the observer will be able to trace the image on paper.

There are other combinations of prism systems like those found in binocular eyepieces for microscopes. Many range finders have a complex

C. P. Goerz System

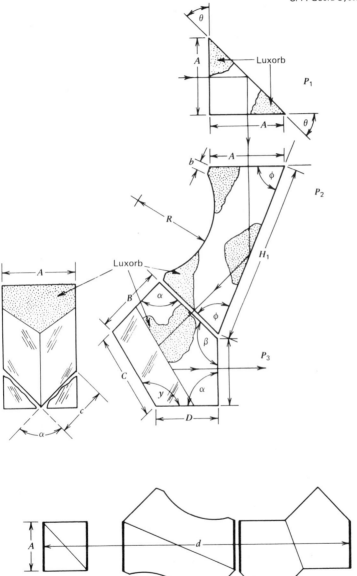

Fig. A1.28. C. P. Goerz system. Given: $A = 1.0$, $\alpha = 90°$, $\theta = 45°$, $\phi = 67°30'$, $\beta = 135°$, $y = 112°30'$, $n = 15170$, $H_1 = 2.4140A = 2.4140$, $C = 1.1620A = 1.1620$, $B = 0.9200A = 0.9200$, $D = 0.8600A = 0.8600$, $c = 0.7071A = 0.7071$, $R = A$.

491

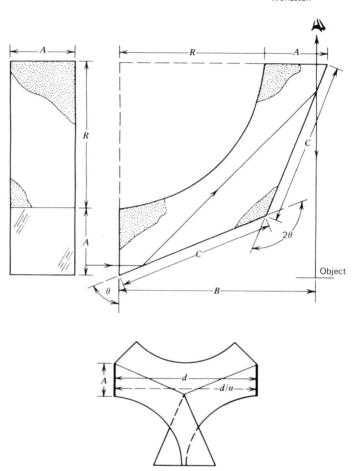

Wollaston

Fig. A1.29. Wollaston prism. Given: $A = 1.00$, $n = 1.517$, $\theta = 67°30'$, $C = 2.6131A = 2.6131$, $B = 3.4142A = 3.4142$, $R = 2.4142A = 2.4142$, $d = 2R = 4.8284A = 4.8284$, $d/n = 3.1829$.

assembly of cemented prisms near the eyepiece for simultaneous viewing of the coincidence images of ranged objects. Careful examination of any of these prism assemblies will show that they consist of rhomboids, Dove, beam-splitters, and other prisms which, when cemented together, yield the desired viewpoint and coincidence of the two telescopic objects.

Appendix 2

Making Small Planoparallels

Small planoparallels have many uses in the optical shop, especially when making prisms. Old reticles can be used because they are well-made plano-parallel elements with optical-quality surfaces. The diameters range from 0.5 in. (12.5 mm) to 0.75 in. (19 mm) for shop use; larger diameters are impractical.

Because these planoparallels are wet-wrung to the surfaces of prisms or other fine-ground areas, they will not change the optical flat surface by the Twyman effect. It is good practice to fine-grind (95 aloxide grit) one surface. To maintain parallelism an autocollimator is set up in an upright position. (See Fig. A2.1.) Shown in the diagram is the preground reticle resting

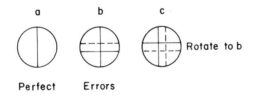

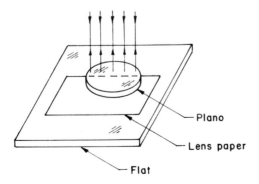

Fig. A2.1. Fine-grinding a small reticle into a planoparallel.

on a sheet of lens paper, with a small rim of its aperture extends beyond its edge. The entire assembly rests on the polished surface of the autocollimator's platform or an old optical shop flat. The autocollimator is set to pick up the autocollimating image from the shop flat and the preground reticle. If the images are superimposed, the reticle is a planoparallel. Generally separated images are observed as illustrated in (b) and (c). Image (c) can be rotated to (b). From position (b) press on an area of the preground reticle where the separated image tends to close up. This is the area in which pressure is applied during fine-grinding to correct the reticle. Work for superimposed images.

Recheck the polished surface for flatness. Often the Twyman effect will warp another polished surface when it is reground. If this happens, the polished surface will have to be retouched on a pitch polisher by hand. The surface quality must be one-half fringe or better. Recheck its parallelism by carefully placing it directly on the aluminized surface of the optical flat, and recheck also for superimposed images it must have to be a planoparallel element. Have it aluminized.

This is perhaps the first glass element that beginners can make.

Appendix 3

Pitch-Cementing Glass Runners and Covers

At times it is desirable to pitch-cement glass runners and cover plates to prisms and other components for easier fabrication. There are several reasons for this:

1. Rhomboid and dove prisms have unequal mass distributions when their entrance and exit faces are worked.
2. Very small prisms [0.5-in. (12.5-mm) base] can be handworked only singly with their aid.
3. Most opticians find their use mainly in correcting and maintaining sharp edges for small and medium-sized prisms during fine-grinding and polishing. There are several factors that govern the makeup of the glass runners: (a) They must be the same glass type and preferrably the same melt. (b) They must be well made with good glass angles and beveled rather strongly because the bevels vanish during grinding. The pitch-cement must have high tensile strength at normal room temperature and a low melting temperature.

1. CEMENTING PITCH

One of the best low-melting pitch mixtures for cementing glass elements to metal holders and glass runners to prisms consists by weight of three parts beeswax, three parts rosin, and two parts montan wax. Carnuba wax can be substituted for the montan wax. Small diameter rods of the pitch can be formed by poring the melted pitch into the recesses in aluminum foil. The cooled pitch can be rolled into single rods on a flat surface. Any entrapped foils can be easily removed. The pitch generally melts at a temperature of 80°C, which is too hot to work without the protection of soft cotton gloves.

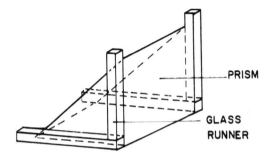

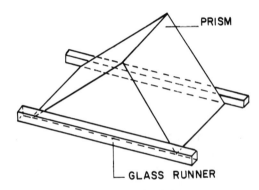

Fig. A3.1.   The pitch-cementing of prefabricated glass runners to a typical prism.

2. MAKING THE GLASS RUNNERS AND COVER PLATES

Glass runners are generally elongated rectangles with square or right-angled corners. The glass runners are approximately 1.5 times longer than the base length of the prism. In Fig. A3.1 note the cementing of a typical (right-angle) prism with two sets of glass runners cemented to its sides: the glass runners are cemented to the hypotenuse side of the lower prism. Lens paper is often placed between the two glass sections to help offset the Twyman effect which causes twisting of a polished surface when another surface is fine-ground.

The glass runners are made with some degree of accuracy, especially those with 45° angular surfaces. These require metal toolholders to hold the glass runners during generation of their 90 and 45° angular surfaces. The number required is governed by the number of prism assemblies to be made.

Generally four assemblies are sufficient because this permits a polished assembly to cool down while another is being worked.

The glass runners are generally cut from the scrap glass of polished glass blanks left over from the presawed prisms. Plate glass can be substituted for the Blanchard-generated prisms with parallel sides. Save any polished sides for mounting to the sides of prisms. The generated surfaces are then beveled strongly 0.02 in. (0.5 mm) with 145 aloxide grit and the corners are faceted by hand on the cylinder grinder or edger. These large bevels are necessary because they are slowly worn away during the grinding of the prism's surfaces. Each fabricated angular surface is a special case, especially prisms tooled for fixtures. The manufacture of tank prisms and beam splitters cut in pairs from one large right angle requires the use of glass runners.

Protective cover plates used on semipolished planoparallels require circular-edged plates. These cover plates are generally made of plate glass or thin window glass and are octagonal before they are circular edged. A glass cutter is used for octagonal shaping. Always blunt all glass cutter's broken edges with a emery stone to protect against glass cuts.

3. CEMENTING GLASS RUNNERS AND COVER PLATES

Prisms with a base length of 1 in. (25.4 mm) or smaller are heated on a electric heating plate. First place white paper toweling on the top plate; this serves two purposes: (a) the protection of polished surfaces; (2) control of the heating of the glass elements; if the paper begins to scorch, the temperature is too high. Larger prisms should be heated in a controlled oven on a asbestos pad covered with white toweling. Prisms with odd masses of glass are subject to breakage when heated unevenly.

Assembling the heated prisms with glass runners requires some care but is easily learned. Strips of lens paper are cut slightly wider and longer than the prisms. The working table is a flat cast-iron plate on which are placed several sheets of paper toweling. Cover one hand with a cotton glove to protect it from the hot prisms and smear them with the rod of pitch while holding them above the flat working surface. Then place the strips on the sides of the prisms with the other hand. The paper strips must not extend below the surface [0.04 in. (1-mm) above is sufficient]. A small wooden stick will do to move the lens paper. In the next step place the glass runners on each side of the prisms, making sure that they are equidistance at the base. Then place the assembly on a newspaper covering a second cast-iron flat surface and press firmly down on the prisms and their cemented runners. Allow the assembly to cool down and press it from time to time. Finally square the ends of the glass runners with a wooden stick to make

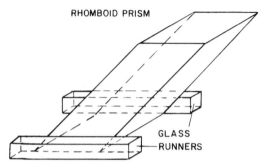

Fig. A3.2. The pitch-cementing of glass runners to a rhomboid prism to offset the overhang mass of the prism.

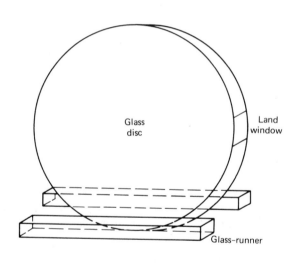

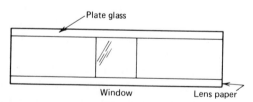

Fig. A3.3. The pitch-cementing of glass runners to a glass disk for fabricating windows for a homogeneity test or glass covers for diamond sawing.

them equal. Press again. When cooled, determine whether the glass runners and surfaces are in the same plane. If there is any misalignment of the two surfaces, they can be reheated slightly and reblocked. After cooling down the bevels and paper are removed with xylene and a toothpick.

As shown in Figs. A3.2 rhomboid and Dove prisms require glass runners cemented to their sides during grinding and polishing. Here it is necessary to have the heavier and more massive glass runners to balance the massive overhang of the assembly. The optician is frequently asked to grind and polish four small windows on the peripheral area of a glass disk for checking the homogeneity. This is shown in Fig. A3.3. Here two sets of parallel glass runners are required. Lens paper is necessary to protect the semi-polished surface of the planoparallel element.

Cover-glass plates are often cemented to the semipolished planoparallel elements for sawing out the predrawn prism shapes. For the heavier massive elements they must be slowly heated in a controlled oven. The glass blank rests on an abestos pad with paper toweling between. The first circular plate-glass plate is cemented to the top surface with the glass blank in the oven. Of course, lens paper is sandwiched between the two glass surfaces. It is turned over and the second circular plate is also cemented to it. All operations for cementing the glass covers require cotton gloves during handling. This is an individual operation because if two opticians are involved there will surely be some conversation. I have seen heated glass blanks broken because saliva was sprayed on one glass during dual handling.

Appendix 4

Interferometric Testing of Deviation Prisms

Interforometric testing of deviation prisms has been described in part by Twyman. This appendix provides other pertinent information required, especially during the correction of the 90° roof during polishing.

Twyman illustrates the placement of the prism in the interferometer. For better and faster prism replacement during the correction of the roof use a third retrodirective flat (m_3) near the prism. This stabilizes the interferometer and permits vertical fringes on (m_2) to be checked. The second mirror (m_2) is blocked off with a cardboard while the prism is tested. In another method use a low-power laser (0.3 mW) and direct the light through the beam splitter, the prism, and the third flat and superimpose each laser spot on itself. Be sure to wear safety goggles and turn off the laser after setting up. (See Fig. A4.1.)

Twyman illustrates the two sets of horizontal fringes which show the 90° roof reflecting angle in error. He does not illustrate the two sets of vertical fringes for a well-corrected 90° roof or the vee-shaped fringes for a 90° roof with a glass error. (See Fig. A4.2.) The slant of the fringes is due to two conditions: (a) because the 90° angle is not perfect; (b) because the fringes are actually segments of a circular pattern caused by the verticle images from m_1 and m_3, one behind the other. [See Figs. A4.2(a)–(c).]

Several methods are available for determining whether the 90° roof is positive or negative. One method is to slightly pull back on m_3 mirror, thereby increasing the light path. If the vee fringes move toward the roof's ridge, the glass angle is obtuse (plus) or greater than 90°. If the vee fringes move away from the ridge, the glass angle is acute (minus) or less than 90°.

Many opticians are confused by positive or negative glass angles; they think the pressure point to use during polishing is on the positive angle; actually it is always on the negative glass angle. An examination of Fig. A4.3 will show this to be true. Glass, can never be added; only computers can do that. Of course, a positive glass angle is always a negative glass angle on a prism.

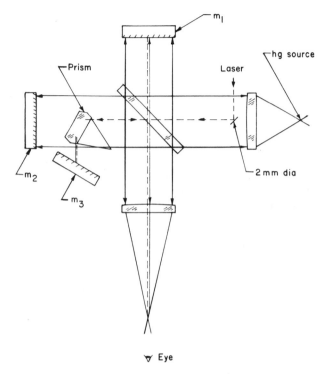

Fig. A4.1. Establishing the interferometer with a third mirror (m_3) near the prism while testing its 90° roof; this makes replacement of the prism easier. The dashed lines show a laser being used to superimpose spots on the optical elements for alignment purposes.

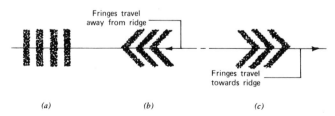

(a)   *(b)*   *(c)*

Fig. A4.2. The two sets of parallel and slant fringes observed in an interferometer while testing the 90° roof of a typical deviation prism.

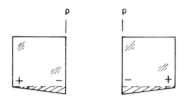

Fig. A4.3. The minus and plus glass angles of a typical prism and the one pressure point on the minus glass-angle area for its necessary correction during polishing.

It is assumed that the deviation prisms have been fabricated by the techniques described in Chapter 13 which shows external and internal autocollimation setups. There may be other variations in the fringe patterns that are caused by nonhomogeneity in the prism's glass and astigmatic polished surfaces. This can be verified by checking the prism's polished surfaces with a proof plane. If they are flat, the OPD fringes (optical path difference) due to the nonhomogeneity in the glass, are there.

Appendix 5

Comparison of Abrasives

Maker	Manufacturers grading	Approximate Micron size	Trade name or other	Type
COARSE ABRASIVES				
Coarse abrasives	8 Mesh	357		Synthetics
	50 Mesh	297		Synthetics
	60 Mesh	250		Synthetics
(All sizes are	76 Mesh	210		Aluminum
standardized by	80 Mesh	177		Oxide and
Bureau of Standards,	100 Mesh	149		Silicon
Carborundum Co.,	120 Mesh	125		Carbide (SiC)
Norton Co., Universal	150 Mesh	100		
Shellac Co., and	180 Mesh	88		
others)	200 Mesh	77		
	230 Mesh	62		
	270 Mesh	50		
FINE ABRASIVES				
Carborundum Co.	F	50	Carborundum	Si C
Universal Shellac	F	50	Unasil	Si C
Norton Co.	F	50	Crystalon	Si C
Carborundum Co.	F	50	Aloxide	Al_2O_3
Universal Shellac	F	50	Unalite or Unalum	Al_2O_3
Carborundum Co.	FF	45	Carborundum	Si C
Norton Co.	FF	45	Crystalon	Si C
Barton Mines	W-0	35	Garnet	Al_2O_3
Bausch & Lomb	500	31	Corundum	Al_2O_3
Universal Shellac	W-0 & W_0A	30	Unalite or Unalum	Al_2O_3
Barton Mines	W-1	30	Garnet	Al_2O_3
Carborundum Co.	320	30	Carborundum	Al_2O_3
Norton Co.	320	30	Crystalon	Si C

Maker	Manufacturers grading	Approximate Micron size	Trade name or other	Type
Universal Shellac	320	30	Unilite or Unalum	Al_2O_3
Carborundum Co.	FF	30	Carborundum	Si C
Norton Co.	FF	30	Alundium	Al_2O_3
Universal Shellac	FF	30	Unsil	Si C
Universal Shellac	FF	30	Unalite or Unalum	Al_2O_3
Universal Shellac	#30	30	Garnet	Al_2O_3
General Abrasive Co.	XF	30	Garnet	Al_2O_3
Universal Shellac	#28	28	Garnet	Al_2O_3
Universal Shellac	W-1 & W_1-A	28	Garnet	Al_2O_3
Carborundum Co.	400	26	Carborundum	Si C
Norton Co.	400	26	Crystalon	Si C
Universal Shellac	400	26	Unsil	Si C
General Abrasive Co.	FFF	26	Garnet	Al_2O_3
Baush & Lomb	600	25	Garnet	Al_2O_3
Universal Shellac	#25	25	Unalite or Unalum	Al_2O_3
Universal Shellac	W-2 or W-A	25	Garnet	Al_2O_3
Universal Shellac	24	24	Unifine	Corundum
Universal Shellac	FFF	24	Unisil	Carborundum
Barton Mines	FFF	23	Garnet	Al_2O_3
King & Malcom	KA	23	Anchor	Al_2O_3
Carborundum Co.	225	22.5	Aloxide	Al_2O_3
American Optical Co.	302	22	A.O. 302	Al_2O_3
Universal Shellac	21	21	Unifine	Al_2O_3
Universal Shellac	21	21	Unalum	Al_2O_3
Universal Shellac	20	20	Unalum	Al_2O_3
Universal Shellac	20	20	Unsil	Si C
Carborundum Co.	FFFF	20	Aloxide	Al_2O_3
Carborundum Co.	FFFF	20	Carborundum	Si C
Universal Shellac	FFFF	20	Unalum	Al_2O_3
Universal Shellac	FFFF	20	Unlite	Al_2O_3
Universal Shellac	20	20	Unsil	Si C
Universal Shellac	W_3, W_2A	20	Garnet	Garnet
Bausch & Lomb	750	20	Corundum	Corundum
Barton Mines	W-3	20	Garnet	Garnet
General Abrasive Co.	24	19		Al_2O_3
Universal Shellac	21	21	Unalite	Garnet
Universal Shellac	21	21	Unafine	Corundum
American Optical	302½	18	A.O.	Al_2O_3
Bausch & Lomb	850	18	Garnet	Al_2O_3
King & Malcom	KC	17½	Anchor	Al_2O_3
Carborundum Co.	600	17½	Carborundum	Si C
Universal Shellac	600	17½	Unsil	Si C

Comparison of Abrasives (Continued)

Maker	Manufacturers grading	Approximate Micron size	Trade name or other	Type
Universal Shellac	600	17½	Unalite	Garnet
General Abrasive Co.	2T	17	2T	Al_2O_3
Titmus Optical Co.	2X	16	2X	Al_2O_3
Baush & Lomb	950	16	A.O. 950	Al_2O_3
Universal Shellac	16	16	Garnet	Garnet
Universal Shellac	W4-W4A	16		Garnet
Barton Mines	W-4	15½		Garnet
Bausch & Lomb	1000	15½		Corundum
Universal Shellac	15	15	Unalum	Al_2O_3
Universal Shellac	15	15	Unlite	Garnet
American Opt.	303	15		Corundum
J. Rhodes	1015	15		Al_2O_3
King & Malcom	KH	14½	Anchor KH	Al_2O_3
Carborundum Co.	145	14½	Aloxide	Al_2O_3
Barton Mines	W-5	14		Garnet
Universal Shellac	14	14		Garnet
Universal Shellac	W-5	14		Washed garnet
Titmus Optical Co.	3X Fine	13½		Al_2O_3
Carborundum Co.	125	12½	Aloxide	Al_2O_3
King & Malcom	KO	12½	Anchor	Al_2O_3
Universal Shellac	12	12	Unalite or	Garnet
Universal Shellac	12	12	Unalum	Al_2O_3
Universal Shellac	W-6	12		Washed garnet
American Optical Co.	303½	11		Corundum

Appendix 6

Test Plate Conversion Fringe Change

Test Plate Conversion Fringe Change
(per Inch and Centimeters)

R/D	Fringes/in.	Fringes/cm
0.7	39,808.91	15,672.8
0.8	25,759.91	10,141.7
0.9	18,768.76	7,389.3
1.0	14,285.71	5,624.3
1.1	11,301.98	4,449.6
1.2	9,197.93	3,621.2
1.3	7,649.93	3,011.8
1.4	6,493.51	2,556.5
1.5	5,555.55	2,187.2
1.6	4,824.39	1,899.4
1.7	4,231.55	1,666.0
1.8	3,743.63	1,473.9
1.9	3,336.89	1,313.8
2.0	2,994.01	1,178.7
2.1	2,702.11	1,063.8
2.2	2,451.46	965.1
2.3	2,234.53	879.7
2.4	2,045.47	805.3
2.5	1,879.69	740.0
2.6	1,722.46	678.1
2.7	1,603.79	631.4
2.8	1,483.27	584.0
2.9	1,383.88	544.8
3.0	1,291.98	508.6
3.1	1,208.19	475.7
3.2	1,132.34	445.8
3.3	1,063.46	418.7
3.4	1,000.72	393.7
3.5	943.39	371.4
3.6	890.88	350.7
3.7	842.65	331.7
3.8	798.26	314.3
3.9	757.30	298.1

Test Plate Conversion Fringe Change
(per Inch and Centimeters)

R/D	Fringes/in.	Fringes/cm
4.0	710.23	279.6
4.1	676.00	266.1
4.2	644.20	253.6
4.3	614.58	242.0
4.4	586.96	231.1
4.5	561.17	220.9
4.6	537.03	211.4
4.7	514.42	202.5
4.8	493.21	194.2
4.9	473.29	186.3
5.0	454.55	178.9
5.2	420.25	165.4
5.4	389.7	153.4
5.6	362.36	142.7
5.8	337.80	133.0
6.0	315.66	124.3
6.2	295.62	116.4
6.4	276.43	108.8
6.6	260.87	102.7
6.8	245.75	96.7
7.0	231.91	91.3
7.2	219.21	86.3
7.4	207.52	81.7
7.6	196.74	77.5
7.8	186.78	73.5
8.0	177.56	69.9
8.2	169.00	66.5
8.4	161.05	63.4
8.6	153.65	60.5
8.8	146.74	57.8
9.0	140.29	55.2
9.2	134.26	52.9
9.4	128.61	50.6
9.6	123.30	48.5
9.8	118.32	46.6
10.0	113.64	44.7
10.2	109.22	43.0
10.4	105.06	41.4
10.6	101.14	39.8
10.8	97.42	38.3
11.0	93.14	36.7
11.2	90.59	35.7
11.4	87.44	34.4
11.6	84.44	33.2

Test Plate Conversion Fringe Change
(per Inch and Centimeters)

R/D	Fringes/in.	Fringes/cm
11.8	81.61	32.1
12.0	78.91	31.1
12.2	76.35	30.0
12.4	73.63	29.0
12.6	71.58	28.2
12.8	69.36	27.3
13.0	67.24	26.5
13.2	65.22	25.7
13.4	63.29	25.0
13.6	61.44	24.2
13.8	59.67	23.5
14.0	57.98	22.8
14.2	56.36	22.2
14.4	54.80	21.6
14.6	53.31	21.0
14.8	51.88	20.4
15.0	50.51	20.0
15.2	48.18	19.0
15.4	47.92	18.9
15.6	46.70	18.4
15.8	45.52	17.9
16.0	44.39	17.5
16.4	42.25	16.6
16.8	40.26	15.8
17.0	39.32	15.5
17.4	37.53	15.0
17.8	35.87	14.0
18.0	34.31	13.5
18.4	33.56	13.2
18.8	32.15	12.7
19.0	31.48	12.4
19.4	30.19	11.9
19.8	28.99	11.4
20.0	28.41	11.2
20.4	27.31	10.7
20.8	26.27	10.3
21.0	25.77	10.1
21.4	24.81	9.8
21.8	24.36	9.6
22.0	23.48	9.2
23.0	21.48	8.5
24.0	19.73	7.7
25.0	18.18	7.2

Test Plate Conversion Fringe Change
(per Inch and Centimeters)

R/D	Fringes/in.	Fringes/cm
26.0	16.81	6.6
27.0	15.59	6.1
28.0	14.49	5.7
*29.0	13.51	5.3
30.0	12.63	5.0
31.0	11.82	4.6
33.0	10.43	4.1
35.0	9.28	3.6
37.0	8.30	3.3
39.0	7.47	2.9
40.0	6.79	2.7
50.0	4.55	1.8
60.0	3.16	1.2
70.0	2.32	0.91
80.0	1.78	0.70
90.0	1.40	0.55
100.0	1.14	0.45

Examples. (1) Test plate: 2.5 in. (63.5-cm); R/D = 2.222; *nominal radius 5.555 ± 0.001 in.; 5.555/2.5 = 2.222; 141.097 ± 0.00254-cm; 141.097/63.5 = 2.222. Therefore 0.001 in. (0.00254 cm). Looking only under R/D 2.2 2,702.11 fringes/in. (965.1 fringes/cm) are found, and by moving the decimal over three places (multiplying by 0.001) 2.45 fringes/in. (0.96 fringes/cm) are found.

* Nominal tolerance given under inches and centimeters can be variously ranged and is always at midpoint. The shop print should point out the number of fringe (often called rings) departures for a given test-plate tolerance.

Appendix 7

Silvering Solution, Amounts, and Order of Mixing

1. 5.67 g NaOH disssolved in 1800 cm³ of distilled water.
2. Dissolve 10 g AgNO₃ in 11 cm³ of distilled water.
3. Add 35 cm³ of (1) to the AgNO₃; stir well slowly.
4. Add to (1) all of (3); stir well slowly.

This results in a silver solution that will keep for many weeks.

Stock. 4.62 cm³ of H_2SO_4 (concentrated sulfuric acid) to 58 cm³ of distilled water.

Reducer. Mix 170 cm³ of distilled water and 30 g of dextrose; add 3.9 cm³ stock.

Thinner. Add two match heads of Stannous Chloride to 1 pt of tap water. This solution should be made up fresh. For silver use 1 liquid oz of reducer to 1 qt of silver solution. Clean all holding and silvering vessels thoroughly by washing in a (hot) detergent solution and follow with a rinse in hot water. Given any break in the surface of the water vessels must be rewashed. Swab with piece of cotton on a stick dipped in cp nitric acid. Rinse with distilled water. Clean all optical surfaces in a similar manner and then wash with the *thinner* solution. Rinse thoroughly in running water and follow with a quick rinse in distilled water (spray bottle) at least twice. Place in a tray of distilled water.

Mix the reducer solution with the silver solution quickly and immediately place the optical element in the silvering solution. Agitate the solution with a wooden stick (tongue depresser) and silver for 2 to 3 min. Remove the optical element and wash well in water. Air dry on a paper towel and then spray with plastic. Remove the plastic from unworked surfaces with acetone.

Rinse all silvering solutions in the sink, and wash all mixing vessels as well. Some firms like to retain used silvering solutions and they are poured into a salt solution. Rinse silvering vessels with a dilute nitric acid solution and rinse in water.

Appendix 8

Optical Figuring of Prisms

Optical figuring of prisms, a technique of localizing or refiguring the optical surface of a prism, is used to arrive at a quality assurance of OPD of one-half wave or better because of the nonhomogeneity in the glass. This is a measurment observed when the prism is place in a Twyman-Green or Hendrix interferometer. Because the interferometer actually gives the double passage of the transmitted fringes of the optical path difference (called OPD), it is necessary to divide it by two. It is assumed through this appendix that the transmitted fringes (OPD) have been divided by two. As a rough calculation, a change in the surface of the prism by one fringe of OPD is four to one. It is important that only *one* face of the prism be localized and its second side kept flat. It is to be expected when prisms are refigured that the deviation also changes. The terms localizing, figuring, and retouching are synonymous.

1. INTERFEROMETRIC TESTING OF PRISMS

A number of interferometers are on the market for checking the OPD of optical elements and lens systems. Most of them are patterned after the Twyman-Green or Hendrix interferometers. The most common interferometer for prism testing is the 102-mm aperture instrument. Interferometers of the Hendrix type have been made as large as 30.5 cm by Davidson Optronics, West Covina, CA. The larger interferometers are valuable because they allow large blocks of glass to be checked for homogeneity in three directions.

Some prisms have small areas and it is difficult to set them up in an interferometer. A low-power laser is helpful. The laser beam is directed through the divider plate on the prism's surface toward the second flat, and a series of laser dots is superimposed on the divider plate. Adjustments of the prism and second flat nearest the prism are required. See Fig. A8.1 for the optical set up of the Hendrix interferometer.

Prisms have often been worked to one-eighth of a fringe on all surfaces, and gonimeter measurements or the autocollimation check may show all glass angles as 2 sec or better. When checked in the interferometer for

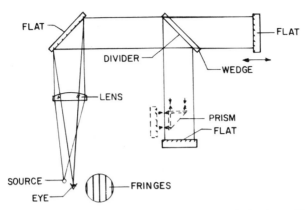

Fig. A8.1. Checking the optical difference (OPD) of a typical prism with a Hendrix interferometer. See also Fig. 8.6.

OPD, there might be two or more fringes of irregularity. Keep in mind that there are two fringes for one wave. Most prism call-outs for OPD are one-quarter fringe or one-half wave.

As noted at the beginning of this appendix only one face of the prism is figured or retouched. It is important to work only one surface. Do not try to balance the refiguring between the entrance and exit faces. Some prisms are worked to 2 sec or better deviation; then, when the prisms are retouched for one-quarter wave OPD, the 2-sec deviation is lost. In this case it must be reworked to the 2-sec deviation. The refigured face must *not* be altered by reworking. What has been accomplished must not be destroyed. The other face of the prism must be corrected; for example, by rough calculation over a 50-mm aperture of a typical prism a one-fringe change amounts to 1 sec of arc. Large prisms such as the penta sometimes have two to three fringes of OPD. In this case refiguring (8-to-12 fringe change of the prism's surface) will be required to meet the one-quarter fringe OPD tolerance.

Astigmatic fringes of the OPD occur 90% of the time. Sometimes more fringes in one plane show a convex or concave wavefront; the second plane might be within one-half fringe of the opposite wavefront. The net results of this figuring is that one astigmatic surface is actually offsetting another. Beginners in optical figuring are amazed by the amount of refiguring that is necessary to make one fringe change for the OPD. Roughly, it takes a four-fringe change in the optical surface for a one-fringe change in the OPD. Opticians call these wavefronts of OPD "irregularities." When placing the prism in the interferometer it is always good practice to check it from corner to corner and in both planes. Figure A8.2(*a*)–(*h*) illustrates a number

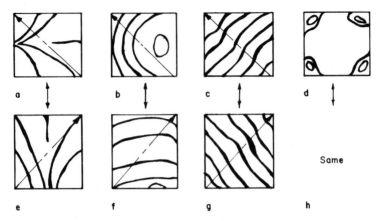

Fig. A8.2. Checking the OPD fringes in two planes to locate the quadrant with the convex contour before figuring.

of fringes of the OPD wavefront for a prism with small variations in the index of refraction of the glass. Figure A8.2(*a*) represents a case that would be impossible to figure.

If the central area in Fig. A8.2((*b*) is a convex or concave wavefront (opticians always say high or low), press lightly on the front of the reflecting flat holder nearest to the prism, thus increasing the light path length between the prism and the flat. If the fringes expand outward, the OPD wavefront is convex or, in the language of the optician, "high." Conversely, if the fringes converge toward the center, the wavefront is concave or "low." Using this method of testing the wavefront is similar to that of using a test plate on optical flat surface.

Summarizing, keep in mind that the OPD fringes observed in the interferometer are transmission fringes and not surface quality fringes. Recognize, when checking the OPD fringes in both planes and astigmatism is observed, that one plane is convex and the other flat or concave or vice versa. It is good practice for beginners to paint the observed fringes with rouge or drafting ink using a camel's hair brush. Only the main contours should be outlined, not the irregularities. It is soon learned that the *convex* wavefront or high area is the plane of the prism that must be figured first. It is *important* to place a piece of masking tape on the figured face of the prism to keep it from being reworked. It is always good practice to place this face toward the autocollimator. This enables the flat surface of the prism to be the required autocollimating flat surface for checking the prism. Of course, the tape can be removed for refiguring.

2. FIGURING, LOCALIZING, OR RETOUCHING

As previously noted, the terms figuring, localizing and figuring, or refiguring are synonymous. These terms are often intermingled in descriptions of optical procedures. Typical double-passage fringes are shown in Fig. A8.3.

Method 1.   This method uses pitch polishers larger than the prism's surfaces. (See Chapter 3.) Channel-cut facets with smaller embossed facets were used by eminent opticians like Harold Hall and Don Hendrix. In this method contoured pressing plates are used with concave or convex surfaces ranging in quality from one to six fringes; for example, one side of the pressing plates can have two fringes concave and two fringes convex on the other. The same is true for the other pressing plates. Only three sets may be required. With a convex pressing plate a concave impression is made on the polisher. Conversely, a concave pressing plate impresses a convex contour. The convex contour polisher is used to change a high area of the OPD, a concave polisher for a low OPD; for example, in Fig. A8.3(*b*) a prism with three fringes convex OPD across one corner and perhaps one fringe concave across the second corner is found. Now, because the interferometer is observing double-passage transmission OPD fringes, it is evident that it is half of three, or 1.5 fringes in one plane and one-half fringe in the second. The general contour of the fringes in the convex plane is given if the polished area is high or low.

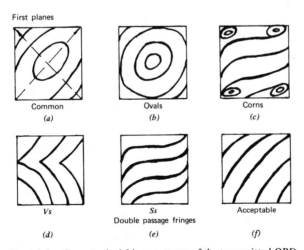

Fig. A8.3.   Some typical fringe patterns of the transmitted OPD.

Compress a polisher with a concave pressing plate with a four-fringe surface quality. Perhaps a three-fringe pressing plate could be used. It depends on many factors: the hardness of the polisher and the pressure applied by the optician. The secret of good figuring is to know that a convex contour polisher polishes the high area down and a concave polisher tends to raise a concave surface toward the convex. In any case, glass is always removed. Polish across the one predetermined convex area (no rotation of prism) observed by the interferometer check. After a wet check the surface of the prism with an optical flat, or in the planointerferometer, and observe how many fringes have been changed from the original surface. Checking with the flat tells whether the optical surface of the prism is progressing in the right direction. Observe an astigmatic surface being polished in the prism (one astigmatic surface is actually offsetting another.) Caution: do not rotate the plane of the prism.

Because it is necessary to make a four-fringe change for a one-fringe change in the OPD, change the surface to approximately six fringes, and recheck in the interferometer. *Caution!* Let the prism rest in the interferometer for a period of time. (Generally it depends on the glass of the prism.) It takes 10 min or more for the latent heat to dissipate. If, when rechecking the prism, it is found that the original orientation is slowly rotating into another plane, it is to be expected. Then find the convex or high area again and work the surface of the prism by perhaps one and one-half fringes and recheck in the interferometer.

As the OPD fringes approach the desired tolerance of 0.25 of a wave, or 0.5 fringe over all the area, there is a tendency for the edges to become turned down or up. Here a critical examination is necessary. In making the high area concave with the convex contour polisher, note that it has reached the entire area and determine whether there is a turned edge. Turned edges often appear as humps of "corns" on the prism during figuring. These humps can be removed with the index or middle finger, but exercise great caution. Do not overdo it. After being localized the prism should be retouched on the correct contour polisher. Do not "paint" the surface of the prism with colored fringes. For smaller irregularities it is better to use smaller pitch laps of a solid type, perhaps as small as 1 cm in diameter. (See Fig. A8.4.)

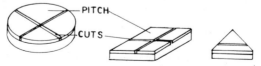

Fig. A8.4. A number of small pitch polishers used during localizing.

The working of a concave OPD wavefront or low central area and high edge still means the removal of glass. Here work the edges of the prism down slowly until they and the central area are nearly the same height. To do this contour the smaller polisher with a convex pressing plate to compress the polisher to a concave contour. With a smaller polisher it means that the pressing plate must have six to eight fringes over the full apperture of the polisher.

The figuring procedures for a concave OPD wavefront or low area are the same as those discussed for the high area. Here the previously polished surface of the prism is changed to a convex configuration that can be checked with a test flat. A concavve area on a prism is harder to figure than a high area because the edges must be reduced.

Method 2. This method uses solid polishers with a generally softer pitch (Swiss pitch No. 64) which are never larger than 6 in. (153-mm) in diameter. In Chapter 3 it is stated that these unfaceted polishers are scratched with a razor blade in certain areas to help control the desired polishing configuration on the prism's surface. A glass burnisher of approximately one-third diameter is used to help break down the polishing compound and to shape the contour of the polisher. Here the polisher's contour is controlled by a combination of pressing plates and glass burnishers and by scratching the polisher in various areas of its aperture.

A greater number of pressing plates is required because of the smaller diameters of the pitch laps, which range from 1 to 10 fringes, both concave and convex contours. A set of five with a 6-in. (152-mm) diameters and concave and convex semipolished surfaces on alternate sides will suffice. It is possible to place a number of smaller laps on the pressing plates for reduction to the desired contours. The pitch laps are dry and must have a small amount of dry polishing compound lightly rubbed over them. The pitch laps will often adhere to the pressing plate after a prolonged period of time, but they can generally be removed by striking heavily around the periphery with a rawhide hammer. Some care must be taken not to break the plates. Another method uses Saran-Wrap; place a sheet between the pitch lap and pressing plate with polishing compound in between.

The optical procedures of figuring the prisms with these pitch laps are generally the same. The only difference is that polishers of smaller diameter lose their compressed contours more rapidly and must be recontoured more often than the faceted polishers. The main advantage of smaller polishers is that they can be contoured quickly. It was mentioned earlier that small diameter polishers are used in conjunction with the larger diameter polishers.

3. CONCLUSIONS

The figuring of the OPD fringes of prisms has been described at length to arrive at the 0.25 wave tolerance for prisms of high quality. Figuring of the other OPD fringe patterns in Fig. A8.2 have also been discussed. It is important therefore that the fringe patterns in both planes of the prism be adjusted to determine whether OPD is present. Caution must be used to prevent the OPD fringes from washing out by adjustment of the interferometer but by keeping about one fringe/0.5 in. (1 cm) of prism area.

In Fig. A8.3(B) are irregular oval fringes (encompassed in the area of the prism) that shows only a small amount of astigmatic pattern. To determine whether the fringe pattern is high or low increase the light path by gently touching the flat mirror nearest the prism. If the fringe enlarges outward, the OPD fringes are convex or high, if the fringes compress inward the OPD area is concave or low.

If the homogeneity of the glass were perfect, OPD fringes would not be present. An unfortunate optician once worked three weeks on a series of 2-in. (51-mm) beam splitters without checking the tank flow slab of glass from which the prisms had been cut. When the prisms were checked in the interferometer, the OPD fringes were hopelessly crisscrossed with fine rippled fringes.

If large masses of glass blocks such as 12-in. (31-cm) sides are checked for homogeneity, it does not mean that all areas are nonhomogeneous. Glass is expensive and almost all parts of the glass block must be used therefore OPD fringes are a vital concern of the optician.

Appendix 9

Optical Contacting

An optician is often confronted with optical contacting of two or more surfaces of components to another optical holding surface during fabrication. Little has been published on the subject. It is easily learned if a few fundamentals are kept in mind. A number of fundamentals are the following: the two matching surfaces to be contacted must be one-fourth fringe or better; surgically meticulous cleanliness must be maintained; an orderly procedure must be followed. A clean-room facility is preferable, but a plastic lucite or glass cabinet may be made up or purchased. This cabinet has a slow air current moving outward through the open doors. (See Fig. A9.1.) It should be fitted with two or more uncoated fluorescent tubular bulbs with a sheet of frosted mylar over them for better light diffusion.

1. EQUIPMENT REQUIRED

1.   A plastic box 1 m square or larger with front hinge doors and a small electric fan that draws the air through a filter into the box.

2.   A portable low-cost monochromatic light source can be obtained from Edmund Scientific Co., Barrington, NJ 08007. Catalog No. 71780 and replacement bulb No. 41920.

3.   A focusing light source. American Optical Co., microscope illuminator.

4.   A binocular eye loupe (watchmatker 2×).

5.   Redistilled acetone stored in two brown bottles.

6.   Six or more linen handkerchiefs or three small chamois washed in detergent and throughly rinsed in distilled water.

7.   A Static Master Camelhair brush (Nuclear Products, Costa Mesa, CA.) The brush is cleaned by dipping it in acetone.

8.   Several orange sticks or lucite rods with a sharpened wedge and point on each end.

9.   A half dozen small finger sols and cottons swabs.

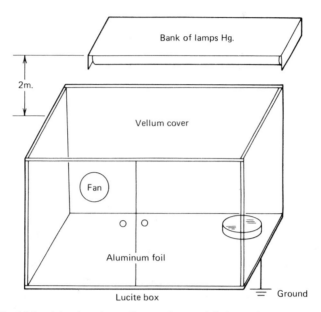

Fig. A9.1. A laminated zero flow test box used during optical contacting.

2. PREPARATION AND CONTACTING

1. Check all bevels and clean off polishing compound and pitch. Use a typewriter eraser and shop razor blades.

2. Wash all bevels twice in xylene using cotton swabs. Wash in a warm detergent solution and rinse in medium hot running water. Dry the optical components with a soft clean cotton towel and wrap in clean lens paper.

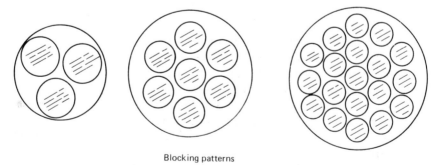

Blocking patterns

Fig. A9.2. A number of configurations used during the optical contacting of components.

3. Put on finger sols on thumb and middle finger to hold prism. Put on loupe.

4. Scrub the component's surface with a cotton swab dipped in distilled acetone. The acetone should be poured from a storage bottle into a small beaker to avoid contamination of the storage bottle.

5. Place the component under the microscope illuminator and with a clean linen handkerchief or chamois wipe off all lint particles. Check for cleanliness by breathing on the surface; a gray surface indicates a clean surface.

6. Remove floating lint particles which have been attracted by the electrically charged glass element with a Static Master brush. Antistatic solutions that prevent the charging of the element are available, but it should be checked because some solutions prevent the antireflecting coatings from adhering. Sema Products, Chicago, Ill., makes an excellent solution.

7. Clean the planoparallel or proof plane that is to be used to contact the components.

8. Take the cleaned element, recheck under the monochromatic light source, and make a parallel contact. Watch for the appearance of fringes. If no fringes or too many appear, the optical surfaces have lint or dust on them. Wipe again with the Static Brush. It is often helpful to touch the charged elements to a piece of grounded aluminum foil which should be available in the box.

9. Watch for any premature contacting; pull away and start anew.

10. After the elements are placed in the planned configuration (see Fig. A9.2), press each component lightly and move the element into position.

11. If the optical contact is ideal, all the color of the expanded fringe will assume an even gray coloration.

12. After the elements are placed in the required configuration, they are vapor-sprayed with acetone. Paint around all edges of the elements with red fingernail polish. Shown in Figs. A9.2 are some of the best circular configurations used in optical fabrication.

Optical contacting is time consuming and rather tedious at the start. Few opticians readily acquire the skill. A meticulous optician with clean equipment acquires the skills faster than one who has little patience.

3. DECONTACTING

1. Remove the nail polish with acetone and wash all surfaces with a warm detergent solution and dry thoroughly.

2. Using an orange stick with a sharpened edge pry under the bevels of each component and watch for spreading fringes to appear. Continue until the component has separated. A shop razor blade is handy for more stubborn elements, but it should have a piece of tape wrapped around its metal top to prevent scratching of the bottom surface. Place the sharp edge under the bevel and pry upward. Stubbornly contacted elements can be heated with a small Bunsen burner. A sharp snap will be heard when the element releases.

3. Often thin crystal windows are optically contacted on a glass planoparallel. Here the heat is applied to the planoparallel and not to the crystal. Some crystal elements will adhere so strongly to glass planoparallel, that it will wrap the plano element. Then polish the contacted crystal element the same number of fringes as on the planoelement.

Appendix 10

Factors and Conversion Tables

LENGTH—INCHES TO METRIC

	Millimeters (mm)	Microns (μm)	Angstroms (Å)
1 Inch (1.00″)	equals 25.4 (exactly)	25,400.	254,000,000
1 Thous. (0.001″)	equals .025 4	25.400	254,000
1 Micro inch μin. (0.000 001″)	equals .000 025 4	.025 4	254

LENGTH—METRIC TO INCHES

	Inches	Thousandths	Micro Inches
1 Millimeter (mm)	equals .039 37*	39.37	39,370.
1 Micron (μm)	equals .000 039 37	.039 37	39.37
1 Angstrom (A)	equals .000 000 003 937	.000 003 937	.003 937

* Carried further—1 mm. = 0.039 370 078 740 in.
This table is based on the exact relationships 1 yard = 0.9144 meters; 1 inch = 25.4 millimeters.

WAVE LENGTH UNITS, IN INCHES, IN KILOMEGACYCLES

Wave length unit	Equivalence in inches	Equivalence in kilomegacycles (KMc)
1 meter	39.37″	0.3
1 decimeter	3.937″	3
1 centimeter	.3937″	30
1 millimeter	.03937″	300
1 micron	.00003937″	300,000
1 milli-micron	.00000003937″	300,000,000
1 angstrom unit	.000000003937″	3,000,000,000

METROLOGY TERMS

Prefix	Symbol	Meaning	Shorthand
Micro	μ	One-millionth	10^{-6}
Milli	m	One-thousandth	10^{-3}
Centi	c	One-hundredth	10^{-2}
Deci	d	One-tenth	10^{-1}
Deka	dk	Ten	10
Hecto	h	One Hundred	10^2
Kilo	koek	One thousand	10^3
Mega	m	One million	10^6
Mega Mega	mm	One million million	10^{12}

CONVERSION—ANGLES TO LINEAR MEASUREMENT

Angles	per inch	per 10 inches	per foot
1 second	.000 005″*	.000 048″	.000 058″
5 seconds	.000 024″	.000 242″	.000 291″
10 seconds	.000 048″	.000 485″	.000 582″
20 seconds	.000 097″	.000 970″	.001 163″
30 seconds	.000 145″	.001 454″	.001 745″
1 minute	.000 291″	.002 909″	.003 491″

CONVERSION—LINEAR MEASUREMENT TO ANGLES

Linear Measure	per inch	per 10 inches	per foot
.000 001″	0.206 seconds**	0.021 seconds	0.017 seconds
.000 025″	5.157 seconds	0.516 seconds	0.430 seconds
.000 05″	10.3 seconds	1.03 seconds	0.86 seconds
.000 1″	20.6 seconds	2.06 seconds	1.72 seconds
.001″	3 min. 26 sec.	20.6 seconds	17.2 seconds
.005″	17 min. 11 sec.	1 min. 43 sec.	1 min. 26 sec.

  * Carried further, this value is .00000 48481
** Carried further, this value is .20626 48062
1 radian = 57° 17′ 44.81″

1 second = .00485 mm per meter
1 mm per meter = 3 min 26 sec
1 mill (exact) = 0.001 radian = 206.26 secs
1 mill (military) = 360°/6400 = 202.50 secs

WAVE LENGTHS OF MONOCHROMATIC RADIATIONS FOR INTERFEROMETRY

Color	λ vac. in mm.	λ air in mm*	$\frac{\lambda}{2}$ Air in inches*	Bands per inch for λ air*
Kr. 86 Orange Red	.000 605 780 21	.000 605 615 73	.000 011 921 569 58	83,881.5722
Cd Red	.000 644 024 91	.000 643 850 37	.000 012 674 219 88	78,900.3197
Hg Yellow 1	.000 579 226 84	.000 579 069 34	.000 011 399 002 76	87,726.9724
Hg Yellow 2	.000 577 119 83	.000 576 962 88	.000 011 357 537 20	88,047.2572
Hg Green	.000 546 227 04	.000 546 078 19	.000 010 749 570 66	93,026.9711
Hg Violet	.000 435 956 21	.000 435 836 05	.000 008 579 449 80	116,557.5900

* λ in air at 20° (C), 760 mm Hg; water vapor pressure 10 mm Hg; CO^2 content, .03% by volume

Appendix 11

On Using a Williams Interferometer for Making a Divider Plate

Making a beam-divider plate for interferometers is an optical specialty. The experience and know-how come only after years of trial and error in optically figuring a number of various sizes and shapes. Most divider plates are finally figured by using the divider plate as an integral part of the interferometer. There are several difficulties even in such a procedure, one being the foreshortening effect in large aperture interferometers because of the thickness and large angles of incidence necessary.

The "in-line interfereometer or series interferometer" by Saunders and Post can prevent these difficulties and is also stable and compact. However, it has too many optical components for the ordinary optical shop, and for a 8-in. (203.2-mm) interferometers it would be costly to make.

The Williams interferometer offers a less expensive method. The two spherical mirrors have 12-in. (30.5-cm) apertures and a 762-in. (300-cm) radius of curvature. This is adequate to check glass plates from 0.5 in. (1.25 cm) to 2.5 in. (6.25-cm) thick. See M.V.R.K. Murty, *Appl. Opt.* **2**, 1337–1340 (1963). The setup and method are described in Chapter 5, and the transmission fringes in Fig. A11.1 are the fringes that are optically figured to straight and equidistance parallel fringes; for example, observe the number of loops of one-half fringe in the fringe contour. This optical difference would meet the optical tolerance for windows but would not be good enough for a divider plate. Still by optical figuring one side this window could be made into a divider plate.

Before any optical figuring is begun on the divider plate check its surface quality on both sides. Of course, this check is done with a planointerferometer or an optical plane of sufficient diameter. For an divider plate both sides ought to have a surface quality of one-tenth or better, free of astigmatism. Because one surface of the element is always convex and other concave (even for a perfect element) place a piece of tape on the concave side, to prevent a mixup during the optical figuring. The convex surface is generally figured. The testing area is important and it must be free of drafts or moving air currents. All the glass holders must be of heavy construction

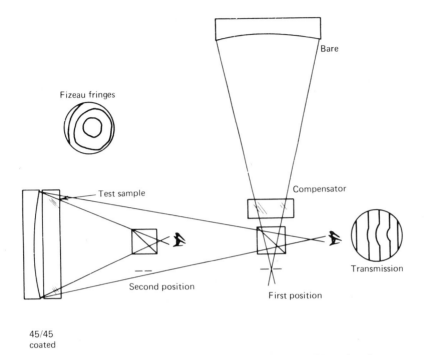

Fig. A11.1. A transmission and reflective fringe pattern with a Williams interferometer.

mounted on a "damped table" with a honeycomb structure. It is good practice always to replace the divider plate in its original position during testing.

The practicing optician would like to know if the area is high or low. If on shorting the optical path ray by pushing one of the mirrors with a stick (thereby increasing the light path) the contour of the fringe area increases, the area is high: conversely, if its contour decreases, the area is low.

It is obvious if the central area is low in the divider plate, this convex surface must be reworked again to make it one or two waves (two to four fringes) more convex. It is soon realized that only high areas can be localized with smaller polishers. The optician must learn to wait during the intervals of optical figuring. It has been said a good optician knows when to quit. Often the last remaining observed fault will be a small hump, and this is generally removed with a small polishing lap or middle finger. Again a period of several hours or overnight should elapse before refiguring. If the figuring is overshot, the surface must be repolished to its original state.

Appendix 12

Laser Illuminated Divergent Ball-Bearing Sources

The use of silverglass beads and mercury globes as divergent point sources has been discussed by Bell, Twyman, Hendrix and Christe. Bell mentioned the use of a silver ball mounted on a big sheet of black cardboard for testing telescope objectives in the sun. Twyman chose small mercury globes for the retrodirective component in the interferometric testing of microscope objectives. Strong described many illustrations of autocollimation setups with silverbead sources. Hendrix and Christie mention their use for making corrector plates for Schmidt cameras. Of course, all ball sources were illuminated by the sun's rays or other bright artificial sources.

It can be shown by ray tracing with Gaussian formulas that small-diameter ball sources as point sources for laser illumination contributes little spherical aberration to the optical setup—less than $1/M$ wave. Actually, the divergent rays originate at the dead center of the ball bearing.

Hendrix and Christe added that the illumination must be centered and concentrated on the face of the silvered bead. I have found that this is not true for laser illumination because the speckled illumination returning is well expanded. In Figs. A12.1 and A12.2 note methods of illuminating the ball bearing. A shield is used to protect off-axis illumination from reaching the observer's eyes. The laser beam can be made to pass through the back of the primary mirror if it is fine-ground by using transparent plastic tape. The steel ball bearing was first polished with a pitch lap to remove some of the irregular crystalline fractures. This gave a bright speckled expanded beam while using a low-power laser of 0.5 mW. Several ball bearings of various diameters ranging from 0.5 and 2 mm were used with the laser. The smaller balls gave a more divergent beam, similar to microscope objectives of 20 or 40×. The 2-mm ball was used for the optical setups illustrated.

In Fig. A12.3 a concave mirror of 40-cm diameter with a wax-ground surface is setup with a small source mounted on a three-axis adjusted holder at the focal point of the $f/1.5$ mirror. The laser beam is directed off-axis but it could be directed through the back of the primary mirror or by using a 3-mm diameter diagonal near the center of the mirror. Each method will give

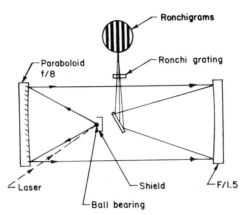

Fig. A12.1. The laser illumination of a ball bearing off-axis placed at the focus of a $f/8$ collimator used to test another paraboloid mirror being figured. Ronchigrams are formed when the paraboloid is tested.

identical results. The concave, which is being fine-ground by hand with flexible tile tools and 95 aloxide grit into a parabolic figure, semireflects the laser beam from its wax-ground surface to a 54-cm diameter $f/8.0$ collimator. Of course, the mirror was first full polished and figured to a near spherical surface. Because the peripheral edge becomes the reference zone, it is not touched during the fine-grinding at the central zones. The laser speckle beam is focused on a small diagonal mirror and the Ronchi grating is placed inside or outside of focus: the Ronchigrams are observed on a frosted screen and become progressively brighter as the concave mirror is polished with hand-held semiflexible pitch polishers as the mirror slowly rotates.

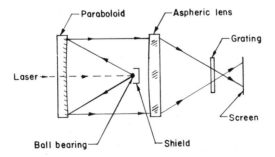

Fig. A12.2. A laser beam passing through the back of a paraboloid collimator; the planoconvex element is being optically figured null.

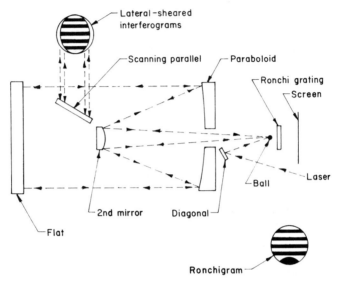

Fig. A12.3. Using a laser beam illuminating ball source for testing a Cassegrainan system with a scanning planoparallel. Both Ronchigrams and lateral sheared interferograms can be observed.

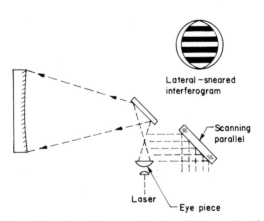

Fig. A12.4. Checking a Newtonian telescope with a laser beam passing through a 0.25-in. eyepiece while testing it with a scanning planoparallel.

530

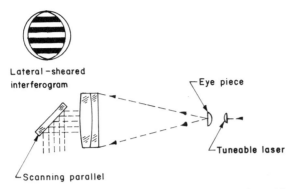

Lateral-sheared
interferogram

Eye piece

Tuneable laser

Scanning parallel

Fig. A12.5. Testing an objective with a tunable for spherochromatism with a scanning planoparallel.

The progress of the optical figuring can be followed by interpretation of the Ronchigram as in Appendix 14 or Fig. A12.6.

Figure 17.2 illustrates a planoconvex lens with an aspheric curve figured on it and a rose-leaf polisher working at the 0.707 zone. The lens is placed in the parallel light beam of the collimator. The laser beam is directed through the semipolished back surface of the paraboloid, which has a 2-mm area of aluminum coating removed, or on a 4-mm aluminzed diagonal near the surface of the mirror. As previously mentioned, if the back surface is not polished, make it transparent by using clear plastic tape.

Figure A12.3 illustrates a method that uses a small aluminized diagonal to direct a laser beam at the focal point of a compound telescope and to

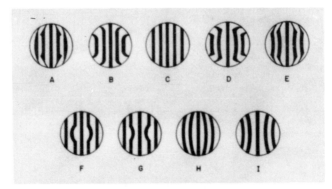

Fig. A12.6. Ronchigram patterns showing the degree of correction (see text).

observe the autocollimated Ronchigrams. Another possibility is to scan the aperture of the telescope with a planoparallel (shown by the dashed lines) and to observe the lateral-sheared interferograms with the Ronchigrams; therefore a qualitative and quantitative correleation is observed.

Check the optical figuring of the Newtonian telescope by simply directing the laser beam through the ¼-inch eyepiece and scanning the aperture with a planoparallel. The lateral sheared interferograms are projected on a frosted screen.

Appendix 13

Angular Settings for Radius Generating

These tables are based on the formula $\sin t = D/2r$, where D equals the cutting edge of the cutter. For concave-surface generation the cutting edge is the peripheral edge and for convex surfaces, the inside edge; r is the required radius. Example: 53.5 in. concave radius. Diameter of cutter 4 in. $\sin t = 4/2 \times 53.5$, $\sin t = 4/107$, or $\sin t = 0.0373$. From a table of natural sine functions 0.0373 equals 2° 9'.

CIRCLE SETTINGS FOR RADIUS GENERATOR

1.0 in. OD Cutter

Concave		Convex	
D = 1.0 in.		ID = 0.750 in.	
Radius required	Circle setting	Radius required	Circle setting
2.00″	14°–29′	2.00″	10°–48′
2.25	12°–50′	2.25	9°–35′
2.50	11°–33′	2.50	8°–38′
2.75	10°–29′	2.75	7°–50′
3.00	9°–36′	3.00	7°–11′
3.25	8°–51′	3.25	6°–37′
3.50	8°–13′	3.50	6°–9′
3.75	7°–40′	3.75	5°–45′
4.00	7°–11′	4.00	5°–23′
4.25	6°–45′	4.25	5°–4′
4.50	6°–23′	4.50	4°–47′
4.75	6°–2′	4.75	4°–32′
5.00	5°–45′	5.00	4°–18′
5.25	5°–28′	5.25	4°–6′
5.50	5°–13′	5.50	3°–54′
5.75	4°–59′	5.75	3°–44′
6.00	4°–47′	6.00	3°–35′
6.25	4°–36′	6.25	3°–27′
6.50	4°–22′	6.50	3°–18′
6.75	4°–15′	6.75	3°–11′
7.00	4°–6′	7.00	3°–4′

CIRCLE SETTINGS FOR RADIUS GENERATOR

2 in. OD		1.75 in. ID	
Concave Radius use OD		Convex Radius use ID	
Radius required	Circle setting	Radius required	Circle setting
7.25	3°–57′	7.25	2°–58′
7.50	3°–49′	7.50	2°–52′
7.75	3°–42′	7.75	2°–46′
8.00	3°–35′	8.00	2°–41′
8.25	3°–28′	8.25	2°–36′
8.50	3°–22′	8.50	2°–31′
8.75	3°–17′	8.75	2°–27′
9.00	3°–11′	9.00	2°–23′
9.25	3°–6′	9.25	2°–19′
9.50	3°–1′	9.50	2°–15′
9.75	2°–56′	9.75	2°–12′
10.00	2°–52′	10.00	2°–9′
10.50	2°–44′	10.50	2°–3′
11.00	2°–36′	11.00	1°–57′
11.50	2°–29′	11.50	1°–52′
12.00	2°–23′	12.00	1°–47′
12.50	2°–18′	12.50	1°–44′
13.00	2°–12′	13.00	1°–39′
13.50	2°–7′	13.50	1°–35′
14.00	2°–3′	14.00	1°–32′
14.50	1°–58′	14.50	1°–29′
15.00	1°–55′	15.00	1°–26′
15.50	1°–51′	15.50	1°–23′
16.00	1°–47′	16.00	1°–20′
16.50	1°–44′	16.50	1°–18′
17.00	1°–41′	17.00	1°–16′
17.50	1°–38′	17.50	1°–14′
18.00	1°–35′	18.00	1°–12′
18.50	1°–33′	18.50	1°–10′
19.00	1°–30′	19.00	1°–8′
19.50	1°–28′	19.50	1°–6′
20.00	1°–26′	20.00	1°–4′
21.00	1°–22′	21.00	1°–1′
22.00	1°–18′	22.00	0°–59′
23.00	1°–14′	23.00	0°–56′
24.00	1°–11′	24.00	0°–54′
25.00	1°–9′	25.00	0°–52′
26.00	1°–6′	26.00	0°–49′
27.00	1°–3′	27.00	0°–47′
28.00	1°–1′	28.00	0°–46′
29.00	0°–59′	29.00	0°–44′
30.00	0°–57′	30.00	0°–43′
35.00	0°–49′	35.00	0°–37′
40.00	0°–43′	40.00	0°–32′

CIRCLE SETTINGS FOR RADIUS GENERATOR

2 in. OD		1.75 in. ID	
Concave Radius use OD		Convex Radius use ID	
Radius required	Circle setting	Radius required	Circle setting
45.00	0°–38'	45.00	0°–28'
50.00	0°–35'	50.00	0°–26'
60.00	0°–28'	60.00	0°–21'

DEGREE SETTINGS FOR OPTICAL GENERATOR (2.00″ OD Cutter)

2.00 in. OD Cutter		1.50 in. ID Cutter	
Concave		Convex	
Radius required	Circle setting	Radius required	Circle setting
1.00″	90°–00'	1.00″	48°–36'
1.25	55°–8'	1.25	36°–53'
1.50	41°–46'	1.50	30°–00'
1.75	35°–50'	1.75	25°–22'
2.00	30°–00'	2.00	22°–2'
2.25	26°–23'	2.25	19°–28'
2.50	23°–35'	2.50	17°–27'
2.75	21°–20'	2.75	15°–49'
3.00	19°–28'	3.00	14°–29'
3.25	17°–53'	3.25	13°–20'
3.50	16°–36'	3.50	12°–22'
3.75	15°–28'	3.75	11°–32'

RADII: From 1.00 to 100″

15' = ¼ of Division
20' = ⅓ of Division
30' = ½ of Division
40' = ⅔ of Division
45' = ¾ of Division

535

DEGREE SETTINGS FOR OPTICAL GENERATOR (2 in. Cutter)

2 in. OD Cutter		1.5 in. ID Cutter	
Concave		Convex	
Radius required	Circle setting	Radius required	Circle setting
4.0″	14°–30′	4.0″	10°–47′
4.5	12°–50′	4.5	9°–34′
5.0	11°–32′	5.0	8°–37′
5.5	10°–28′	5.5	7°–50′
6.0	9°–35′	6.0	7°–10′
6.5	8°–50′	6.5	6°–38′
7.0	8°–13′	7.0	6°–8′
7.5	7°–40′	7.5	5°–45′
8.0	7°–10′	8.0	5°–22′
8.5	6°–35′	8.5	5°–4′
9.0	6°–22′	9.0	4°–47′
9.5	6°–3′	9.5	4°–32′
10.0	5°–44′	10.0	4°–18′
10.5	5°–28′	10.5	4°–6′
11.0	5°–13′	11.0	3°–54′
11.5	4°–59′	11.5	3°–44′
12.0	4°–47′	12.0	3°–35′
12.5	4°–35′	12.5	3°–23′
13.0	4°–24′	13.0	3°–18′
13.5	4°–15′	13.5	3°–12′
14.0	4°–6′	14.0	3°–4′
14.5	3°–57′	14.5	2°–58′
15.0	3°–49′	15.0	2°–52′
15.5	3°–42′	15.5	2°–46′
16.0	3°–35′	16.0	2°–41′
16.5	3°–28′	16.5	2°–36′
17.0	3°–22′	17.0	2°–32′
17.5	3°–16′	17.5	2°–28′
18.0	3°–11′	18.0	2°–23′
18.5	3°–6′	18.5	2°–19′
19.0	3°–0′	19.0	2°–16′
19.5	2°–56′	19.5	2°–12′
20.0	2°–52′	20.0	2°–9′
25.0	2°–17′	25.0	1°–40′
30.0	1°–50′	30.0	1°–20′
30.5	1°–53′	30.5	1°–26′
31.0	1°–51′	31.0	1°–24′
31.5	1°–49′	31.5	1°–22′
32.0	1°–48′	32.0	1°–21′
32.5	1°–46′	32.5	1°–20′
33.0	1°–44′	33.0	1°–18′
33.5	1°–42′	33.5	1°–17′
34.0	1°–41′	34.0	1°–16′

DEGREE SETTINGS FOR OPTICAL GENERATOR (2 in. Cutter)

2 in. OD Cutter		1.5 in. ID Cutter	
Concave		Convex	
Radius required	Circle setting	Radius required	Circle setting
34.5	1°–39′	34.5	1°–15′
35.0	1°–38′	35.0	1°–14′
35.5	1°–37′	35.5	1°–13′
36.0	1°–35′	36.0	1°–12′
36.5	1°–35	36.5	1°–11′
37.0	1°–33′	37.0	1°–10′
37.5	1°–32′	37.5	1°–9′
38.0	1°–31′	38.0	1°–8′
38.5	1°–29′	38.5	1°–7′
39.0	1°–28′	39.0	1°–6′
39.5	1°–27′	39.5	1°–5′
40.0	1°–26′	40.0	1°–4′
40.5	1°–25′	40.5	1°–3′
41.0	1°–24′	41.0	1°–2′
41.5	1°–23′	41.5	1°–1′
42.0	1°–22′	42.0	1°–0′
42.5	1°–21	42.5	1°–0′
43.0	1°–20′	43.0	0°–59′
43.5	1°–19′	43.5	0°–58′
44.0	1°–18′	44.0	0°–58′
44.5	1°–17′	44.5	0°–57′
45.0	1°–16′	45.0	0°–57′
45.5	1°–15′	45.5	0°–56′
46.0	1°–14′	46.0	0°–56′
46.5	1°–14′	46.5	0°–55
47.0	1°–13′	47.0	0°–55′
47.5	1°–13′	47.5	0°–54′
48.0	1°–12′	48.0	0°–54′
48.5	1°–11′	48.5	0°–53′
49.0	1°–10′	49.0	0°–53′
49.5	1°–9′	49.5	0°–52
50.0	1°–9′	50.0	0°–52
50.5	1°–8′	50.5	0°–51′
51.0	1°–8′	51.0	0°–51′
51.5	1°–7′	51.5	0°–50′
52.0	1°–7′	52.0	0°–50′
52.5	1°–6′	52.5	0°–49′
53.0	1°–5′	53.0	0°–49′
53.5	1°–4′	53.5	0°–48′
54.0	1°–3′	54.0	0°–48′
54.5	1°–3′	54.5	0°–47′
55.0	1°–2′	55.0	0°–47′
55.5	1°–2′	55.5	0°–47′

DEGREE SETTINGS FOR OPTICAL GENERATOR (2 in. Cutter)

2 in. OD Cutter		1.5 in. ID Cutter	
Concave		Convex	
Radius required	Circle setting	Radius required	Circle setting
56.0	1°–1′	56.0	0°–46′
56.5	1°–0′	56.5	0°–46′
57.0	1°–0′	57.0	0°–46′
57.5	0°–59	57.5	0°–45′
58.0	0°–59′	58.0	0°–45′
58.5	0°–58′	58.5	0°–44′
59.0	0°–58′	59.0	0°–44′
59.5	0°–57′	59.5	0°–43′
60.0	0°–57′	60.0	0°–43′
61.0	0°–56′	61.0	0°–42′
62.0	0°–55′	62.0	0°–42′
63.0	0°–54′	63.0	0°–41′
64.0	0°–54′	64.0	0°–40′
65.0	0°–53′	65.0	0°–39′
66.0	0°–52′	66.0	0°–39′
67.0	0°–51′	67.0	0°–38′
68.0	0°–51′	68.0	0°–38′
69.0	0°–50′	69.0	0°–37′
70.0	0°–49′	70.0	0°–37′
71.0	0°–49′	71.0	0°–36′
72.0	0°–48′	72.0	0°–36′
73.0	0°–47′	73.0	0°–35′
74.0	0°–47′	74.0	0°–35′
75.0	0°–46′	75.0	0°–34′
76.0	0°–45′	76.0	0°–34′
77.0	0°–45′	77.0	0°–33′
78.0	0°–44′	78.0	0°–33′
79.0	0°–44′	79.0	0°–32′
80.0	0°–43′	80.0	0°–32′
81.0	0°–43′	81.0	0°–31′
82.0	0°–42′	82.0	0°–31′
83.0	0°–42′	83.0	0°–30′
84.0	0°–41′	84.0	0°–30′
85.0	0°–41′	85.0	0°–30′
86.0	0°–40′	86.0	0°–30′
87.0	0°–40′	87.0	0°–29′
88.0	0°–39′	88.0	0°–29′
89.0	0°–39′	89.0	0°–29′
90.0	0°–38′	90.0	0°–29′
91.0	0°–38′	91.0	0°–28′
92.0	0°–37′	92.0	0°–28′
93.0	0°–37′	93.0	0°–28′
94.0	0°–37′	94.0	0°–27′

DEGREE SETTINGS FOR OPTICAL GENERATOR (2 in. Cutter)

2 in. OD Cutter		1.5 in. ID Cutter	
Concave		Convex	
Radius required	Circle setting	Radius required	Circle setting
95.0	0°–36′	95.0	0°–27′
96.0	0°–36′	96.0	0°–27′
97.0	0°–35′	97.0	0°–27′
98.0	0°–35′	98.0	0°–26′
99.0	0°–34′	99.0	0°–26′
100.0	0°–34′	100.0	0°–26′

DEGREE SETTINGS FOR RADIUS GENERATOR

2 in. OD		1.75 in. ID	
Concave Radius use OD		Convex Radius use ID	
Radius required	Circle setting	Radius required	Circle setting
1.00″	89°–55′	1.00″	61°–3′
1.25	55°–8′	1.25	44°–26′
1.50	41°–48′	1.50	35°–41′
1.75	34°–51′	1.75	30°–00′
2.00	30°–00′	2.00	25°–57′
2.25	26°–23′	2.25	22°–53′
2.50	23°–35′	2.50	20°–30′
2.75	21°–19′	2.75	18°–33′
3.00	19°–28′	3.00	16°–57′
3.25	17°–55′	3.25	15°–37′
3.50	16°–36′	3.50	14°–29′
3.75	15°–28′	3.75	13°–29′
4.00	14°–29′	4.00	12°–38′
4.25	13°–36′	4.25	11°–53′
4.50	12°–50′	4.50	11°–13′
4.75	12°–9′	4.75	10°–37′
5.00	11°–33′	5.00	10°–5′
5.25	10°–59′	5.25	9°–35′
5.50	10°–29′	5.50	9°–9′
5.75	10°–1′	5.75	8°–45′
6.00	9°–35′	6.00	8°–23′
6.25	9°–13′	6.25	8°–3′
6.50	8°–51′	6.50	7°–44′
6.75	8°–31′	6.75	7°–27′
7.00	8°–12′	7.00	7°–11′
7.25	7°–55′	7.25	6°–56′

539

DEGREE SETTINGS FOR RADIUS GENERATOR

2 in. OD		1.75 in. ID	
Concave Radius use OD		Convex Radius use ID	
Radius required	Circle setting	Radius required	Circle setting
7.50	7°–40′	7.50	6°–42′
7.75	7°–25′	7.75	6°–29′
8.00	7°–11′	8.00	6°–17′
8.25	6°–58′	8.25	6°–5′
8.50	6°–45′	8.50	5°–54′
8.75	6°–34′	8.75	5°–45′
9.00	6°–23′	9.00	5°–35′
9.25	6°–12′	9.25	5°–26′
9.50	6°–2′	9.50	5°–17′
9.75	5°–53′	9.75	5°–9′
10.00	5°–45′	10.00	5°–1′
10.25	5°–36′	10.25	4°–54′
10.50	5°–28′	10.50	4°–47′
10.75	5°–20′	10.75	4°–40′
11.00	5°–13′	11.00	4°–33′
11.25	5°–6′	11.25	4°–27′
11.50	4°–59′	11.50	4°–22′
11.75	4°–53′	11.75	4°–16′
12.00	4°–47′	12.00	4°–11′
12.25	4°–41′	12.25	4°–6′
12.50	4°–36′	12.50	4°–1′
12.75	4°–30′	12.75	3°–56′
13.00	4°–25′	13.00	3°–52′
13.25	4°–20′	13.25	3°–47′
13.50	4°–15′	13.50	3°–43′
13.75	4°–10′	13.75	3°–39′
14.00	4°–6′	14.00	3°–35′
14.25	4°–1′	14.25	3°–31′
14.50	3°–57′	14.50	3°–27′
14.75	3°–53′	14.75	3°–24′
15.00	3°–49′	15.00	3°–20′
15.25	3°–45′	15.25	3°–17′
15.50	3°–42′	15.50	3°–14′
15.75	3°–38′	15.75	3°–11′
16.00	3°–35′	16.00	3°–8′
16.25	3°–31′	16.25	3°–5′
16.50	3°–28′	16.50	3°–2′
16.75	3°–25′	16.75	3°–0′
17.00	3°–22′	17.00	2°–57′
17.25	3°–19′	17.25	2°–54′
17.50	3°–16′	17.50	2°–52′
17.75	3°–14′	17.75	2°–49′
18.00	3°–11′	18.00	2°–47′
18.25	3°–8′	18.25	2°–45′

DEGREE SETTINGS FOR RADIUS GENERATOR

2 in. OD		1.75 in. ID	
Concave Radius use OD		Convex Radius use ID	
Radius required	Circle setting	Radius required	Circle setting
18.50	3°–6'	18.50	2°–41'
18.75	3°–3'	18.75	2°–39'
19.00	3°–1'	19.00	2°–38'
19.25	2°–58'	19.25	2°–36'
19.50	2°–56'	19.50	2°–34'
19.75	2°–54'	19.75	2°–32'
20.00	2°–52'	20.00	2°–30'
20.25	2°–50'	20.25	2°–29'
20.50	2°–48'	20.50	2°–27'
20.75	2°–46'	20.75	2°–25'
21.00	2°–44'	21.00	2°–23'
21.25	2°–42'	21.25	2°–21'
21.50	2°–40'	21.50	2°–19'
21.75	2°–38'	21.75	2°–18'
22.00	2°–36'	22.00	2°–17'
22.25	2°–34'	22.25	2°–15'
22.50	2°–33'	22.50	2°–13'
22.75	2°–31'	22.75	2°–12'
23.00	2°–29'	23.00	2°–11'
23.25	2°–28'	23.25	2°–9'
23.50	2°–26'	23.50	2°–7'
23.75	2°–25'	23.75	2°–6'
24.00	2°–23'	24.00	2°–5'
24.25	2°–22'	24.25	2°–3'
24.50	2°–20'	24.50	2°–2'
24.75	2°–19'	24.75	2°–1'
25.00	2°–18'	25.00	2°–0'
25.25	2°–16'	25.25	1°–59'
25.50	2°–15'	25.50	1°–58'
25.75	2°–14'	25.75	1°–57'
26.00	2°–12'	26.00	1°–56'
26.25	2°–11'	26.25	1°–55'
26.50	2°–10'	26.50	1°–54'
26.75	2°–8'	26.75	1°–53'
27.00	2°–7'	27.00	1°–52'
27.25	2°–6'	27.25	1°–51'
27.50	2°–5'	27.50	1°–50'
27.75	2°–4'	27.75	1°–48'
28.00	2°–3'	28.00	1°–47'
28.25	2°–2'	28.25	1°–46'
28.50	2°–0'	28.50	1°–45'
28.75	1°–59'	28.75	1°–44'
29.00	1°–58'	29.00	1°–43'
29.25	1°–57'	29.25	1°–43'

DEGREE SETTINGS FOR RADIUS GENERATOR

2 in. OD		1.75 in. ID	
Concave Radius use OD		Convex Radius use ID	
Radius required	Circle setting	Radius required	Circle setting
29.50	1°–56′	29.50	1°–42′
29.75	1°–56′	29.75	1°–41′
30.00	1°–55′	30.00	1°–40′
30.25	1°–54′	30.25	1°–39′
30.50	1°–53′	30.50	1°–38′
30.75	1°–52′	30.75	1°–38′
31.00	1°–51′	31.00	1°–37′
31.25	1°–50′	31.25	1°–36′
31.50	1°–49′	31.50	1°–36′
31.75	1°–48′	31.75	1°–35′
32.00	1°–47′	32.00	1°–34′
32.25	1°–46′	32.25	1°–33′
32.50	1°–45′	32.50	1°–32′
32.75	1°–45′	32.75	1°–31′
33.00	1°–44′	33.00	1°–31′
33.25	1°–43′	33.25	1°–30′
33.50	1°–42′	33.50	1°–29′
33.75	1°–41′	33.75	1°–28′
34.00	1°–41′	34.00	1°–28′
34.25	1°–40′	34.25	1°–27′
34.50	1°–39′	34.50	1°–27′
34.75	1°–38′	34.75	1°–26′
35.00	1°–38′	35.00	1°–26′
35.25	1°–37′	35.25	1°–25′
35.50	1°–36′	35.50	1°–25′
35.75	1°–35′	35.75	1°–24′
36.00	1°–35′	36.00	1°–24′
36.25	1°–35′	36.25	1°–23′
36.50	1°–34′	36.50	1°–23′
36.75	1°–34′	36.75	1°–22′
37.00	1°–33′	37.00	1°–21′
37.25	1°–32′	37.25	1°–21′
37.50	1°–31′	37.50	1°–20′
37.75	1°–31′	37.75	1°–20′
38.00	1°–30′	38.00	1°–19′
38.25	1°–29′	38.25	1°–19′
38.50	1°–29′	38.50	1°–18′
38.75	1°–28′	38.75	1°–18′
39.00	1°–28′	39.00	1°–17′
39.25	1°–27′	39.25	1°–17′
39.50	1°–27′	39.50	1°–16′
39.75	1°–26′	39.75	1°–16′
40.00	1°–26′	40.00	1°–15′

DEGREE SETTINGS FOR RADIUS GENERATOR

2 in. OD		1.75 in. ID	
Concave Radius use OD		Convex Radius use ID	
Radius required	Circle setting	Radius required	Circle setting
41.00	1°–24′	41.00	1°–13′
42.00	1°–22′	42.00	1°–12′
43.00	1°–20′	43.00	1°–10′
44.00	1°–18′	44.00	1°–8′
45.00	1°–17′	45.00	1°–7′
46.00	1°–15′	46.00	1°–5′
47.00	1°–14′	47.00	1°–4′
48.00	1°–12′	48.00	1°–3′
49.00	1°–10′	49.00	1°–1′
50.00	1°–9′	50.00	1°–0′
51.00	1°–8′	51.00	0°–59′
52.00	1°–6′	52.00	0°–58′
53.00	1°–5′	53.00	0°–57′
54.00	1°–4′	54.00	0°–56′
55.00	1°–2′	55.00	0°–55′
56.00	1°–1′	56.00	0°–54′
57.00	1°–0′	57.00	0°–53′
58.00	0°–59′	58.00	0°–52′
59.00	0°–58′	59.00	0°–51′
60.00	0°–57′	60.00	0°–50′
65.00	0°–53′	65.00	0°–46′
70.00	0°–49′	70.00	0°–43′
75.00	0°–46′	75.00	0°–40′
80.00	0°–43′	80.00	0°–37′
85.00	0°–40′	85.00	0°–35′
90.00	0°–38′	90.00	0°–33′
95.00	0°–36′	95.00	0°–31′
100.00	0°–35′	100.00	0°–30′
110.00	0°–31′	110.00	0°–27′
120.00	0°–29′	120.00	0°–25′
130.00	0°–26′	130.00	0°–23′
140.00	0°–25′	140.00	0°–21′
150.00	0°–23′	150.00	0°–20′
160.00	0°–22′	160.00	0°–18′
170.00	0°–20′	170.00	0°–18′
180.00	0°–19′	180.00	0°–17′
190.00	0°–18′	190.00	0°–16′
200.00	0°–17′	200.00	0°–15′

DEGREE SETTINGS FOR OPTICAL GENERATOR

4 in. OD Cutter		3.75 in. ID Cutter	
Concave		Convex	
Radius required	Circle setting	Radius required	Circle setting
4.0″	30°–0′	4.0″	27°–57′
4.5	26°–25′	4.5	24°–37′
5.0	23°–35′	5.0	22°–2′
5.5	21°–20′	5.5	19°–56′
6.0	19°–30′	6.0	18°–13′
6.5	17°–50′	6.5	16°–46′
7.0	16°–38′	7.0	15°–32′
7.5	15°–30′	7.5	14°–29′
8.0	14°–25′	8.0	13°–33′
8.5	13°–38′	8.5	12°–44′
9.0	12°–52′	9.0	12°–1′
9.5	12°–10′	9.5	11°–23′
10.0	11°–30′	10.0	10°–48′
10.5	11°–0′	10.5	10°–17
11.0	10°–26′	11.0	9°–49′
11.5	10°–0′	11.5	9°–23′
12.0	9°–40′	12.0	8°–59′
12.5	9°–13′	12.5	8°–38′
13.0	8°–52′	13.0	8°–17′
13.5	8°–31′	13.5	7°–59′
14.0	8°–13′	14.0	7°–42′
14.5	7°–56′	14.5	7°–26′
15.0	7°–40′	15.0	7°–11′
15.5	7°–25′	15.5	6°–56′
16.0	7°–11′	16.0	6°–44′
16.5	6°–57′	16.5	6°–31′
17.0	6°–46′	17.0	6°–19′
17.5	6°–33′	17.5	6°–9′
18.0	6°–23′	18.0	5°–58′
18.5	6°–12′	18.5	5°–49′
19.0	6°–3′	19.0	5°–39′
19.5	5°–53′	19.5	5°–31′
20.0	5°–45′	20.0	5°–22′
25.0	5°–36′	25.5	4°–18′
30.0	3°–57′	30.0	3°–35′
30.5	3°–45′	30.5	2°–35′
31.0	3°–41′	31.0	2°–32′
31.5	3°–38′	31.5	2°–30′
32.0	3°–35′	32.0	2°–28′
32.5	3°–32′	32.5	2°–26′
33.0	3°–28′	33.0	2°–24′
33.5	3°–25′	33.5	2°–22′

DEGREE SETTINGS FOR OPTICAL GENERATOR

4 in. OD Cutter		3.75 in. ID Cutter	
Concave		Convex	
Radius required	Circle setting	Radius required	Circle setting
34.0	3°–22′	34.0	2°–20′
34.5	3°–19′	34.5	2°–18′
35.0	3°–16′	35.0	2°–15′
35.5	3°–13′	35.5	2°–13′
36.0	3°–11′	36.0	2°–11′
36.5	3°–8′	36.5	2°–9′
37.0	3°–6′	37.0	2°–7′
37.5	3°–3′	37.5	2°–5′
38.0	3°–1′	38.0	2°–4′
38.5	2°–58′	38.5	2°–2′
39.0	2°–56′	39.0	2°–0′
39.5	2°–54′	39.5	1°–59′
40.0	2°–52′	40.0	1°–58′
40.5	2°–49′	40.5	2°–56′
41.0	2°–47′	41.0	1°–54′
41.5	2°–45′	41.5	1°–53
42.0	2°–42′	42.0	1°–52′
42.5	2°–41′	42.5	1°–50′
43.0	2°–40′	43.0	1°–49′
43.5	2°–38′	43.5	1°–47′
44.0	2°–36′	44.0	1°–46′
44.5	2°–34′	44.5	1°–45′
45.0	2°–32′	45.0	1°–44′
45.5	2°–31′	45.5	1°–42′
46.0	2°–29′	46.0	1°–41′
46.5	2°–27′	46.5	1°–41′
47.0	2°–26′	47.0	1°–40′
47.5	2°–25′	47.5	1°–39′
48.0	2°–23′	48.0	1°–38′
48.5	2°–21′	48.5	1°–37′
49.0	2°–20′	49.0	1°–36′
49.5	2°–19′	49.5	1°–35′
50.0	2°–18′	50.0	1°–34′
50.5	2°–16′	50.5	1°–34′
51.0	2°–15′	51.0	1°–33′
51.5	2°–13′	51.5	1°–32′
52.0	2°–12′	52.0	1°–31′
52.5	2°–11′	52.5	1°–30′
53.0	2°–10′	53.0	1°–29′
53.5	2°–9′	53.5	1°–28′
54.0	2°–7′	54.0	1°–27′
54.5	2°–6′	54.5	1°–26′

DEGREE SETTINGS FOR OPTICAL GENERATOR

4 in. OD Cutter		3.75 in. ID Cutter	
Concave		Convex	
Radius required	Circle setting	Radius required	Circle setting
55.0	2°–5′	55.0	1°–25′
55.5	2°–4′	55.5	1°–24′
56.0	2°–3′	56.0	1°–23′
56.5	2°–2′	56.5	1°–22′
57.0	2°–1′	57.0	1°–22′
57.5	2°–1′	57.5	1°–21′
58.0	1°–59′	58.0	1°–21′
58.5	1°–58′	58.5	1°–20′
59.0	1°–56′	59.0	1°–19′
59.5	1°–55′	59.5	1°–18′
60.0	1°–55′	60.0	1°–18′
61.0	1°–53′	61.0	1°–17′
62.0	1°–51′	62.0	1°–16′
63.0	1°–49′	63.0	1°–15′
64.0	1°–47′	64.0	1°–14′
65.0	1°–45′	65.0	1°–13′
66.0	1°–43′	66.0	1°–12′
67.0	1°–41′	67.0	1°–11′
68.0	1°–40′	68.0	1°–10′
69.0	1°–39′	69.0	1°–9′
70.0	1°–38′	70.0	1°–7′
71.0	1°–37′	71.0	1°–7′
72.0	1°–36′	72.0	1°–6′
73.0	1°–34′	73.0	1°–5′
74.0	1°–33′	74.0	1°–4′
75.0	1°–32′	75.0	1°–3′
76.0	1°–31′	76.0	1°–2′
77.0	1°–29′	77.0	1°–1′
78.0	1°–28′	78.0	1°–1′
79.0	1°–27′	79.0	1°–0′
80.0	1°–26′	80.0	1°–0′
81.0	1°–25′	81.0	0°–59′
82.0	1°–24′	82.0	0°–58′
83.0	1°–23′	83.0	0°–57′
84.0	1°–22′	84.0	0°–56′
85.0	1°–21′	85.0	0°–55′
86.0	1°–20′	86.0	0°–55′
87.0	1°–19′	87.0	0°–54′
88.0	1°–18′	88.0	0°–53′
89.0	1°–17′	89.0	0°–52′
90.0	1°–16′	90.0	0°–52′
91.0	1°–15′	91.0	0°–51′
92.0	1°–14′	92.0	0°–51

DEGREE SETTINGS FOR OPTICAL GENERATOR

| 4 in. OD Cutter | | 3.75 in. ID Cutter | |
| Concave | | Convex | |
Radius required	Circle setting	Radius required	Circle setting
93.0	1°–13′	93.0	0°–50′
94.0	1°–12′	94.0	0°–49′
95.0	1°–12′	95.0	0°–49′
96.0	1°–11′	96.0	0°–48′
.97.0	1°–10′	97.0	0°–48′
98.0	1°–10′	98.0	0°–47′
99.0	1°–9′	99.0	0°–47′
100.0	1°–9′	100.0	0°–47′

Appendix 14

Profiling Pitch Polishers

In optical figuring pitch polishers with cutouts or scratched profiles are often used. Much time and effort are lost in trying to figure the optical element's surface with previously made polishers. In this Appendix we develop a correcting polisher from the test patterns slopes.

In Chapter 14 the correlation of Ronchigrams and focograms is established for typical refracting and reflecting elements in an autocollimating setup. In Figs. A14.1 and A14.2 observe the slope patterns (in box) of the high or low regions from the focograms and Ronchigrams. These box sections contain other information than that of the contour or configuration of the profiling pitch polisher required to change by optical figuring any Figs. A14.1 or A14.2 to a null pattern (a well corrected system free of spherical aberration); for example, optical figuring a concave spherical mirror into a paraboloid with an optical flat gives an oblate spheroid focogram or Ronchigram (under correction). (See Figs. A14.1 and A14.2.) Because this is a reflecting element rotate Fig. A14.2 180° and find the pincushion pattern of the inside-of-focus Ronchigram in quadrant 1. Similarly Fig. A14.2 is turned upside down, and pattern 1, 1 or 1, 2, depending on the f/ratio of the concave mirror being figured, is found. The slope pattern with a high center and high peripheral area is illustrated in either box section. The desired profile pitch polisher would be a sixpoint star with small triangles of pitch near the peripheral edge. Allowing for some translation of the polisher across the glass element the small triangles of pitch would be slightly inside the edge. Another type is the third-diameter polisher. Polish across the central and highest areas of the 0.7 zone.

This star-type polisher with triangles of pitch must be used with care when the full correction is nearly completed or a turned down edge will result. The one-third diameter polisher invariably causes a turned up edge as the central area becomes fully corrected. This can be offset by lengthening the translating stroke and causing it to pass well over the edge. Then, again, for small diameter mirrors [12-in. (30 cm)] a solid polisher slightly larger than the mirror, held stationary on the rotating element, can be used. This polisher is valuable because it irons out any astigmatic surfaces developed

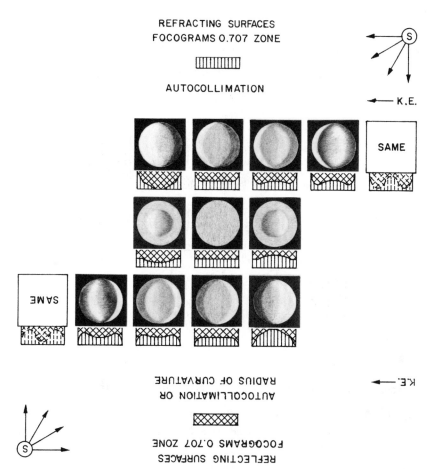

Fig. A14.1. Focograms for typical refracting and reflecting elements in an autocollimation setup. The slope patterns (in box) show high or low areas.

by the smaller type polishers. For larger diameter mirrors a small diameter polisher is hand-rotated while being held in position by a stationary arm with a pin.

From Figs. A14.1 and A14.2 in an upside-down position look at the top box with parallel lines and develop a profile of the polisher for minimum glass removal. It becomes a rose-leaf polisher working at the 0.707 zone for the developed profiles of the pitch polishers. (See Fig. A14.3.)

Another example of a refracting element is a corrector plate for a Schmidt-Cassegrain system. Again the pattern is an oblate spheroid that

RONCHIGRAMS

Autocollimation (inside of focus). Parallel sections (box) refracting surfaces

Autocollimation (outside of focus). 45° sections (box) reflecting surfaces

Y AXIS

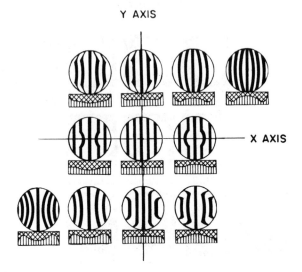

X AXIS

Autocollimation (inside of focus). 45° sections (box) reflecting surfaces

Autocollimation (outside of focus), Parallel sections (box) refracting surfaces

Fig. A14.2. Ronchigrams for typical reflecting and refracting elements in an autocollimated setup. The slope patterns (in box) show high or low areas.

results from the optical setup in the autocollimation test of the system. This is the pincushion pattern in the third quadrant (either -1, -1 or -1, -2, depending on the eccentricity value of the system). Here the rose-leaf profile polisher can be developed by the parallel line section of the box below the patterns. A. Cornejo in Chapter 9 of *Optical Shop Testing* (Wiley, New York, 1978) p 299, describes Ronchigrams similar to those shown in Fig. A14.2 but these diagrams are a little too difficult for the beginning optician to understand.

The optical figuring of a Schmidt-Cassegrain or other reflecting mirror systems with or without refracting elements involves several aspects of optical testing. Consider a Schmidt-Cassegrain system with equal radii on both the primary and secondary mirrors separated at a parametric distance for an 1.72 amplification. This system is described in Chapter 23.

Set the two reflective mirrors of the system in an autocollimation arrangement with a flat. (See Fig. A14.4.) Two optical tests can be used alternately if a slit source is incorporated in the knife-edge apparatus: then the knife-edge can be used or the Ronchigrams observed when a Ronchi-

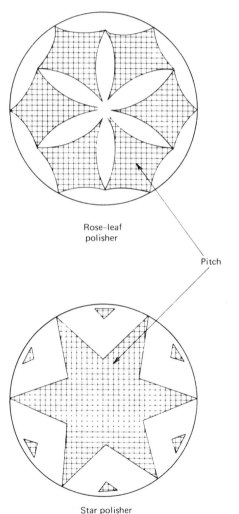

Rose-leaf
polisher

Pitch

Star polisher

Fig. A14.3. The profiling pitch polishers of the rose-leaf and star patterns for correcting undercorrected lens system to a null pattern. An inverse configuration would be pitch within the white areas of the polisher.

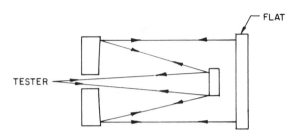

Fig. A14.4. The autocollimation of two reflective mirrors—one primary and one secondary—with spherical surfaces that would give a test pattern of an oblate spheroid.

grating is placed over the lighted slit with some extension for viewing. The oblate spheroid pattern of undercorrection is shown in Fig. A14.4; *Question:* What test pattern is observed in Fig. A14.5 if the convex secondary mirror is mounted on or near the corrector plate? *Answer:* It would still be an oblate spheroid with undercorrection. This is not a contradiction. It is not until one surface of the corrector plate has four or more waves aspheric that the refractive test patterns are observed. Two flate surfaces of one to two fringes on each side of the planocorrector plate would not cause a change.

Another puzzling condition that can occur at the onset of the optical figuring concerns the step patterns and ring zones observed at the edge of the corrector plate, one of which is shown in Fig. A14.6. Note the well-corrected central zone figured by the rose-leaf polisher and the badly turned down edge. Because the polisher has barely worked on its edge, what we

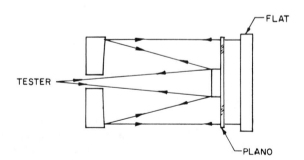

Fig. A14.5. An identical setup of the two reflective mirrors with the secondary mirror attached to a refractive planoparallel with near flat surfaces. The same oblate spheroid pattern will be observed (see text).

—STEP ZONE

Fig. A14.6. The inside-of-focus Ronchigram showing a well-corrected central area with a badly turned down edge. This step zone shows that full correction has not reached the edge.

observe here is the step zone described by Thomas Johnson in *Optical Shop Notebook* (Optical Society of America, Washington, D.C., 1976), pp. 92–95. The polishing should be continued, even to a point at which some slight overcorrection is observed at the central and peripheral areas. At this point refigure the secondary mirror with an inverse star polisher, making the system null in its autocollimation setup. This improves the off-axis image of the compound system. Here the optical figuring of the reflector element moves four times faster than when figuring the corrector (refractive) element.

Summarizing, the optical figuring of a lens system is always dependent on the optimum design of the system; for example, the optician would generally aspherize the central area of a spheroid to form a paraboloid mirror during polishing. For a Schmidt-Cassegrain correcting plate lower the 0.86 zone first with a rose-leaf polisher to a slight overcorrection and then retouch the secondary mirror. Again, it is interesting to estimate the apparent focal plane while optical testing, as projected backward or forward to any major optical component surface, and then figure that surface. Perhaps it is best to consider figuring a smaller surface—one with its high amplification factor (as for a secondary mirror) and eccentricity value. In figuring a Maksutov compound telescope system less glass is removed by optically figuring the primary mirror; optical figuring of the corrector lens requires the removal of four times more glass. For optimum performance of a well-designed lens system the optician should be told what surface or surfaces need to be retouched to arrive at a well-corrected system.

1. FORMING THE CUTOUT PATTERNS

The cutout patterns can be cut from heavy wrapping or mylar drafting paper, which is expensive but it has one advantage that it is stronger and the wrapping paper is easily torn. By using either paper as raw material it is possible to form radially symmetric cutout patterns that serve to contour the pitch laps by compressing the pitch for low areas and by exposing and elevating the pitch in areas in which the desired polishing cutting action is to take place.

Figure A14.7 demonstrated how the profiles are cut out. Using this cutout technique a dozen or more different polishers can be made up on one working day. Soft pitch similar to Zobal soft or Swiss (Gugolz) # 64 (U.S.

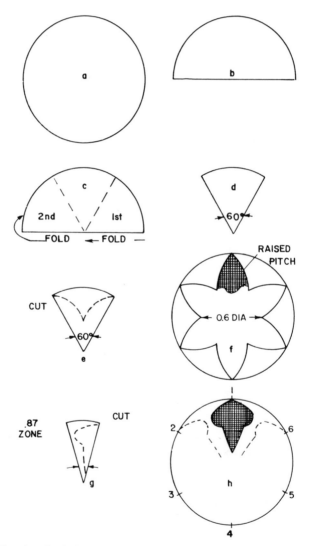

Fig. A14.7. Directions for folding and cutting out the configurations of the patterns to profile the pitch polisher.

distributor A. Meller, Providence, R.I.), which clearly polishes many types of glass and flows readily when heated in warm water or under heating infrared lamps. It is well to emboss the polisher with plastic screen door netting so that good impressions are formed by the cutout sheeting made from either of the papers used. The exposed embossed areas make excellent contact during polishing.

Let us summarize the known optical tests: (1) knife-edge test qualitatively gives the contour errors in the glass of a lens system; (2) Ronchi grating provides information about focal plane errors (the thickness of the intermediate area between the fringe patterns is the magnitude of the slope change of the glass errors); (3) the interferometer gives glass contours errors and therefore is differentiated to arrive at focal plane errors. *Question:* Is a concave reflecting mirror, tested at its radius of curvature, a self-autocollimated device? *Answer:* The pattern is null for a truly spherical surface.

These focograms are formed by a knife-edge apparatus when testing a reflecting or refracting surface of a lens system in autocollimation or parallel light. The focograms depict the systems with or without spherical aberration. The light source is on the *right-hand* side and the focograms are formed by making the cutoff of the knife-edge from *right-to-left* when a lens system is tested in parallel (parallel light from a lens collimator setup or formed by using an autocollimating flat in conjunction with the lens system being figured). To view the focograms for the autocollimation (see Fig. A14.2) or radius of curvature (autocollimating setup) of a reflective mirror turn Fig. A14.2 upside down and make the knife-edge blade cutoff from *left to right*. It is important to remember several aspects while testing: all slope departures are illuminated by the light source's position which for the knife-edge apparatus is on the right-hand side. Some confusion can exist because Texereau has a stationary light source and he makes it a rule that it be on the *left* of the axis facing the mirror, and that the *knife-edge* be inserted *right-to-left*. Porter reverses the whole setup: stationary light source on the left, knife-edge on the right. Both Texereau and Porter make this rule: *assume* that the light always comes from the side opposite the knife-edge (in testing a refractive *lens* or transmitting element the rule is reversed). Figure A14.1 as shown is correct because the knife-edge is following the sun. In any case, all slopes *away* from the light will be *dark*. Also, the diameter of mirror or lens is the dividing line of lights and shadows, and a bright area on one half will be a dark area on the opposite half.

As already described the box sections below the patterns contain other information; that is, the contour or configuration of the profiling pitch polisher would be required to change by optical figuring any Ronchigram or focogram to a null pattern (well corrected). For the autocollimation or radius of curvature testing of the reflecting element the 45° cross-sectional lines show the apparent slope of the optical surface or reflecting surface of the element. Summarizing, any intermediate zonal departures observed from the null pattern or well-corrected system as shown in the Ronchigrams or focograms can be considered as an eccentricity value change (positive or negative). The autocollimation or radius of curvature testing have purposely been made zero for the paraboloid or spheroid.

Appendix 15

Using a Street Lamp in Place of an Optical Flat

Amateur telescope makers Jeff Schrodeer and Charles C. Chinizi of Los Angeles have used distance street lamps in place of optical flats. Since the advent of modern mercury and sodium street lamps which give a monochromatic light source this has been possible. Chinizi has used this setup to figure Cassegrain secondary (hyperbolic) mirrors optically for Cassegrainian telescopes. Of course, a telescope unit built of plywood and mounts for the primary and secondary mirrors must be available. Only nights when the atmospheric conditions are reasonably stable can this test be made. The mercury street lamps (blue-white) and sodium (bright yellow) several miles distant offer a point source that is very bright. Chinizi and I have made use of a Newtonian setup of plywood and tested parabolic primary mirrors with these street lights. I now have a 12 in. optical flat with a 1/20 fringe surface.

Jeff Schrodeer's method of using these street lamps several miles distant is unique. When he made his 11-in. achromatic objective with little experience in making telescope mirrors, he first made his mount for testing the degree of correction for spherical aberration. It is well known that even using the best available design information there is always residual spherical aberration of several waves. This must be removed by figuring optically as described in chapter 27. Here the front crown element is figured optically with an autocollimating flat to form parallel rays. Schroeder's method is unique in that he figured the concave surface of the flint element that he could test at the radius of curvature for which he had much experience in making telescope mirrors. He learned that the figure he must have was a oblate conic surface. Since Schroeder was using the Ronchi grating test he could have a feeling for the desired Ronchi pattern. His article in *Sky and Telescope*, March 1978, describes the testing and optical figuring of a large achromatic objective telescope.

Edward Lumley's method of drawing a rough template for the desired conic surface could also be used. Here Schroeder could have drawn the patterns of undercorrection and overcorrection because he undoubtedly soon learned while testing his telescope with the newly corrected concave surface on the flint element. The desired template could be approximately drawn from the previous pair of slope departures. The templates can be made from carboard by first drawing the Ronchigram of the slope with a marking pen. Here the identical Ronchigram setup, must be used. Set the Ronchigram with three fringe patterns on each side of the middle aperture of the lens system at its radius of curvature test set. Next draw the second contour of the fringe pattern with a marking pen and then place a piece of mylar drafting paper over the outline and draw the same slope departure. It is important that the Ronchi grating be precisely inside-of-focus by the same distance each time for the radius of curvature test. The prior mylor slope pattern is drawn with marking pencil each time the test is made to determine whether Ronchi fringes are shallower or longer radius than the drawn Rochi pattern after each figuring and polishing phase. Lumley's paper appeared in the *Sky and Telescope* November 1961, p. 298–299. This test should be used more often by the amateur telescope maker for making conic aspheric mirrors. It is well known that a coninc surface is always a small correction of a parabola. Of course, the parabolic surface eccentricity is -1 but this is squared and to becomes $+1$. (See Chapter 28.)

Lumley has even made secondary mirrors for Cassegrainian telesecopes by testing the secondary in a reversed manner making it a concave test at the radius of curvature. Here he ray traces a prescribed template for the Ronchigram's contour.

Appendix 16

Figuring Large Paraboloids

At Mt. Hamilton in California the time is nearing when the Lick Observatory will put into operation the second largest optical telescope in the world (in 1979 the fourth). The giant 120-in. mirror, which will probe the depths of the space-scanning objects a billion and more light years away, has reached the stage at which final correction must be done with small hand-controlled polishers. In the picture reproduced here the late Don Hendrix and Howard Cowan are seen during the painstaking task of figuring a small area of the mirror (Fig. A16.1).

Since 1953, when grinding began, about 900 lb of glass have been removed from the original four-ton, 10-ft disk of pyrex glass, but only a few thousandths of a ounce will be polished away in the final figuring to within one-tenth wavelength of light (1/400,000 in.).

This delicate phase was carried out under the personal direction of Don Hendrix, who was head of the optical shop at the Mount Wilson and Palomar Observatories, with the assistance of Howard Cowan, a retired Lick Observatory optician. Henrix had the experience needed for he did the final figuring on the 200-in mirror in precisely the same manner.

Hartmann and knife-edge tests are being used to find the areas of the mirror that are not perfect. For both it is necessary to have an excellent night for observation, to have the mirror in place, collecting the light of a suitable bright star. The mirror is not yet aluminized, but its five percent reflectivity is sufficient for the tests. Either Hendrix or Dr. N. U. Mayall, Lick astronomer, rides in the observer's cage at the top of the telescope tube to take photographs of the test patterns.

In the Hartmann test an opaque screen with regularly spaced holes is placed a few feet below the observer's cage and an out-of-focus photograph is taken. Points of light on the negative show the positions of the holes; if one appears out of position, it indicates a distortion in the slope of the surface at a particular on the mirror.

The knife-edge test is carried on in much the same way that an amateur telescope maker might test a lens or mirror at focus by autocollimation, using an optical flat or parallel light from another optical system. There is no large flat mirror of sufficient size and precision to make such tests on the

Fig. A16.1. The last stages of figuring the surface of the 120-in. telescope mirror are the slowest and delicate. Howard Cowan watches as Don O. Hendrix presses a small hand polisher into contact. Photograph by Robert E. Watson.

120-inch mirror. Nature's own source of parallel light—the stars—must be used, although with the limitations imposed by atmospheric turbulence.

Enlargements of some of the knife-edge focograms are shown in the picture, scattered above on the mirror surface. These prints, marked to correspond to the points of compass labeled on the edge of the disk, tell the worker where to apply the necessary corrections. A wooden ruler is necessary for measuring distances on the mirror itself, and the working area is outlined with the grease glass-marking pencil lying in front of the white box in the picture.

The hand-polishing tools are 6 in. in diameter or less, with a backing of pressed wood and a facing of pitch. Optical rouge is used to work the area slowly and lightly. Each spell of polishing is followed by further Hartmann and knife-edge tests to determine the results.

Hendrix is seen pressing a hand polisher to ensure perfect contact on the glass before correcting begins: the area is located about a foot inside the edge of the mirror in its northeast quadrant. To avoid scratches cleanliness is of great importance. Both workers wear washable trousers with rolled-up cuffs, and washable footgear. The side of the mirror, as well as the base of the grinding and polishing machine on which it rests, is covered with sheets of polyethylene plastic held to the mirror's edge with waterproof adhesive tape.

The supporting cores of the mirror's floating system and the weight-reducing honeycomb structure of the mirror back are easily seen through the highly polished surface. The central hole is only 8 in. in diameter, unexpectedly small for a mirror of this size.

The work goes very slowly. After a test is made, two or three days are needed to calculate where imperfections are located on the surface. A cloudy night can cause delays. One hour's work requires about a week of testing. The acutal time of completion depends on the occurrence of nights of excellent viewing which at this time of the year are rather infrequent on Mt. Hamilton.

The 120-in. mirror has already required 300 hr for its grinding and polishing. The entire instrument, built at a cost of 2.5 million dollars to the State of California, has been 10 years in the making. A recent portraite of this optical giant and other photographs may be seen in *Sky and Telescope* for December 1956, pp. 61 to 63.

ADDENDUM

Consider briefly the history of the final phases of the optical figuring of the 200-in. mirror and some of the personnel involved.

The final testing of the 200-in. paraboloid at the optical shop in Pasadena, CA, made use of a null lens system designed by Dr. F. E. Ross in conjunction with the knife-edge testing at the radius of curvature. The chief optician was the late Marcus T. Brown under the supervision of the late Dr. J. A. Anderson of the California Institute of Technology who was in charge of the grinding and figuring. Brown is the author of Chapter 26 on the Foucault knife-edge or Schlieren test.

Here the mirror could be tilted while on the polishing machine during the testing at its radius of curvature with the null lens and the knife-edge. This is not the best method to hold a very large mirror and was undoubtedly the cause of the flexure or raised zone discovered afterwards. Another factor that contributed was that very little experimentation had been done on cellular recessed mirrors. One small 16-in. Model had been worked to a spheroid and then mounted as reversed Dall-Kirham (actually Schmidt made these compound telescopes in Europe) to test out the mounting and so on. I knew all the personnel involved and it is not the intent by hindsight to discredit anyone of them—only to report the history and speculate on what had happened.

When the aluminized mirror was mounted in the cell, attached to the telescope and star tested, it was found to be unacceptable. Much time and effort was made to adjust the 200-in. mirror in its cell. Because the floating system was similar to the mounting of the 100-in. telescope on Mt. Wilson, they finally found the problem to be a high raised zone midway between the 0.707 zone and the periphery. Floating metallic systems to hold large telescope mirrors is not used today; instead, the cellular structures has automatic controlled air bags, which support approximately 95% of the mass of the mirror, the other 5% being held by holding pads. The mirror is held in position around its periphery by a mercury-filled tube. This prevents the lateral shifting of the mirror while in various tilting positions.

Finally it was decided that the raised mirror zone had been due to flexure while being figured and tested in a vertical position. Dr. Ira Bowen along with Don Hendrix decided to optically figure the large mirror in the observatory dome! The plan was to leave the mirror in its cell (now called marrying the mirror to the cell) while it rested on the polishing machine. The star test would be used to test the mirror by photographing a zenith star of sufficient brightness. Here the knife-edge makes a cutoff at the focus of the mirror. Perhaps Fig. A16.2 better illustrates how this raised zone would appear in a photograph with parallel light.

Bowen describes the technique in an earlier paper (see Bibliography). This publication appeared after the final optical figuring and the 200-in. telescope was being used.

Consider the technique used by Don Hendrix. The location of the faulty zone (raised) was midway between the 0.707 zone and was estimated to be 1.5 wave high (private communication with Hendrix). Here he used a 6-in.

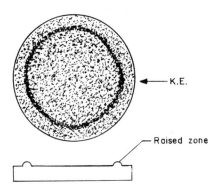

Raised zone

K.E.

Fig. A16.2. The focogram of a raised zone in a paraboloid when tested with parallel light from a star and a knife-edge.

diameter polisher with semiflexibility that was placed on the crest of the raised zone. The small polisher was hand-rotated on the crest of the high zone with an eccentric throw of ⅜ to ¼ in. The large mirror rotated at perhaps 3 rpm. Here the rotation can be stopped as desired to see if the polisher is hogging in. All small polishers have cut facets in them to allow for the pitch to flow and be kept moist. This type of polishing during optical figuring is called "rapid-rotation, no stroke."

The type of polishing machine with its swinging arm and rotating spindle can be observed in the photograph. [See Fig. A16.1) (*upper right*).] This is often used as above.

When Don Hendrix went ot Mt. Hamilton (Lick Observatory) he actually had the dome and telescope mounting constructed while the 120-in. mirror was being fabricated. From this basis modern-day telescopes are now tested in tower buildings; for instance, at Parsons and Grubb and Optical Center at Arizona University. The large mirror is married to the cell during its grinding and polishing.

In the description above, there was a small local zone in the northeast quadrant that was being localized by hand polishing. These are hard stops that are not randomily distributed over the surface of the mirror. Generally they are around one-eighth to one-tenth wave.

It is known that the largest telescope mirror in Russia (250-in.) has a number of hard and soft stops that have been blacked out so it can be used. A new mirror is being cast for future use.

Those interested in the Hartmann test should read Chapter 10 of *Optical Shop Testing*, D. Malacara, Ed. Wiley, New York, 1978.

Dr A. E. Whitford, former director of Lick Observatory, gives this information on the fabrication of the 120-in. mirror. There was an underground test tunnel that had sufficient length to test the mirror at its radius of curvature. The test tunnel was used to bring the mirror to a spherical surface.

An aspheric correcting lens, made in the optical shop, made possible a Foucault cutoff test of the parabolized mirror. Howard Cowan thinks the lens was to a design by I. S. Bowen (more likely by F. E. Ross) because Bowen was director at Mt. Palomar. Much of the parabolizing was done in the tunnel, and there was a problem with distortion of the mirror by the edge-arcs that supported it when it was tipped up (vertically) from its horizontal position on the polishing machine. The same problem occurred on the 200-in. testing to a vertical position.

Is the dome of sufficient height to be used for radius of curvature testing was one question asked. According to Dr. Whitfford, the optical shop is at the lowest level of the northeast part of the dome foundations and the clear space overhead was not great enough for any testing along a vertical line of sight. When an approximate paraboloid had been produced, the mirror was transferred on its support system to the telescope for direct testing on the stars. Here the both the Foucault cutoff and the Hartmann screen were used. A description of the procedures is given by Mayall and Vasilevskis in *Astronom. J*, **65**, 304–317, (1960).

The swing arm shown in the picture had a central spindle sturdy enough to handle a full-size grinding and polishing tool. Polishing tools of all sizes down to 12 in. were used in the figuring process. The final touches were done by hand. The photographs in front of Don Hendrix are Foucault cutoffs in various directions made in the prime focus cage.

BIBLIOGRAPHY

Bowen, Ira S. "Optical Problems at the Palomar Observatory." *J. Opt. Soc. Am.*, **42**, 795 (1952).

Hendrix, Don, (private communication).

Appendix 17

Scanning A Murty Interferometer for Optical Testing

A Murty's interferometer setup can be designed for scanning polished or wax-ground polished aspheric surfaces. Use of Penta's prisms to scan optical surfaces has been described by Ransom, Hendrix, Hooker, and Hochgraf. I have outlined the placement of small planoparallels with or without a slight wedge in the parallel beam of the Murty interferometer. In this section the use of a pair of planoparallels cut from a larger one or one planoparallel for optical scanning, both of which will give interferograms projected on a viewing screen, is discussed.

Some of the earliest scanning by a penta prism was done by Ransom. He scanned with a pair of penta prisms split from on larger prism or by a single prism for scanning aspheric mirror surfaces or telescope systems. Hendrix's publication describes and early use of the penta prism in scanning the optical figure of a corrector plate in an assembled Schmidt camera. Hooker states in a recently published abstract that the angular deviation of the reflective beam from an aspheric surface of a mirror can be sensed to better than 2 μm by a scanning penta. Hochgraft in a recent abstract showed that the zonal error across the (40-cm) aspheric for a typical $f/1.25$ paraboloid was 1 μ. It is well known that the angular deviation can be converted to surface deformations and plotted as surface glass to convert the best-fit curve into an aspheric surface.

A typical design patterned after the scanning prism can be used for converting the Murty interferometer setup for interferometer setup for interferometric testing. (See Figs. A17.1, 2.) The large concave mirror is resting on the polishing machine abd is being optically figured. The scanning unit mounted in the parallel beam consists of a pair of 153-mm planoparallels which are fastened to a steel ribbon running around two pulleys. (Do not use planoparallels that are too small because of the foreshortening effect caused by the 45° scan which leaves only a small area of the aspheric surface.) The pair of planoparallels scan the aperture of the mirror as they are translated across the aperture of the mirror in opposite directions by

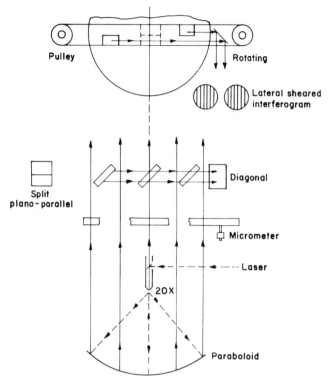

Fig. A17.1. A split planoparallel (Murty's interferometer) for scanning an aspheric surface that can be wax-ground and buffed to form interferograms of the optical figuring. A laser and 20 X and pinhole are used.

rotating one of the pulleys. A laser source consisting of a 10-mm aluminized diagonal inside the microscope tube with its diverging objective is mounted on a lever arm that can be raised or lowered by a differential micrometer. This focusing arm and the micrometer measure the reflective deviations at each zone scanned by a pair of planoparallels. The two projected reflective interferorgrams are viewed on a frosted viewing screen. The rotating viewing screen has two sets of 12.5-mm width parallel lines ruled 90° to each other. In a like manner but not shown in Fig. A17.1, a single 153 or larger planoparallel can be used instead of a pair and only one interferogram will be projected.

The use of the diverging laser (0.5 mW) source allows the optician to alternately grind and wax-grind polish the surface aspheric mirrors or lenses during optical figuring. The application of wax-to-grind surfaces to stimulate polished surfaces is described by Moreau and Hopkins. The optical figuring is generally carried out with small rotating flexible tiled tools and flexible (nonrotating) polishers. (See also Chapter 20.) My experimentation with wax-ground polished surfaces indicated the following: a 30-cm $f/6.0$ mirror was wax-ground polished with a 95 aloxide grit, and the simulated polish allowed the mirror to be tested at its radius of curvature with a Ronchi tester. The best wax to use is a paste rich in Brazilian carnauba wax (Trewax Brand). This is lightly smeared on the fine-ground surface and buffed strongly with a nylon cloth ball. The interferograms which are lateral-sheared (similar to defocused Ronchigrams) could be observed on a screen in a darkened room. (See Appendixes 12 and 14.)

In conclusion, a number of aspects are to be considered: The two projected interferograms must be properly orientated by rotating the viewing screen with its ruled lines, thus keeping all the interferograms evenly

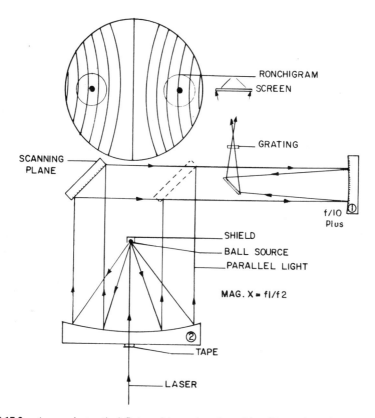

Fig. A17.2. A scanning optical flat used in conjunction with a Newtonian telescope to view the Ronchigrams of the paraboloid being optically figured. A wax-ground surface that is then buffed can be used with a laser source to illuminate a small polished ball bearing.

spaced. Remember during optical figuring that the edge zone becomes the reference zone (see Chapter 20) and to change a polished spheroid to a paraboloid, the glass is removed from the central zone or near the 0.707 zone, although in many aspheric lenses or corrector plates glass is removed only near the 0.707 zone. The reference zone can pivot to any zone after some preliminary optical figuring. (See Chapter 26.) The use of spatial filter pinholes with a more powerful laser source to help improve the contrast of the lateral-sheared interferograms is necessary in the last stages of pitch polishing. Again, a ball bearing laser source can be used if a hollow spindle with a drilled hole in the supporting plate permits the laser beam to illuminate the ball. (See Appendix 12.) It is apparent, but not discussed, that it is possible to scan aspheric lenses on a rotating spindle with a tunable laser while they are being wax-ground polished and optically figured.

Another favorite optical setup in scanning is an aluminized optical flat and a Newtonian telescope. The Ronchigrams are observed on a frosted screen. Here the inside-of-focus Ronchigram is observed for an undercorrected concave mirror. This setup is shown in Fig. A17.2.

Appendix 18

Using a Murty Interferometer to Test the Homogeneity of Samples of Optical Materials

Many optical shops make prisms, lens elements, etc., as a rapid method of checking the homogeneity of crystals and glasses. One of the least expensive interferometers of large aperture is the Murty interferometer. A typical setup is shown in Fig. A18.1(a), which illustrates a 30-cm paraboloid of $f/4.0$ ratio in a Newtonian collimator. A low-power visible laser (about 0.5 mW) and a $10\times$ spatial pinhole completes the collimator. (See also Figs. A12.2 and A12.4.) When a parallel plate with or without a slight wedge is placed in the collimated beam, it will project lateral-sheared interferograms of the correction of the spherical aberration of the $f/4.0$ paraboloid. This is a versatile interferometer because finished lenses (for which a tunable laser illumination is required), aspheric elements, and planoparallels can be interferometrically tested by a slight modification and rearrangement of the components making up the collimator. [See Fig. A18.1(a).]

To observe the transmission interferograms $[(2nt - 1)t]$ the prepared test samples of the optical materials made as planoparallels (semipolished) with or without a slight wedge are placed in the collimated beam at an approximate 45°. The lateral-sheared interferogram is photographed or projected on a frosted screen. It is also possible to observe the reflection Fizeau interferogram $(2nt)$ of the test sample by placing it in a tilted position near the paraboloid mirror. The interferograms can be observed by projecting them either on a frosted screen or photographed to a scaled diameter. [See Fig. A12.1(b).]

Make use of this interferometer for qualitative observation or quantitative measurements of the homogeneity of the variation of the refractive index. For general optical fabrication the qualitative observation of the location of the nonhomogeneity will suffice; for example, look at a sketch from a photograph of the 1.414 enlargement (remember this is a 45° angle foreshortened photograph) of the transmission fringes in Fig. A18.2(a) of a BSC-2 or BK-7 glass disk for prisms. Observe the break in the parallel

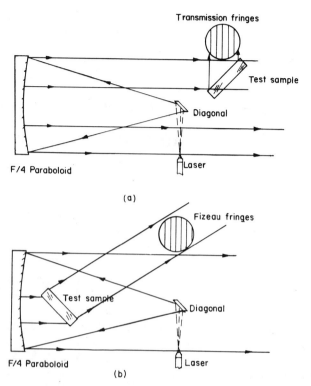

(a)

(b)

Fig. A18.1. (*a*) A Murty interferometer with a laser source and a single planoparallel plate of optical transmitting material that forms lateral-sheared interferograms of the transmission homogeneity quality of the material. (*b*) The Fizeau interferogram formed by the optical transmitting material when placed in the laser beam near the paraboloid mirror.

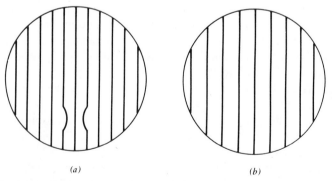

(*a*)   (*b*)

Fig. A18.2. (*a*) Lateral-sheared transmission interferogram as observed in Fig. A18.1*a*. (*b*) Fizeau interferogram as observed in Fig. A18.1*b*. Here the BSC-2 glass test sample is 150 mm in aperture and 50 mm thick.

transmission fringes, which indicates an area of nonhomogeneity. It would be necessary to cut or drill out this area for prisms but for an optical window or even a divider plate this area could be localized or refigured by polishing to match the overall parallel fringes. It is apparent that the homogeneity of optical materials governs what particular use is made of it.

The quantitative measurement of the variations of the refractive index can be made by the Twyman-Perry method. The method makes use of two properly enlarged drawings (determined by the 45° angular displacement when photographed) as shown in Fig. A18.2. The sketch on mylar drafting paper of the transmission fringes in Fig. A18.2(a) is superimposed over the sketch of the Fizeau fringes in Fig. A18.2(b). From this information develop a contour map of the variation of the refractive index of the optical materials.

Often in testing the homogeneity of the materials in test samples, some very small irregularities in the fringe patterns of the transmission fringes will not move when the test samples are translated in the collimated laser beam. These come from the uncorrected zones of spherical aberration in the paraboloid mirror. Any small nonhomogeneous area in the test samples remain fixed in their own areas while the optical materials are translated. It is of interest that the interferometer can test a window 1.4 times its own diameter because of its placement of 45° in the collimated beam.

Appendix 19

Near Perfect Optical Square

A perfect optical square is a glass cube that has all eight corners and 12 90° angles perfect. The optical square described is one in which all the tolerances for the 90° glass angle are ±2 sec. This means that for the cube to close around all sides it would have to be less than 1.5 sec at the first corner *a*, as shown in Fig. A19.1(*a*). Figure 19.1(*b*) is the finished optical square with the measured listed values.

1. The 50.8-mm BK-7 glass cube was generated and fine-ground around all six sides with corner *a* as the reference. A precision 90° square was used to check the six right-angle corners The next step is to semipolish (25%) all surfaces of the cube; this eliminates the Twyman effect when another pre-ground surface is being polished. With this rapid method it may be necessary to regrind one or more of the 90° glass angles with fine emery.

2. There are several methods of determining whether the 90° glass angles are positive or negative. (See Chapter 6.) The 4× amplification factor external autocollimation mode may be preferred because it permits the location of one pressure point for correcting 90° glass-angle errors. Here, as shown in Fig. A19.2 use the touch method to locate the one pressure area for correcting the 90° glass angles by touching an area that causes the separated images to draw together. It is obvious that the touch method cannot be used to test the frozen images in the internal autocollimation mode.

Using one surface at corner *a*, polish to one-eighth fringe surface quality. This becomes the reference corner for working the two remaining 90° glass angles; it should be marked with tape to prevent mixup. The first 90° glass angle around *a* is now checked with the setup in Fig. A19.2. If the separated images are too far apart, regrinding will be necessary. Always face the polished reference side toward the collimator and place the surface to be corrected on the three prewaxed nylon balls. This surface of the 90° glass angle must be one-eighth fringe surface quality with a well corrected 90° angle.

Both surfaces are then aluminized and their 90° angular surface is measured on a Hilger-Watts spectrometer. Nine circular readings (using the four reading microscopes) are required to measure the air-angle between the

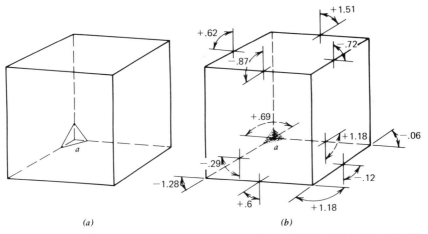

(a)   (b)

Fig. A19.1. (*A*) Glass cube with a reference corner (*a*) used during fabrication. (*B*) The measured values in seconds of arc are listed at (6) 90° glass angles of the optical square. The values at the corner (*a*) were −.29 in., +.69 in. and +1.18 in.

90° coated surfaces. This air-angle measured on the cube is not the glass angle. (See Fig. A19.3.) Pressure is always applied on the negative glass-angle side to correct the glass-angle error. (See Chapter 7.) The first 90° glass angle is finished when it measures 1.5 sec or less.

The third 90° around corner *a* is similarly finished and corrected for both 90° angles that are now involved. Table 7.1 shows the one pressure point required for one or two glass-angular corrections. This third 90° glass angle is measured on the spectrometer. There are two 90° angular surfaces at corner *a* to be measured and the required tolerance is 1.5 sec. All aluminized surfaces should be retained as they are used in (3).

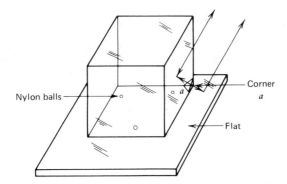

Fig. A19.2. The external autocollimation of the 90° polished reference side of the glass cube.

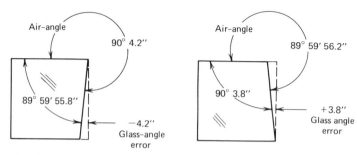

Fig. A19.3. The difference between a measured air angle and a glass angle: if the air angle is plus the glass-angle error is minus; conversely, if the air angle is minus the glass-angle error is plus.

3.   The final finishing of the three remaining sides of cube around corner *a* is accomplished by each surface rendering in turn. This final parallel is made using an interferoscope setup. The interferoscope has an isotope mercury source. (A 0.3-mW laser source can be used to project the interferogram on a frosted screen.) Any lack of parallelism is shown by a large number of parallel fringes or off-centered circular fringes (small errors). To determine the wedge angle in the cube touch one of the fringes with one finger and it will actually point toward the thinner edge of the cube. This is shown in Fig. A19.4. Parallelism of both surfaces is found when circular fringes are dead center on the cube and the polished surface is one-eighth

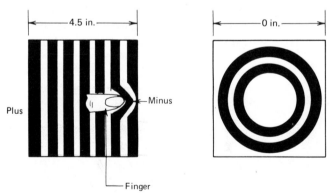

Fig. A19.4. The Fizeau interferogram of parallelism and the method of determining the wedge angle existing in the glass cube. Parallelism between the two surfaces of the cube in a Fizeau interferometer is dead-center circular fringes.

fringe surface quality. Then finish the optical square without measuring the 90° glass angles.

The final quality assurance using the values measured on the spectrometer requires 27 readings for each air-angle. As shown in Fig. A19.3 the glass-angle error always has the opposite value of the air-angle error. The method of least squares is involved to justify the 87% probability that the measured 90° glass angles are valid. The method of least squares is described in William Chauvenet, *A Manual of Spherical and Practical Astronomy*, Vol. 2, Dover New York, pp. 469–566.

Appendix 20

Testing Glass Reflecting Angles of Prisms

In Chapters 6 and 7 the testing of a prism's glass reflecting angles while in an internal or external autocollimation set were discussed. In this Appendix further aspects of the testing of prism angles in an external and internal autocollimation of air-angles are covered and a critical examination of a near perfect glass reflecting angle in an external or internal autocollimation mode is made.

Figure 6.8 illustrates how one collimator can autocollimate a wide range of angular settings of the collimator from 72° in the vertical plane and 4° in the horizontal, for a total range of 82° over which the air-angle can still be externally autocollimated. It should be obvious that in any preset angle in this range the angle in question can be checked in one position of the auto-collimator and then rotated to a second angular position, thus both the glass reflecting angle and its 90° side angle (pyramidal control) can be tested with one setting of the autocollimator. This is shown in Fig. 6.8. Note that the prism is resting on three nylon balls which are tightly waxed (lab wax) to an aluminized flat. The direct comparison of the prescribed angular setting previously setup for the prism being made is a 2× amplification factor. The chosen reference side of the prism is either semipolished or made reflective by a small aluminized plano wet-wrung to the fine-ground surface. Here there is a 4× amplification factor of the 90° air-angle of the prism.

Consider first air angles: there are several methods of determining where the external autocollimation of the 90° reference side and the flat are obtuse or acute (See Fig. A20.1.)

1. Defocus the eyepiece 1 mm *in* toward the prism and observe at the same time. If the images tend to separate further, the air-angle is plus or greater than 90°.
2. Defocus the eyepiece as above. If the images tend to draw together, the air-angle is minus or less than 90°.

Verify these findings by defocusing the eyepiece in the outside direction (out). Then the opposite effect takes place: a positive air-angle will draw the

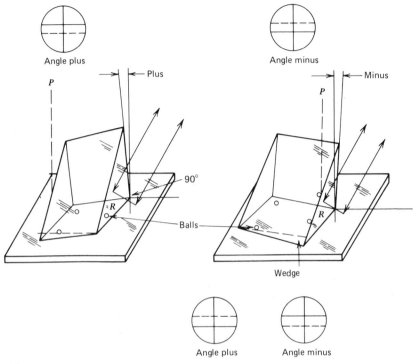

Fig. A20.1.   The external autocollimation mode of test-air angles formed by a prism's 90° side angle (pyramidal control) and the aluminized flat while the prism rests on three balls.

horizontally separated images together and, conversely, the negative air-angle will separate the images further.

When air-angles are involved, as in prisms in which a direct comparison of the prescribed setting by a master angle block and the 90° reference side is being externally autocollimated, use the touch method. Here the prism is touched by a finger on an area which causes the separated images to draw together. (See Fig. A20.2.) This area becomes one of the pressure points when pressure is applied in fine-grinding or polishing (ultrafine-grinding) during the correction of the glass angles of the prism. It is obvious that this touch method cannot be used to test the frozen images in the internal auto-collimation.

Consider the glass angle as discussed in Chapter 5, Rule 1, and compare with (1) above. Note that they are identical. It follows that Rule 2 and (2) are also identical. Enlarge on Rule 1 and (1) in either case, because this results in misinterpretation by beginners in making prisms. This comes

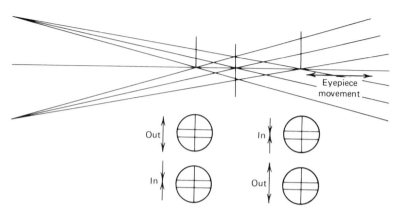

Fig. A20.2. The ray trace of plus and minus reflecting-angles of either an internal or external autocollimation setup of a prism. Note that the movement of the horizonal images on defocusing the eyepiece.

about when there are superimposed images for a well-corrected prism glass reflecting angle and the beginner defocuses the eyepiece. Here the images will again separate, and the beginner will falsely interpret the prism's glass angle to be slightly positive (obtuse), which is incorrect. What the beginner should do to verify his observation is to defocus 1 mm in the opposite direction with the eyepiece. Then the image will again separate a like amount. This is analogous to star testing a well corrected astronomical objective which gives equal diameter with equal intensity in the ring system on each side of the focal plane. This is illustrated in Fig. A20.2 *out* images.

To summarize the testing of externally or internally autocollimated prisms for glass reflecting angular errors, use the arrangement in Fig. A20.2. Here the movement of the images, by defocusing the eyepiece *in* and *out* where there are reflecting glass angular errors is illustrated. Always focus the eyepiece on the sharpest image possible and then defocus the eyepiece 1 mm. It must be kept in mind that this checks only one angle of the compound reflecting glass angle; therefore we must rotate the prism and make a comparison check of the prescribed glass angle. When two glass angles are involved (as in controlling a compound angle), the final deciding pressure point is midway between two predetermined points. This is due to three angular conditions that can exist: positive or negative glass angles and glass angles without errors.

Appendix 21

Interpreting Wavefront and Glass-Error Slopes from an Interferogram

Who has not seen an interferogram and wondered what information it contains? There is a wealth of data to be found especially when coma, astigmatism, and spherical aberrations are present. In this appendix the interest is primarily in residual spherical aberration and a novel method of detecting wavefront, glass-error slopes in an interferogram.

Consider a catadioptric Maksutov system tested interferometrically in a Twyman-Green or lateral-sheared interferometer. The resulting interferogram is shown in Fig. A21.1. Make a quick survey of the location of the high or low slopes in this interferogram if four things are known: (a) the type of image-forming system—all transmitting glass elements, all reflective, or catadioptric with both types of glass elements; (b) some identification marking (such as a piece of tape) on the system so that a developed print can be properly oriented; (c) whether the lateral-sheared interferograms are defocused inside or outside of focus; (d) focal-plane Ronchigrams for inside and outside of focus for slope departures. See Appendix 14. Note that a catadioptric system was chosen because of the two types of glass surface present.

An optician can optically figure the corrector lens's concave surface with quasi-Ronchigrams. The reference 0.707 zone is reasonably correct, the central zone is high out to the 0.707 zone, the peripheral area is reasonably correct, and the extreme edge is badly turned down (low). Note that this information is derived from the inside-of-focus quasi-Ronchigrams for light-transmitting elements in parallel light. (See Appendix 14.) The outside-of-focus quasi-Ronchigrams show the same glass-slope departures. (See Fig. A21.1.)

Question. Can the optician figure the reflective primary mirror's surface to a null pattern? *Answer.* He can, but he must know that there is a reversal of the quasi-Ronchigram, as Fig. A21.1 shows. Because these quasi-Ronchigrams are reversible for reflective optical surfaces, when optically figuring them one considers the outside-of-focus as the *inside-of-focus* quasi-Ronchi-

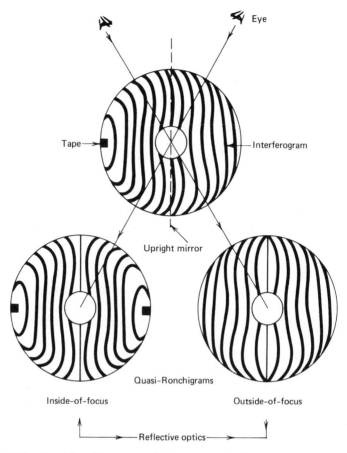

Fig. A21.1. Quasi-Ronchigrams as reflective symmetries observed in an interferogram.

grams and vise versa. Reinterpretation of the zones is as follows: the central area is high out to the 0.707 zone, the 0.707 zone is reasonably correct, the peripheral area is reasonbly correct, and the extreme edge is badly turned down (low). Note that the glass-slope zonal departures are interpreted as described for light-transmitting elements. Appendix 14 shows that a star-pattern pitch polisher can be used for optically correcting the concave surfaces of either lens or mirror.

In *Optical Shop Note Book*. J. B. Houston, Jr., Ed., Optical Society of American, Washington, D.C., 1974–1975, T. J. Johnson of Celestron summarizes the known optical tests: (a) interferograms actually give a contour map of the glass errors; therefore they must be differentiated to arrive at focal-plane errors; (b) Ronchigrams give information about focal-plane errors (the thickness of the Ronchiband between the fringe patterns is the magnitude of the slope change of the glass errors; (c) the focograms give the slope contours in the glass.

With quasi-Ronchigrams differentiate the interferogram into focal-plane errors that an optician can use to figure optically any lens system to a null pattern. After some practice the user need not draw the quasi-Ronchigrams from an interferogram if the right-hand portion is first covered with a slip of paper and it is assumed that it is the contour slope of the glass transmitting wavefront. Next cover the left-hand portion of the interferogram and assume that it is the contour slope of the reflective surface and its wavefront. The optician must keep in mind that these are *inside-of-focus* quasi-Ronchigram interpretations.

BIBLIOGRAPHY

A. S. De Vany, *Appl. Opt.*, **9**, 1944 (1970).

A. S. De Vany, *Appl. Opt.*, **17**, 3023 (1978).

T. J. Johnson, in *Optical Shop Note Book*, Vol. 1, J. B. Houston, Jr., Ed. Optical Society of America, Washington, D.C., 1974–1975, p. 92.

Appendix 22

Resin Polishers, Holders, and Tools

In this appendix a number of experiments with plastic polishers, plastic-ceramic grinding tools, and holders for lenses during fabrication are discussed.

When fabricating a limited number of lenses it is often desirable to make plastic-ceramic grinding tools because they can be produced more cheaply than with expensive glass or cast iron. It is also a good idea to use resin plastic holders for the lens during fabrication. It is well known that polyurethane loaded with cerium or zirconium oxide is being used extensively in the rapid fabrication of glass components, but a number of disadvantages are apparent; for example, excessive loss of material during the flycutting of sharp radii polishers; also, the polishers must be stored in water to avoid drying out and hardening. The plastic polisher has none of these disadvantages, but a foaming agent has not been found to give a microcellular structure for carrying the polishing agent. D. F. Horne, in *Optical Production Technology*, Crane, Russak & Company, New York, 1953. Sec. 8.9, p. 235, discusses a number of foaming agents for forming the microcellular structure. The use of Freon in a pressure vessel while mixing is probably one method to try that will cause small bubbles to form at air pressure. Pressurized Freon cans are available on the market.

The best and most rapid resin polishers made of eponresin compounds were polishers fabricated from material provided by Minnesota Mining and Manufacturing Company, St. Paul, Minn, 55101. Their Lot 83 was used. The *procedure* is as follows:

1. Prepare the mixture (see Table A22.1) by weighing each compound.

2. Rub silicon grease thoroughly over the fine-ground surface of the lens component that has been checked with an electronic (Strasbaugh) spherometer.

3. Rub grease well into the bevels and 1–2 mm down the peripheral edges.

4. Wrap several layers of masking tape around the periphery of the previously machined radius holders, to hold the lens during polishing and the

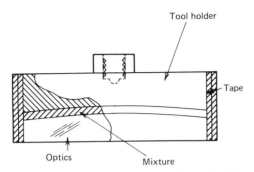

Fig. A22.1. Casting the resin plastic holders and polishers.

polishing mixture: the concave and convex holders and polishers must have a 2-mm radius that is longer or shorter than the surface of the lens radius in order to have an equal amount of mixture between the surface. The tape forms a dike around the periphery to hold the mixture. No cerium or zirconium oxide polishing compound is required in the mixture for lens surface holders.

5. Pour the well-stirred mixture evenly over the surface to a depth of approximately 2 mm. Let it stand for several minutes on a leveling stand. (See Fig. A22.1.)

6. After several hours check by pressing on the tape to see if the mixture has hardened.

7. The lens can be removed by cutting the hardened mixture around the bevels with a shop razor blade.

8. Cut a series of channel facets in the polishers with a shop razor blade. These are important for if the polisher goes dry it will quickly grasp the surface of the lens and pull it off the holder.

Consider the following:

Corrector Plate, 25-cm Diameter.   Three pliable flat surfaces were cast on full-sized flat aluminum holders: one for holding the corrector plate, the other two for figuring the corrector plate. The thickness of the plastic mixture was approximately 2 mm, and the pliable nature of the mixture for holding the optical elements and for preventing astigmatic surfaces from forming far exceeds that of rubberized felt. The two patterns, cut out with a shop razor blade, used a previously developed rose-leaf pattern for marking the desired profile; see Fig. A22.2 and note that one profile is the inverse of the other. The top figure shows the 0.87 zone polisher used to begin the

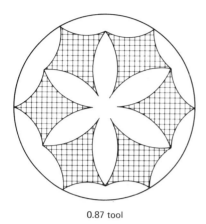

0.87 tool

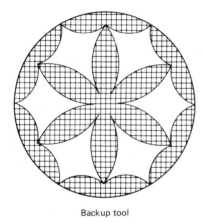

Backup tool

Fig. A22.2. Two types of resin plastic polisher with cutout profiles for figuring an aspheric surface. The upper part shows a rose-leaf type for correcting an undercorrected lens system. The lower part shows a backup tool for removing overcorrection.

optical figuring to remove the spherical aberration of the Schmidt-Cassegrain system. The bottom figure is the inverse profile of the top and is used to remove overcorrection of the spherical aberration. (It was never used, because the interest lay in obtaining 95% at the 0.87 zone (three wave depth); the other 5% was obtained by figuring the secondary mirror null in an autocollimating mode.

The polishing is best carried out on a milling machine. The corrector plate is mounted on its pliable holder and zeroed out on a turntable. The turntable rotates in opposition at 2 rpm, and a small eccentric drive moves the precut polisher over the corrector plate with an approximate 0.5-cm stroke at 50 rpm.

Front surface optical flats. Two 127-mm optical flats—one of Pyrex the other of optical quartz previously finely generated with a 1500-diamond grit

TABLE 22.1

Resin[a]	Mixture (g)	Description	Handwork	Machine work	Remarks
EC 1614 A EC 1614 B	50 50	Rock hard			Add 50 g 95 aloxide for grinding tools
EC 1614 A EC 1614 B	60 40	Slightly hard	Sleeks	Good if weighted: excellent holders	50 g cerium oxide: one wave surface, good memory
EC 1614 A EC 1614 B	65 35	Like pitch	Works to 0.75 wave	Good for aspherics	Add 50 g cerium oxide; good memory
EC 1614 A EC 1614 B	70 30	Rubbery	Works to 2–3 waves	Works to 2 waves	Add 50 g cerium oxide; poor memory

[a] Scotch Weld Brand Lot 83. Polishing compound: Cerium oxide (40 h milled). Lens holders do not add some cerium oxide, but no harm results.

wheel were polished with the plastic polisher impregnated with 40-hr milled cerium oxide. Polishing time was 4 hr using 2 g/1 cm² circular weight on top of the polisher. The best surface quality was about 1.5 fringes overall. No better quality could be obtained by handwork, or by reversing the polisher and work.

Test Plates. A pair of 127-mm aperture test plates, with a prescription radius of 241.5 mm were polished to a test plate fit of 1.5 fringes concave. They could not be further improved by changing the profile of the polisher or working them by hand, although they maintained excellent spheroidicity.

Duplication of a scratched lens surface. A badly scratched front surface of a 25.4-cm field camera required reworking. The lens was removed from its cell and a polisher and holder were cast on the surfaces of the lens. The grinding tool had 50 g of 96 aloxide grit added to the mixture. (See Table 22.1.) The lens holder was a pliable mixture but the polisher was a conventional pitch type. The radius of the original surface was maintained by checking with an electronic spherometer reading to 10^{-4} mm. The lens was checked during polishing by testing through its back and convex surface to form a concave reflective surface; a knife edge was used for testing before and after grinding and polishing. The figure before and after polishing was a near spheroid. The tested assembled lens has shown no noticeable change from its previous test pattern.

Flexible aspheric tools. A corrector plate for 16-cm, $f/0.7$ Schmidt camera was made of quartz, against the original one of crown glass. The slope departures at the 0.87 zone are approximately 0.2 mm in depth. The fabrication of the corrector plate followed the method used by the Cox brothers, as described in "The Journal of the British Astronomical Association, Volume 50 (61–68), December 1939. A reprint appears in *Amateur Telescope Making, Book Three, Sci. Am.* **3**, 349–354 (1953). Here the corrector is made as a planoconvex element, the 0.87 zone of the convex surface is lowered by the facet tile tool, with the pitch-cemented tiles in a rose-leaf configuration on a pliable flat holder. The pliable rose-leaf grinding tool (see Fig. A22.2) as a polisher is much better than using sponge rubber with or without thin microscope cover-glass supports under the tile or using pitch facets during the fabrication of the steep aspheric surfaces.

The corrector plate's surface was brought to an excellent, well-corrected figure with a screen at the focal plane in the fine-grinding phase. This is described in Book Three as the Cox's brothers method of fabrication. Here the corrector is made semitransparent by slightly oiling it and wiping it nearly dry and then placing it in an autocollimation apparatus. From this finished aspheric a cast pliable polisher with cerium oxide was formed. (A precast polisher would not work well on the aspheric surface because it tends to grasp the surface of the corrector plate.) Pitch buttons were then formed by dropping pitch on the pliable polisher with a small soldering iron. This facilitates changing the profile of the polishers as the optical figuring proceeds.

The following *conclusions* have been drawn: These resin-plastic polishers with impregnated polishing compounds work well in bowl-feed slurries for flat and spherical surfaces. Polishers should not be permitted to dry because orange peel polished surfaces then occur. (Hand polishing of optical surfaces is not always successful.) The preformed polishers have excellent memories because the contours do not change during polishing. Microcellular structures formed with Freon or carbon dioxide gas under pressure during the final mixing of the two compounds might help.

BIBLIOGRAPHY

D. F. Horne, *Optical Production Technology*, (Crane, Russak & Company, Inc., New York, 1972), Sec. 8.9 p. 235.

H. W. Cox and L. A. Cox, *Amateur Telescope Making, Book Three* (Scientific American, Inc., New York, 1953), Vol. 3, pp. 349–354.

References

Appendix 1

Department of the Army. "Design of Fire-Control Optics," ORDM-2-1, 68-103, (1940).

Department of Defense. "Military Standardization Handbook," Mil HBK-141, 13-1-13-52 (October 5, 1962).

Office of Naval Research, Department of the Navy. "Handbook of Infra-Red Technology," 307–309 (1965).

Smith, W. J. "Modern Optical Engineering," McGraw-Hill, New York, 1966, pp. 84–95.

Appendix 2

De Vany, A. S. "Reduplication of a Penta Prism Angle Using Master Angle Prisms and a Plano-Interferometer," *Appl. Opt.*, **10**, 1372–1375 (1971).

Appendix 3

De Vany, A. S. "Making Right-Angle and Dove Prisms," *Appl. Opt.*, **7**, 1086 (1968).

Appendix 4

De Vany, A. S. "Testing a Prism's Compound Angle," *Appl. Opt.*, **14**, 279–292 (1975).

De Vany, A. S. "Deviation Prisms: Controlling the 90° Roof-Angle Ridge," *Appl. Opt.*, **15**, 1104 (1976).

Twyman, F. "Prism and Lens Making," Hilger-Watts, Ltd., London, 412–413, 426 (1952).

Appendix 6

De Vany, A. S. Northrop Memo (1964).

Appendix 8

De Vany, A. S. "Reduplication of a Penta Prism Angle Using Master Angle Prisms and a Plano-Interferometer," *Appl. Opt.*, **10**, 1375 (1971).

De Vany, A. S. "Optical Figuring of Prisms," *Appl. Opt.*, **16**, 1459–1460 (1977).

Johnson, B. K. "Optics and Optical Instruments," Dover, New York, 1960, pp. 200–202.

Twyman, F. "Prism and Lens Making," Hilger & Watts, Ltd., London, 428–430 (1952).

Appendix 9

Obreimov, I. V. and Trekhov, E. S. "Optical Contacting of Polished Surfaces," *Soviet Physics JEPT*, 5, 235–242 (1957).

Paul, H. E. "Amateur Telescope Making" Book 3, Scientific American, New York, 393–395 (1953).

Appendix 10

Davidson Incorp, "Optronics," West Covina, CA.

Appendix 11

Saunders, J. B. "In-Line Interferometer," *J. Opt. Soc. Am.*, 44, 241 (1953).

Post, D. "Characteristics of the Series Interferometer," *J. Opt. Soc. Am.*, 44, 243 (1953).

Appendix 12

Bell, C. *The Telescope*, McGraw-Hill, New York, 1923, p. 203.

De Vany, A. S. "Laser Illuninated Divergent Ball Bearing Sources," **Appl. Opt.**, 13, 457–459 (1974).

Hendrix, D. O. and Christe, W. H. "Amateur Telescope Making," Book 3, Scientific American, New York, 354–365 (1953).

Strong, J. *Procedures in Experimental Physics*, Prentice-Hall, Englewood Cliffs, NJ, 1943, p. 70.

Twyman, F. *Trans. Opt. Soc.* 24, 188 (1922–1923).

Appendix 13 Angular Settings for Radius Generator

Northrop Corporation Memo (1963). M. Brown.

Appendix 14

De Vany, A. S. "Optigami—A Profiling Tool for Pitch Polishers," *Appl. Opt.*, 10, 661–668 (1971).

De Vany, A. S. "Profiling Pitch Polishers," *Appl. Opt.*, 17, 3022–3024 (1978).

De Vany, A. S. and De Vany, A. S., Jr. "Patterns of Correlation between Focograms and Ronchigrams," *J. Bol. Inst. Tonantizintla*, 1(5) (1975).

Malacara, D. and Cornejo, A. "Relating the Ronchi and Lateral Shearing Interferometer Tests," *Opt. Spectra*, 8, 54 (1974a).

Appendix 15

Chinizi, C. C. Private communication (1965).

Schroeder, J. "Making and Testing an 11 Inch Objective," Sky & Telesc., 178–180 (March 1978).

Appendix 16

Sky & Telesc., 69 (December 1957).

Appendix 17

De Vany, A. S. "Scanning Murty Interferometer for Optical Testing," *Appl. Opt.*, **11**, 1467–1468 (1972).

Hendrix, D. O. and Christe, W. H. "Some Applications of the Schmidt Principle in Optical Design," Amateur Telescope Making, Book 3, Scientific American, New York 365 (1953).

Hochgraf, N. H. "Abstract," *J. Opt. Soc. Am.*, 655-A (1971).

Hooker, R. B. "Abstract," *J. Opt. Soc. Am.*, **61**, 655-A (1971).

Moreau, B. G. and Hopkins, R. E. "Application of Wax to Fine Ground Surface to Simulate Polish," *Appl. Opt.*, **8**, 2150 (1969).

Murty, M. V. R. K. "The Use of a Single Plane Parallel as a Lateral Shearing Interferometer with a Visible Laser Source," *Appl. Opt.*, **10**, 531 (1964).

Ransom, J. Private communication (1951).

Appendix 18

De Vany, A. S. "Using a Murty Interferometer for Testing the Homogeneity of Test Samples of Optical Materials," *Appl. Opt.*, **13**, 10, 1459 (1971).

Murty, M. V. R. K. "A Note on the Testing of Homogeneity of Large Aperture Parallel Plates of Glass," *Appl. Opt.*, **3**, 784 (1964).

Twyman, F. "Prism and Lens Making," Hilger & Watts, Ltd., London, 504–508 (1952).

Appendix 19

De Vany, A. S., "Correcting Glass Angles of Prisms," *Appl. Opt.*, **16**, 2019 (1977).

De Vany, A. S. "Testing Reflecting Angles of Prisms," *Appl. Opt.*, **17**, 1661 (1978).

De Vany, A. S. "Near Perfect Optical Square," *Appl. Opt.*, **18**, 1284–1286 (1979).

Howell, J. and De Vany, A. S. "A Near Perfect Cube," Northrop Shop Manual (1955).

Appendix 20

De Vany, A. S. "Testing a Prism's Compound Reflecting Angles," *Appl. Opt.*, **14**, 2791 (1975).

De Vany, A. S. "Correcting Glass Angles of Prisms," *Appl. Opt.*, **16**, 2019 (1977).

Appendix 21

De Vany, A. S. "Quasi-Ronchigrams as Mirror Transitive Images of Interferograms," *Appl. Opt.*, **9**, 1944 (1970).

De Vany, A. S. "Profiling Pitch Polishers," *Appl. Opt.*, **17**, (1978).

T. J. Johnson, in *Optical Shop Note Book*, J. B. Houston, Jr., Ed. (Optical Society of America, Washington, D.C., 1974–75), Vol. 1, p. 92.

Appendix 22

1. D. F. Horne, *Optical Production Technology* (Crane, Russak & Company, Inc., New York, 1972), Sec. 8.9 p. 235.

2. H. W. Cox and L. A. Cox, *Amatuer Telescope Making, Book Three* (Scientific American, Inc., New York, 1953), Vol. 3, pp. 349–354.

Index